Research and Development in Intelligent Systems XXX

Incorporating Applications and Innovations in Intelligent Systems XXI

Max Bramer · Miltos Petridis
Editors

Research and Development in Intelligent Systems XXX

Incorporating Applications and Innovations in Intelligent Systems XXI

Proceedings of AI-2013, The Thirty-third SGAI International Conference on Innovative Techniques and Applications of Artificial Intelligence

Editors
Max Bramer
School of Computing
University of Portsmouth
Portsmouth
UK

Miltos Petridis
School of Computing, Engineering and Mathematics
University of Brighton
Brighton
UK

ISBN 978-3-319-02620-6 ISBN 978-3-319-02621-3 (eBook)
DOI 10.1007/978-3-319-02621-3
Springer Cham Heidelberg New York Dordrecht London

Library of Congress Control Number: 2013950355

Printed on acid-free paper

Springer is part of Springer Science+Business Media (www.springer.com)

Programme Chairs' Introduction

This volume comprises the refereed papers presented at AI-2013, the Thirty-third SGAI International Conference on Innovative Techniques and Applications of Artificial Intelligence, held in Cambridge in December 2013 in both the technical and the application streams. The conference was organised by SGAI, the British Computer Society Specialist Group on Artificial Intelligence.

The technical papers included new and innovative developments in the field, divided into sections on Knowledge Discovery and Data Mining I, Knowledge Discovery and Data Mining II, Intelligent Agents, Representation and Reasoning, and Machine Learning and Constraint Programming. This year's Donald Michie Memorial Award for the best refereed technical paper was won by a paper entitled "Pattern Graphs: Combining Multivariate Time Series and Labelled Interval Sequences for Classification" by S. Peter (University of Konstanz, Germany), F. Höppner (Ostfalia University of Applied Sciences, Wolfenbüttel, Germany) and M. R. Berthold (University of Konstanz, Germany).

The application papers included present innovative applications of AI techniques in a number of subject domains. This year, the papers are divided into sections on Medical Applications, Applications in Education and Information Science, and AI Applications. This year's Rob Milne Memorial Award for the best refereed application paper was won by a paper entitled "Knowledge formalisation for hydrometallurgical gold ore processing" by Christian Severin Sauer (University of West London, UK), Lotta Rintala (Aalto University, Finland) and Thomas Roth-Berghofer (University of West London).

The volume also includes the text of short papers presented as posters at the conference.

On behalf of the conference organising committee we would like to thank all those who contributed to the organisation of this year's programme, in particular the programme committee members, the executive programme committees and our administrators Mandy Bauer and Bryony Bramer.

Max Bramer, Technical Programme Chair, AI-2013.

Miltos Petridis, Application Programme Chair, AI-2013.

Acknowledgments

AI-2013 Conference Committee

Conference Chair

Prof. Max Bramer University of Portsmouth

Technical Programme Chair

Prof. Max Bramer University of Portsmouth

Deputy Technical Programme Chair

Prof. Daniel Neagu University of Bradford

Application Programme Chair

Prof. Miltos Petridis University of Brighton

Deputy Application Programme Chair

Dr. Jixin Ma University of Greenwich

Workshop Organiser

Prof. Adrian Hopgood Sheffield Hallam University

Treasurer

Rosemary Gilligan University of Hertfordshire

Poster Session Organiser

Dr. Nirmalie Wiratunga The Robert Gordon University

FAIRS 2013

Dr. Paul Trundle	University of Bradford
Giovanna Martinez	Nottingham Trent University

UK CBR Organiser

Prof. Miltos Petridis University of Brighton

Publicity Officer

Dr. Ariadne Tampion

Conference Administrator

Mandy Bauer BCS

Paper Administrator

Bryony Bramer

Technical Executive Programme Committee

Prof. Max Bramer	University of Portsmouth (Chair)
Prof. Frans Coenen	University of Liverpool
Dr. John Kingston	Be Informed
Dr. Peter Lucas	University of Nijmegen, The Netherlands
Prof. Daniel Neagu	University of Bradford (Vice-Chair)
Prof. Thomas Roth-Berghofer	University of West London

Applications Executive Programme Committee

Prof. Miltos Petridis	University of Brighton (Chair)
Mr. Richard Ellis	Helyx SIS Ltd.
Ms. Rosemary Gilligan	University of Hertfordshire
Dr. Jixin Ma	University of Greenwich (Vice-Chair)
Dr. Richard Wheeler	University of Edinburgh

Technical Programme Committee

Ali Orhan Aydin	Istanbul Gelisim University
Yaxin Bi	University of Ulster
Mirko Boettcher	University of Magdeburg, Germany
Max Bramer	University of Portsmouth
Krysia Broda	Imperial College, University of London
Ken Brown	University College Cork
Frans Coenen	University of Liverpool
Madalina Croitoru	University of Montpellier, France
Bertrand Cuissart	Universite de Caen
Ireneusz Czarnowski	Gdynia Maritime University, Poland
John Debenham	University of Technology, Sydney
Nicolas Durand	University of Aix-Marseille
Frank Eichinger	SAP AG, Karlsruhe, Gemany
Sandra Garcia Esparza	University College Dublin, Ireland
Adriana Giret	Universidad Politécnica de Valencia
Nadim Haque	Thunderhead.com
Arjen Hommersom	University of Nijmegen, The Netherlands
John Kingston	Be Informed
Konstantinos Kotis	VTT Technical Research Centre of Finland
Ivan Koychev	Bulgarian Academy of Science
Fernando Lopes	LNEG-National Research Institute, Portugal
Peter Lucas	Radboud University Nijmegen
Stephen G. Matthews	De Montfort University, UK
Roberto Micalizio	Universita' di Torino
Dan Neagu	University of Bradford
Lars Nolle	Nottingham Trent University
Dan O'Leary	University of Southern California
Juan Jose Rodriguez	University of Burgos
María Dolores Rodríguez-Moreno	Universidad de Alcalá
Thomas Roth-Berghofer	University of West London
Fernando Sáenz-Pérez	Universidad Complutense de Madrid
Miguel A. Salido	Universidad Politécnica de Valencia
Rainer Schmidt	University of Rostock, Germany
Frederic Stahl	University of Reading
Jiao Tao	Oracle USA
Simon Thompson	BT Innovate
Andrew Tuson	City University London
Graham Winstanley	University of Brighton

Application Programme Committee

Contents

Applications in Education and Information Science

AI Applications

Short Papers

Research and Development in Intelligent Systems XXX

Best Technical Paper

Pattern Graphs: Combining Multivariate Time Series and Labelled Interval Sequences for Classification

Sebastian Peter, Frank Höppner and Michael R. Berthold

Abstract Classifying multivariate time series is often dealt with by transforming the numeric series into labelled intervals, because many pattern representations exist to deal with labelled intervals. Finding the right preprocessing is not only time consuming but also critical for the success of the learning algorithms. In this paper we show how pattern graphs, a powerful pattern language for temporal classification rules, can be extended in order to handle labelled intervals in combination with the raw time series. We thereby reduce dependence on the quality of the preprocessing and at the same time increase performance. These benefits are demonstrated experimentally on 10 different data sets.

1 Introduction

In recent years the development of cheaper sensors and bigger storage capacities has led to an increase in the amount of data gathered periodically. Companies are now able to use (mobile and/or wireless) sensor networks more efficiently in many different domains (e.g. health care, climate, traffic, business processes to name a few) to collect data with usually various dimensions. By analysing temporal data, companies try to gather more insight into their processes and are thereby able to draw

S. Peter (✉) · M. R. Berthold
Nycomed-Chair for Bioinformatics and Information Mining, University of Konstanz,
Box 712, D-78457 Konstanz, Germany
e-mail: sebastian.peter@uni-konstanz.de

M. R. Berthold
e-mail: michael.berthold@uni-konstanz.de

F. Höppner
Department of Computer Science, Ostfalia University of Applied Sciences,
D-38302 Wolfenbüttel, Germany
e-mail: f.hoeppner@ostfalia.de

M. Bramer and M. Petridis (eds.), *Research and Development in Intelligent Systems XXX*,
DOI: 10.1007/978-3-319-02621-3_1,

conclusions, enabling them for example, to predict the market for the next week or optimise the output by improving the production process.

One important aspect during the analysis step is often finding *typical* or *characteristic* situations. To grasp or encompass these situations, various notions of *multivariate temporal patterns* are described in literature. Example applications for multivariate temporal patterns include the discovery of dependencies in wireless sensor networks [1], the exploration of typical (business) work flows [3] or the classification of electronic health records [2]. Temporal patterns are often applied to labelled interval data, as the resulting patterns are easy to understand for the experts and also allow us to deal with multivariate data. To incorporate numerical time series in the patterns, they are discretized and their behaviour is described by a linguistic term ('low revolutions', 'slowly accelerating') that holds over a given period of time, hence the term 'labelled (temporal) interval'. The effectiveness of such patterns depend strongly on this discretization step. In this paper we extend the powerful concept of pattern graphs (see Fig. 1 as an example) enabling us to deal directly with time series data and overcome the sensitivity of the preprocessing phase.

The paper is outlined as follows: The next section reviews related work and further motivations for our work. We then give an introduction to pattern graphs (Sect. 3) and the matching and learning algorithms (Sect. 4) [10, 11]. In Sect. 5, we contribute the necessary changes to incorporate numeric time series. Section 6 presents the experimental results, and we conclude the paper in Sect. 7.

2 Motivation and Related Work

In this paper we concentrate on multivariate temporal patterns to characterise the evolution of multiple variables over time. These patterns are used in the antecedents of classification rules. The data consists of labelled temporal intervals; the labels may address categorical (e.g. 'gear-shift' in Fig. 1) or numerical features (e.g. 'low revolutions' in Fig. 1). These labelled intervals and their relationships are combined to form *temporal patterns*, for example by specifying the relationships between all observed intervals like '*A before B*', '*A overlaps C*' and '*C overlaps B*' [2, 5]. This notation is quite strict and somewhat ambiguous [7], because the qualitative

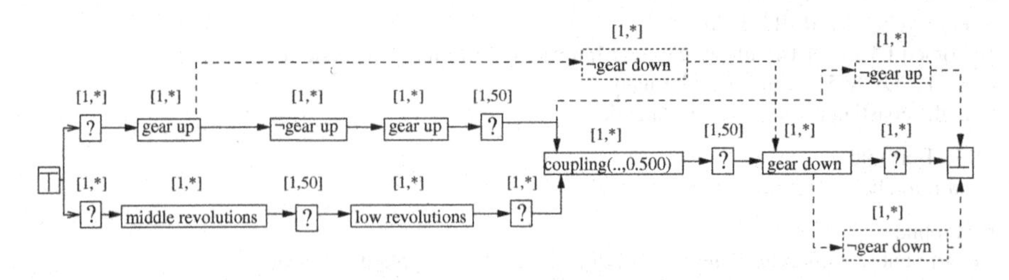

Fig. 1 Example of a pattern graph describing a driving cycle (learned from data, see [10])

relationship does not carry quantitative information about the degree of overlap or size of a gap. Other approaches contain such information [3], but consider only those events that do not include *durations* and thus offer no means to express concurrency as in '*A and B must co-occur for 5–10 time units*'. To be more robust against noise, some approaches allow to address parts of the intervals only [4, 9]. The recently proposed pattern graphs [11] satisfy most of the shortcomings and will be used in this paper (and will be introduced below in more detail).

Regardless of the pattern language, when the recorded data is numeric in nature, this leads to the problem of having to convert the numeric data into labelled intervals. This is usually done by applying thresholds, clustering methods or methods dedicated to finding stable intervals (e.g. 'Persist' [8]). This step is time consuming: multiple iterations and manual inspections are needed for a suitable discretization as a bad discretization can render all efforts to retrieve good patterns useless. An example is shown in Fig. 2, where the values of two time series (a) and (b) are discretized using the threshold 5, leading to the same sequence of labelled intervals (with labels [low: $y \leq 5$] and [high: $y > 5$]) in Fig. 2c. In this case the sequences are not distinguishable anymore, which is undesired if both series belong to different classes and we look for a temporal pattern that distinguishes both classes from each other. Furthermore the one perfect discretization may not exist in a situation where in class *(a)* the threshold needs to be 5 whereas for class *(b)* 6 and for class *(c)* the threshold of 7 would be perfect. To overcome this problem, the selection of optimal thresholds may be postponed in the learning algorithm itself instead of leaving it as a preprocessing step.

3 Pattern Graphs

This section reviews pattern graphs, which were first introduced in [11]. We consider m (categorical or numeric) attributes with value range D_j, composed of multivariate observations $\mathbf{x} \in D$ with $D = (D_1 \times \cdots \times D_m)$.

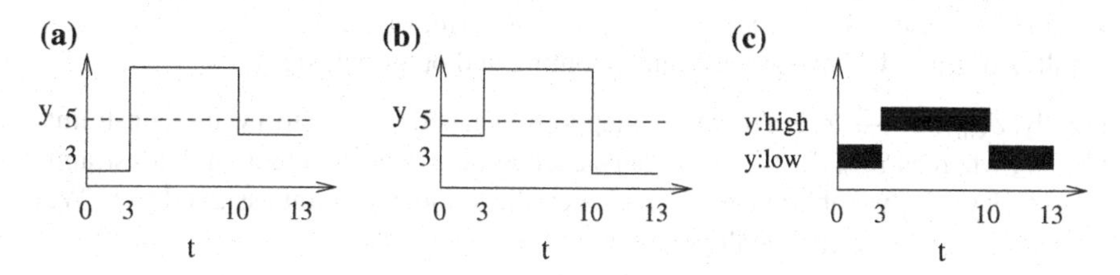

Fig. 2 Two time series: positive class (**a**) and negativie class (**b**) discretized to the interval sequence (**c**) by using the threshold 5 (*dotted line*), thereby loosing information to distinguish them from each other

Definition 1 ((sub)sequence). A sequence S consists of an arbitrary number of observations $(\mathbf{x}_1, \ldots, \mathbf{x}_n) \in \mathscr{S}$ with $\mathscr{S} = \bigcup_{i=1}^{\infty} D^i$. Let $|S| = n$ denote the length of the sequence S. A subsequence from index a to b of S is denoted by $S|_{[a,b]}$.

To describe those parts of the series that are relevant for the classification task, we apply (local) constraints to subsequences:

Definition 2 (set of constraints for (sub)sequences). Let $\mathscr{C} = \{C \mid C : \mathscr{S} \to \mathbb{B}\}$ denote the set of all (possible) constraints on (sub)sequences. We distinguish between value-constraints, restricting the acceptable values of the (sub)sequence, and temporal constraints, which limit their duration. For a sequence $S = (\mathbf{s}_1, \ldots, \mathbf{s}_k)$, examples of value-constraints are:

- $C(S) = true$ ("don't care": is always satisfied)
- $C(S) = true \Leftrightarrow \forall i : 1 \leq i \leq k : s_{i,j} \in C_j$ with $C_j \subseteq D_j$ for all $1 \leq j \leq m$. This constraint limits the range of accepted values for the sequence.

In this paper we consider only one type of temporal constraint:

- Given $t \in T$, $T = \{(a, b) | 1 \leq a \leq b\} \subseteq \mathbb{N}^2$, a temporal constraint is defined as $C(S) = true \Leftrightarrow a \leq |S| \leq b$. Therefore a temporal constraint is represented by an interval $[a, b]$ and restricts the duration of the (sub)sequence S to lie within these bounds. Here a is considered the minimal and b the maximal temporal constraint.

Up to now, pattern graphs have only been used for interval sequences, that is, a condition (described by the interval label) either holds or not ($D_j = \{0, 1\}$). We thus have three different value-constraints: $C_j \subseteq \{0\}$ (absent), $C_j \subseteq \{1\}$ (present) and $C_j \subseteq \{0, 1\}$ (don't care). A pattern graph defines a partial order of constraints:

Definition 3 (pattern graph). A tuple $M = (V, E, \mathscr{C}_{val}, \mathscr{C}_{temp})$ is a pattern graph, iff (V, E) is an acyclic directed graph with exactly one source ($\top$), one sink ($\bot$), a finite node set $V \subseteq \mathbb{N} \cup \{\top, \bot\}$ and an edge set $E \subseteq (V \times V)$, for which the following properties hold with $V' = V \backslash \{\top, \bot\}$:

- $\forall (v_1, v_2) \in E : v_2 \neq \top$ (no incoming edge to $\top$)
- $\forall (v_1, v_2) \in E : v_1 \neq \bot$ (no outgoing edge from $\bot$)
- $\forall v \in V' : (\exists w \in V : (v, w) \in E) \wedge (\exists w \in V : (w, v) \in E)$ (all nodes $v \in V'$ have at least one incoming and outgoing edge)

Finally, $\mathscr{C}_{val} : V' \to \mathscr{C}$ is a function, mapping each node $v \in V'$ to a value-constraint $C \in \mathscr{C}$, whereas $\mathscr{C}_{temp} : V' \to \mathscr{C}$ maps each node $v \in V'$ to a temporal constraint $C \in \mathscr{C}$. By $\mathscr{C}_{val}^{v}$ we abbreviate $\mathscr{C}_{val}(v)$, i.e., the value constraint assigned to v. We define $\mathscr{C}_{temp}^{v} := \mathscr{C}_{temp}(v)$ analogously.

Definition 4 (mapping). A mapping B for a sequence S and a pattern graph $M = (V, E, \mathscr{C}_{val}, \mathscr{C}_{temp})$ assigns each node $v \in V \backslash \{\top, \bot\}$ a continuous subsequence $S|_{[a,b]}$. $\top$ is assigned the fictitious subsequence $S|_{[0,0]}$ and $\bot$ the subsequence $S|_{[|S|+1,|S|+1]}$. $B(v) = [a, b]$ denotes the start and end index of the subsequence of S assigned to node v.

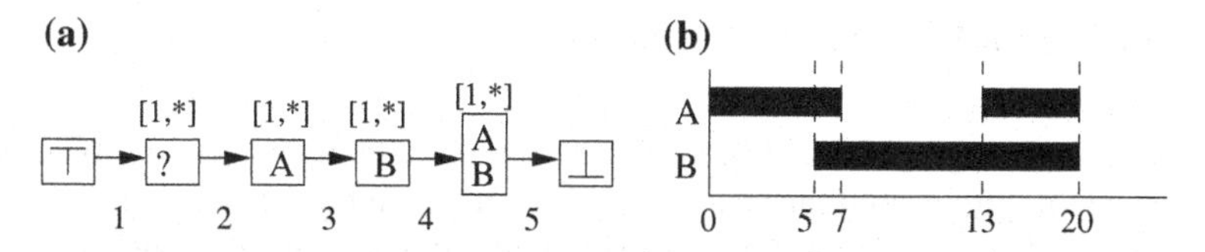

Fig. 3 Example for multiple mapping candidates on a given graph (**a**) and sequence (**b**)

Definition 5 (match, valid mapping). A valid mapping B for pattern graph $M = (V, E, \mathscr{C}_{val}, \mathscr{C}_{temp})$ and a sequence S of length n is a mapping with the following additional properties: with $V' = V \backslash \{\top, \bot\}$

$$\forall (v_1, v_2) \in E : B(v_1) = [a, b] \wedge B(v_2) = [c, d] \Rightarrow b + 1 = c \quad \text{(no gaps)} \tag{1}$$

$$\forall i : 1 \leq i \leq n : \exists v \in V' : i \in B(v) \text{ (each index is assigned at least once)} \tag{2}$$

$$\forall v \in V' : \mathscr{C}^{v}_{val}(S|_{B(v)}) = \text{true} \quad \text{(value-constraint holds)} \tag{3}$$

$$\forall v \in V' : \mathscr{C}^{v}_{temp}(S|_{B(v)}) = \text{true} \quad \text{(temporal constraint holds)} \tag{4}$$

Having defined the pattern graph in detail we will now give an example to illustrate the semantics of the pattern graph. Figure 3a shows an example of a pattern graph with one path, which is read as follows: The temporal constraint of a node is depicted above the node. A star represents an unlimited duration. The value-constraint of a node is shown inside the node. We have two kinds of value-constraints for attribute A: A means that the attribute is active ($D_A = \{1\}$) and $\neg A$ requires its absence ($D_A = \{0\}$). A node labelled '?' (*don't care*) is unconstrained. Please note that if the node states A the behaviour of the other attributes is unconstrained.

Figure 3b shows a sequence where the vertical axis reveals two attributes A and B, which hold over certain periods of time (black bars, time on horizontal axis). We now discuss whether these sequence match the pattern graph in Fig. 3a. As this is a simple graph, it contains only one path from source to sink. For this graph the sequence has to be divided into four contiguous parts, so that the first part satisfies the '*don't care*' constraint; during the second part the property A has to hold; the property B must hold in the third part and both A and B have to hold during the last part. All of these four parts require a duration of at least one time unit (but have no upper bound on the duration) except the A node with a minimum duration of 3. The sequence shown in Fig. 3a can be mapped to the graph, because we can clearly see that A is active until B begins and is active until the end and during the last part A becomes active again through to the end of the sequence. Actually the pattern graph has more than one valid mapping (discussed later in Sect. 4). A more complex and expressive pattern graph is found in Fig. 1, which describes a driving cycle derived from real data [10].

4 Matching and Learning Pattern Graphs

Matching a pattern graph to a sequence is essentially a combinatorial problem, an efficient matching algorithm can be found in [11]. Often multiple matches are possible and for each edge $e = (u, v) \in E$ the algorithm provides a set of valid edge positions $p(e)$, i.e., a set of positions t that satisfy all value-constraints of node u for $t' < t$ and all value-constraints of node v for $t' \geq t$. For the graph in Fig. 3a and the sequence in Fig. 3b we have a set of valid locations $p(e) = \{0\}$ for the edge e from node $\top$ to $\bot$, but for the edge e' from B to AB we have $p(e') = \{13, 14, \ldots, 19\}$. These edge positions are organized in so called *mapping candidates*, which map each edge of the graph to one contiguous interval of valid edge positions. So a *mapping candidate* C may be considered as a precursor of a *valid mapping* B in the following sense (by $(\cdot, v)$ we denote any edge leading to v):

$$\forall v \in V : B(v) = [s, e] \quad \Rightarrow \quad s \in C(\cdot, v) \wedge e \in C(v, \cdot)$$

Multiple mapping candidates exist, from which one or more valid mappings may be derived. Consider the graph in Fig. 3. The '?' node is valid during [0,20], the 'A' node is satisfied during [0,7] and [13,20], 'B' during [5,20] and finally 'AB' during [5,7] and [13,20]. Out of these sets of valid positions the matching algorithm derives two mapping candidates C_1 and C_2, assigning each edge its admissible positions. Three valid mappings B_1-B_3, obtained from C_1 and C_2, are shown below (many more are possible).

C_1 1:[0,0], 2:[1,4], 3:[5,7], 4:[13,19], 5:[20,20]

C_2 1:[0,0], 2:[13,15], 3:[16,18] 4:[17,19] 5:[20,20]

B_1 [0,1] - [1,4] - [4,13] - [13,20]
B_2 [0,2] - [2,7] - [7,16] - [16,20]
B_3 [0,13] - [13,17] - [17,18] - [18,20]

While all edge positions from the mapping candidates ensure that the value constraints hold, the positions must also fulfil the temporal constraint of the node. For instance, if the temporal constraint for the node labelled 'A B' was [5,*], the mapping B_2 and B_3 would no longer be valid (because the sequence assigned to this node has only a length of ≤ 4).

A two-phased learning algorithm for pattern graphs has been introduced in [10], where a general pattern, matching all instances of a class, is learned in the first phase. The second phase implements a beam-search, where in each iteration, the k-best pattern graphs are refined further by special refinement operators, which add new nodes or edges to the graph, modify temporal constraints or add value constraints to nodes in order to improve some measure of interestingness (we apply the J-measure [12]).

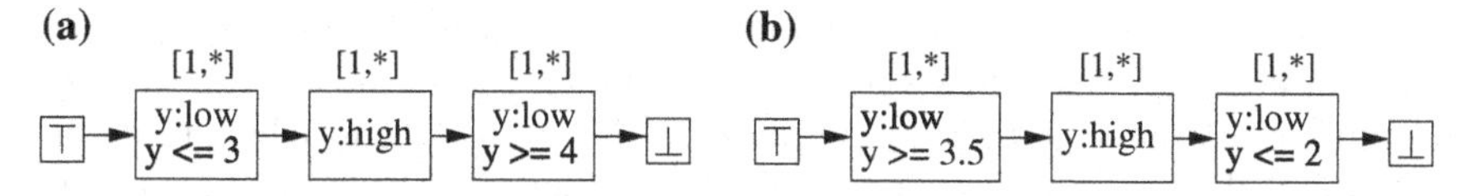

Fig. 4 Two pattern graphs with a new constraint to distinguish the positive between the negative sequences from Fig. 2

5 Extending Pattern Graphs to Time Series

In order to enable pattern graphs to deal with a numeric range $D_j \subseteq \mathbb{R}$ (cf. Def. 1) we introduce new value constraints called *series constraints*, where $C_j = \{x|x \leq \sigma\} \subseteq D_j$ or $C_j = \{x|x \geq \sigma\} \subseteq D_j$ for some threshold σ (cf. Def. 2). With these additional constraints we enable pattern graphs to overcome the obstacle of finding the best discretization to convert time series to labelled intervals, as every node may now use its own threshold σ instead of relying on the predefined intervals alone. This enables us to use different thresholds for different classes and it also allows us to use different series constraints for the same series within the same class (local constraints in different nodes). For example we can create the pattern graphs shown in Fig. 4a and b which are able to separate the sequences shown in Fig. 2a and b nevertheless they have the same interval representation as shown in Fig. 2.

To learn such constraints automatically from data, we have to extend the beam search operators. While it is quite easy to check whether a given assignment of subsequences to graph nodes represents a valid mapping, it is much more complicated to derive new conditions that are in some way 'optimal' for the set of all possible mappings. This is due to the fact that a graph node with constraint '*y:low*' may match an interval $[s, e]$ with this label in many different ways: any sub-interval $[s', e'] \subseteq [s, e]$ satisfies the node constraints; such a constraint still leaves many possibilities for valid mappings. In order to refine (or introduce new) node constraints, we have to consider *all* of these potential mappings at the same time, in order to calculate what would be the best additional constraint to distinguish good from bad cases. Enumeration of all possible mappings is not feasible because of their large number. Thus we operate directly on the *mapping candidates* of the matching algorithm.

Without loss of generality we will consider only the $x \leq \sigma$ constraint in the following. The new operator is instantiated for each individual graph node $v \in V$. As with all the other operators, it receives all mapping candidates (that already reflect the value constraints of that node), the temporal constraint and the data sequence. The objective is to derive a threshold σ on one (numeric) variable x of the sequence, such that the additional node constraint $x \leq \sigma$ improves the discriminative power of the pattern. Expressed more formally: if P is a pattern graph, let $m_P(s) = 1$ denote that P has a valid mapping to $s \in S$ (0, else). Let G' denote the resulting pattern if P is extended by the constraint $x \leq \sigma$ in node v. Then a confusion matrix from $m_{G'}(S) \in \{0, 1\}$ and a class $k/\neg k$ is created for G' to evaluate its utility.

The naive approach to find the best refinement is to extend P with every possible series constraint for x and then match all of the resulting graphs to all sequences. It is sufficient to examine only those thresholds σ that change the matching result of some $s \in S$, that is, for σ we obtain $m_{G'}(s) = 1$ but for $\sigma - \varepsilon$ we have $m_{G'}(s) = 0$, because it is only at these thresholds that the confusion matrices of the rule change. Thus, we need to determine only as many confusion matrices as we have sequences. How do we find the threshold σ for a given sequence s? If a subsequence, mapped to node v, shall satisfy the constraint, we have to choose σ as the maximum of all x. However we do not know this subsequence in advance, but have to consider all possible subsequences that may be obtained from the mapping candidates. To let all subsequences satisfy the constraint, we have to pick the smallest of all maximum values of all possible subsequences. If this value were reduced only slightly ($-\epsilon$), there would be at least one subsequence for node v that would not match anymore with the result that no valid mapping exists anymore. Lemma 1 shows that it is sufficient to inspect only the shortest possible subsequences rather than all possible subsequences.

Lemma 1. *By* $\max S$ *we denote the maximum of the* x*-values in a (sub) sequence* S*. Let* $\mathcal{Q}$ *be the set of all subsequences (that may occur in a valid mapping to node* v*) and* $\mathcal{Q}'$ *the set of shortest subsequences.*[1] *Then* $\min \max_{S \in \mathcal{Q}} S = \min \max_{S \in \mathcal{Q}'} S$ *holds.*

Proof. Let $S \in \mathcal{Q}$. Without loss of generality let us assume that $S \notin \mathcal{Q}'$. Thus, S is not among the shortest subsequences and therefore we find a $T \in \mathcal{Q}' \subseteq \mathcal{Q}$ such that T is a subsequence of S. All values of T are contained in S, but S contains additional entries, therefore we have $s := \max S \geq \max T =: t$. Thus, we know that $\min \max_{S \in \mathcal{Q}'} S \leq t \leq s$. This means, that $s = \max S$ can be ignored safely in the calculation of $\min \max_{S \in \mathcal{Q}} S$.

Algorithm 1 findBestSeriesSmallerThanThresholdConstraint

Require: S: all sequences
Require: v: node to refine
Require: v_{min}: minimal length of the node
Ensure: best refinement

1: **for** $s \in S$ **do**
2: find mapping candidates C_M
3: $values \leftarrow \bigcup_{c \in C_m} getMinMaxValueForMappingCandidate(s, c, v_{min}, v)$
4: $\sigma \leftarrow \min_{values}$
5: collect all thresholds σ in set Σ
6: **end for**
7: sort all thresholds in Σ in ascending order
8: create and evaluate confusion matrices for all found thresholds.
9: add the threshold σ with highest measure to the node v as the series constraint
10: **return** refined pattern graph.

---

[1] shortest in the following sense: $\forall s' \in Q' : \neg\exists s \in Q : s \subset s'$.

Algorithm 2 getMinMaxValueForMappingCandidate

Require: s: sequence
Require: c: mapping candidate
Require: v_{min}: minimal length of the node
Require: v: node to refine
Ensure: smallest maximum value of the subsequences contained in c

1: $pl \leftarrow$ latest start position $\in c((\cdot, v))$
2: $pe \leftarrow$ earliest end position $\in c((v, \cdot))$
3: $S_m \leftarrow \emptyset$
4: $begin \leftarrow pe - v_{min}$.
5: **if** $begin > pl$ **then**
6: **return** maximum value contained in $s|_{[pl,pe]}$
7: **else**
8: **while** $begin \leq pl$ **do**
9: $S_m \leftarrow S_m \cup s|_{[begin,begin+v_{min}]}$
10: $begin \leftarrow begin + 1$
11: **end while**
12: **end if**
13: **return** smallest value out of the maximum values from the subsequences $\in S_m$

The outline of the refinement operator to find the best $x \leq \sigma$ is shown in Algorithm 1. In the lines 1–6 the algorithm computes the threshold as defined by the Lemma 1. It utilises Algorithm 2 to find the maximum value of all shortest subsequences for a given *mapping candidate*. v_{min} denotes either the minimal temporal constraint of node v, or is a greater value if the graph structure requires longer sequences in order to satisfy the temporal constraints of other nodes (for example due to parallel paths).

We find the best refinement by evaluating all possible confusion matrices and picking the one with the highest interestingness measure. In order to avoid overfitting, the series constraint with the best measure will be relaxed similarly to the binary split operator in decision tree learning: We search for the next greater value and use the mean of both. This doesn't change the prediction of the new pattern on the training set, but is less restrictive for new instances. The refinement is completed in line 9 by adding the series constraint with the computed value to the node.

From Algorithm 2 we can see that the shortest subsequences for a single mapping candidate are always subsequences with the same length, shifted by one time unit. This allows us to use a priority queue, in order to extract the constraint value efficiently.

Overfitting. An important step to avoid overfitting is to prevent nodes with the minimum temporal constraints 1 to be refined with a series constraint. This would allow the pattern graph to focus on one single time point and would thus stimulate overfitting. We therefore enforce a minimal length of a node to be refined with the new series constraint. If a node has a minimal duration of 1 during refinement, the minimal length will be set to this minimal length (lower bound of v_{min}). This has the consequence for step 1 that only subsequences with the minimal length, which are mappable to the node have do be analysed. Additionally we have added a likelihood ratio [6] test after every refinement and keep only those graphs with statistically significant improvements to avoid overfitting (which is a problem common to all rule learners).

6 Experimental Evaluation

The experimental evaluation is divided into two different experimental setups. In the first experiment we show that the series constraints help to overcome the preprocessing problem discussed earlier. Whereas in the second experiment we show that the new approach is able to perform better, even if a good discretization is applied beforehand.

6.1 Robustness Against Preprocessing Errors

To show that series constraints could help dealing with sub-optimal preprocessing of the data we took nine data sets from the UCR time series repository.[2] This repository already supplies training and tests partitions for each data set in a common format. All of these data sets consist of a raw univariate time series which requires some preprocessing: a moving average smoothing was applied to the series and we also extracted an additional slope series. Thereby we artificially converted the data into a multivariate time series (original and slope time series). In the second step this preprocessed time series had to be converted into a labelled interval series by applying 3-quantile discretization. To achieve different discretization the quantile boundaries are selected randomly for each iteration. We are aware of the fact that for the given data sets algorithms exists that perform better, but most of these approaches could not deal with multivariate data. These approaches often utilize 1-nearest neighbor classification (1NN) with Euclidean distance or dynamic time warping, whereas the pattern graphs rely on simple elements only (like intervals with a value $\geq \sigma$) and thus highlight structural differences. These simple elements keep pattern graphs interpretable even in the case of complex multivariate data (see Sect. 6.2). Therefore 1NN-approaches are not the real competitors. To show the improvement of the learned graphs with series constraints and allow future comparison for follow up work we decided to use these data sets. Table 1 displays the results obtained by applying the beam with and without the series constraints for 30 iterations per class and dataset. The parameter for the minimum sequence length was set to 10, but the results obtained by using additional operators with 5, 15 and 20 as minimal length led to nearly the same results. The first row names the dataset, the second row displays the class (for which the pattern graphs was learned for). The left side represents the search without

[2] Keogh, E., Zhu, Q., Hu, B., Hao. Y., Xi, X., Wei, L. & Ratanamahatana, C. A. (2011). The UCR Time Series Classification/Clustering Homepage: www.cs.ucr.edu/~eamonn/time_series_data/.

Table 1 Results on the data when the thresholds vary

Data set	Gun Point				Waver				Yoga				Coffee			
Class	1		2		-1		1		1		2		1		2	
Accuracy	67,5	79	61,3	75,1	16,4	91	58,6	77,5	48,2	56,7	51,4	52,7	55,8	60,6	59,5	71
Std. dev.	12,3	5,7	14,2	4,8	21,3	2,7	10,6	0	4,1	2,7	4,5	3,2	13,1	1,1	10,3	1,8

Data set	Lightning 2				Synthetic Control											
Class	-1		1		1		2		3		4		5		6	
Accuracy	53,5	54,1	51,6	56,1	62,5	89,3	88	90,1	93,3	87,7	92,2	93,2	91,2	92,2	87,4	94,1
Std. dev.	8,1	0	8,2	5,3	34	2,1	19,9	5,6	3	5,6	2,9	3,6	3,1	3,3	15	1,4

Data set	Two Patterns								CBF					
Class	1		2		3		4		1		2		3	
Accuracy	34,8	96,8	27,79	91,1	27,4	93,2	29,8	97,5	76,3	95,9	76,1	95,8	67	77,2
Std. dev.	20,6	0	13,5	2,9	13,6	1,7	18,1	0,4	13,5	0	11,4	0,2	15,1	8,5

Data set	Fish													
Class	1		2		3		4		5		6		7	
Accuracy	33,9	85,7	37,1	89,7	42,5	91,8	33,7	24,9	71,4	91,4	28,2	86,3	64,7	90,4
Std. dev.	31,1	1,1	34,8	0,1	36,1	2,1	28,6	22,1	34,4	3,3	26,3	1,2	30,2	1,2

series constraint the right side represents the search with series constraints. The third row indicates accuracy and the fourth contains standard deviation.

For most of the classes the learned pattern graphs with the series constraints are able to perform significantly better in terms of accuracy. We can also see that in most cases standard deviation has decreased, showing that the suboptimal discretization has been compensated. For four classes only small improvements in terms of accuracy and standard deviation occurred. For two classes the performance with series constraints deteriorates: for the class # 3 from the Synthetic Control data set, standard deviation increases while accuracy drops. In these cases the series constraint led the beam search into a local maximum (the interestingness measure is also lower on the training set). This leaves room for further improvements of this approach, because without the limitations of the beam we should obtain at least the same performance as before.

6.2 Improved Accuracy on Data with Good Discretization

We obtained a data set from a German company producing, amongst others, power tools. This data set consists of 8 different classes, where each class describes a different screwing process: screwing in and out using different gears of the power tool and different screws. Each of these 564 instances are described by five time series, e.g. voltage/current at the battery or engine etc. In a first step we manually discretized each of this series to labelled intervals in an interactive manner until we

Table 2 Results on the power tool data set

Power tools																
Class	1		2		3		4		5		6		7		8	
Approach	a	b	a	b	a	b	a	b	a	b	a	b	a	b	a	b
Accuracy	98.7	98.9	98.1	99.5	98.8	98.9	98.7	98.2	97.4	99.1	96.8	97.3	98.4	98.4	95.9	99.0
Std. Dev	0.7	0.9	0.9	0.6	0.9	0.9	0.9	1.3	1.4	1.1	1.4	1.4	1.1	1.6	2.3	2.7
Avg Imp.	0.3		1.4		0.01		−0.5		1.7		0.6		0		3.1	

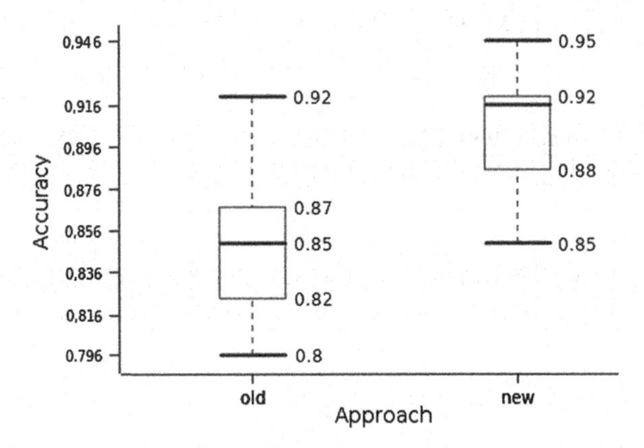

Fig. 5 Box plots showing the results of the complete classifier on the power tool data set

were satisfied with the results. Therefore we may safely assume that discretization is good and it would be hard to achieve better discretization. We applied the beam search (minimal length: 10) to this data set 30 times with and without the series constraint. In each iteration we randomly partitioned the data set into 80 % training and 20 % test. As a result of the good discretization we assume that the accuracy results would be nearly the same, but may be improved by applying different thresholds to one class or in between classes.

Table 2 shows the result. If we sum up the average improvements for all the individual classes, the new graphs performed 6.6 % better. Most of the pattern graphs learned with series constraints perform better (up to 3 %) and are only slightly worse for one class #4 (−0.5 %). However improvements 'per-class' are not significant. So far, each rule has predicted just one class. Next we combine the individual rules to a single, multi-class classifier: we only classify an instance if and only if one pattern graph has a valid mapping on the instance. In case none or more than one pattern graph matches we predict "unknown". The box plot in Fig. 5 and Table 3 shows the result of this classifier, where we use the same pattern graphs as in Table 2, thus the results origin from the same 30 runs with random training and test sets.

We can see that by using series constraints the overall accuracy has improved by an average of 5.8 %. It is also interesting to note that in all 30 iterations, the lowest accuracy of the new approach is at least as good as the average result without the series

Table 3 Mean accuracy and standard deviation of the complete classifier for the powertool dataset

Approach	Old	New
Accuracy	84.7	90.5
Std. Dev	3.2	2.8

constraints. Additionally the mean accuracy using the new approach is nearly equal to the best result obtained without the series constraints ($-1.5\,\%$). By inspecting the learned pattern graphs, we observed that one or two additional series constraints were derived per class. Depending on the class, the thresholds were slightly different, which explains the improved accuracy as the number of false positives was able to be reduced, without increasing the number of false negatives.

7 Conclusion

In this paper we have shown, that the results of pattern learning algorithms for labelled sequences rely heavily on discretization of the source time series. The quality of the learned patterns varies considerably when the discretization changes. In order to overcome the problem of finding good discretization, which is time consuming and not always possible, we introduced an algorithm capable of mining labelled intervals together with the corresponding time series. The first experiment has shown that, in comparison to the approach without the series constraint, the quality of the patterns is higher resulting in and allowing for a more robust approach compared to a priori discretization. Furthermore we have shown that even in situations, where discretization already performs well, the quality of the patterns may be increased, because different levels of discretization for different classes and even different thresholds within one class may be utilized.

For future work the synergies of labelled intervals and numeric time series may be improved further as, so far, we have only used simple constraints ($\leq$, $\geq$). But it is possible to use more sophisticated constraints on mean values or standard deviation. This kind of constraint may provide further insight into the patterns.

Acknowledgments We would like to thank Stefan Mock from the Robert Bosch GmbH for kindly providing the data.

References

1. Basile, T.M.A., Mauro, N.D., Ferilli, S., Esposito, F.: Relational temporal data mining for wireless sensor networks (2009).
2. Batal, I., Valizadegan, H., Cooper, G.F., Hauskrecht, M.: A pattern mining approach for classifying multivariate temporal data. In: Bioinformatics and Biomedicine (BIBM), 2011 IEEE International Conference on, pp. 358–365. IEEE (2011).

3. Berlingerio, M., Pinelli, F., Nanni, M., Giannotti, F.: Temporal mining for interactive workflow data analysis. In: Proceedings of the 15th ACM SIGKDD international conference on Knowledge discovery and data mining, KDD '09, pp. 109–118. ACM, New York, NY, USA (2009).
4. Chen, Y.C., Jiang, J.C., Peng, W.C., Lee, S.Y.: An efficient algorithm for mining time interval-based patterns in large database. In: Proc. Int. Conf. Inf. Knowl. Mngmt., pp. 49–58. ACM (2010).
5. Höppner, F.: Discovery of temporal patterns - learning rules about the qualitative behaviour of time series. 2168, pp. 192–203. Freiburg, Germany (2001).
6. Kalbfleisch, J.G.: Probability and statistical inference: probability, vol. 2. Springer-Verlag (1985).
7. Mörchen, F.: Unsupervised pattern mining from symbolic temporal data. SIGKDD Explor. Newsl.**9**(1), 41–55 (2007).
8. Mörchen, F., Ultsch, A.: Optimizing time series discretization for knowledge discovery. In: Proc. Int. Conf. Knowl. Disc. and Data Mining, pp. 660–665. ACM (2005).
9. Mörchen, F., Ultsch, A.: Efficient mining of understandable patterns from multivariate interval time series. pp. 181–215. Springer (2007).
10. Peter, S., Höppner, F., Berthold, M.R.: Learning pattern graphs for multivariate temporal pattern retrieval. In: Proc Int Symp Intel. Data, Analysis (2012).
11. Peter, S., Höppner, F., Berthold, M.R.: Pattern graphs: A knowledge-based tool for multivariate temporal pattern retrieval. In: 6th IEEE International Conference on Intelligent Systems (IS' 12) (2012).
12. Smyth, P., Goodman, R.M.: An information theoretic approach to rule induction from databases. IEEE Trans. Knowledge Discovery and Engineering **4**(4), 301–316 (1992).

Knowledge Discovery and Data Mining I

Vertex Unique Labelled Subgraph Mining

Wen Yu, Frans Coenen, Michele Zito and Subhieh El Salhi

Abstract With the successful development of efficient algorithms for Frequent Subgraph Mining (FSM), this paper extends the scope of subgraph mining by proposing Vertex Unique labelled Subgraph Mining (VULSM). VULSM has a focus on the local properties of a graph and does not require external parameters such as the support threshold used in frequent pattern mining. There are many applications where the mining of VULS is significant, the application considered in this paper is error prediction with respect to sheet metal forming. More specifically this paper presents a formalism for VULSM and an algorithm, the Right-most Extension VULS Mining (REVULSM) algorithm, which identifies all VULS in a given graph. The performance of REVULSM is evaluated using a real world sheet metal forming application. The experimental results demonstrate that all VULS (Vertex Unique Labelled Subgraphs) can be effectively identified.

1 Introduction

A novel research theme in the context of graph mining [7, 8, 15, 16], Vertex Unique labelled Subgraph Mining (VULSM), is proposed in this paper. Given a particular sub-graph g in a single input graph G; this subgraph will have a specific structure, and edge and vertex labelling associated with it. If we consider only the structure

W. Yu (✉) · F. Coenen · M. Zito · S. E. Salhi
Department of Computer Science, The University of Liverpool Ashton Building, Ashton Street, L69 3BX Liverpool, UK
e-mail: yuwen@liverpool.ac.uk

F. Coenen
e-mail: Coenen@liverpool.ac.uk

M. Zito
e-mail: Zito@liverpool.ac.uk

S. E. Salhi
e-mail: hsselsal@liverpool.ac.uk

M. Bramer and M. Petridis (eds.), *Research and Development in Intelligent Systems XXX*, DOI: 10.1007/978-3-319-02621-3_2,

and edge labelling there may be a number of different compatible vertex labellings with respect to G. A Vertex Unique Labelled Subgraph (VULS) is a subgraph with a specific structure and edge labelling that has a unique vertex labelling associated with it. This paper proposes the Right-most Extension Vertex Unique Labelled Subgraph Mining Algorithm (REVULSM) to identify all VULS. REVULSM generates subgraphs (potential VULS candidates) using Right Most Extension [3], in a DFS manner, as first proposed in the context of gSpan [14]; and then identifies all VULS using a level-wise approach (first proposed by Agrawal and Srikant in the context of frequent item set mining [1, 2, 9]). VULSM is applicable to various types of graph; however, in this paper we focus on undirected graphs.

VULSM has relevance with respect to a number of domains. The application domain used to illustrate the work described in this paper is error prediction in sheet metal forming. More specifically error prediction in Asymmetric Incremental Sheet Forming (AISF) [4, 6, 10, 12, 13]. In this scenario the piece to be manufactured is represented as a grid, each grid centre point is defined by a Euclidean (X-Y-Z) coordinate scheme. The grid can then be conceptualised as a graph (lattice) such that each vertex represents a grid point. Each vertex (except at the edges and corners) can then be connected to its four neighbours by a sequence of edges, which in turn can be labelled with "slope" values. An issue with sheet metal forming processes, such as AISF, is that distortions are introduced as a result of the application of the process. These distortions are non-uniform across the "shape", but tend to be related to local geometries. The proposed graph representation captures such geometries in terms of sub-graphs, particular sub-graphs are associated with particular local geometries (and by extension distortion/error patterns). Given before and after shapes we can create a training set by deriving the error associated with each vertex in the grid. This training data, in turn, can then be used to train a predictor or classifier of some sort. There are various ways that such a classifier can be generated; but one mechanism is to apply VULSM, as proposed in this paper, to identify sub-graphs that have unique error patterns associated with them that can then be used for error prediction purposes (some form of mitigating error correction can then be formulated).

A simple example grid and corresponding graph are given in Fig. 1. The grid (lefthand side of Fig. 1) comprises six grid squares. Each grid centre is defined by a X-Y-Z coordinate tuple. Each grid centre point is associated with a vertex within the graph (right-hand side of Fig. 1). The edges, as noted above, are labelled with "slope" values, the difference in the Z coordinate values associated with the two end vertices. Each vertex will be labelled with an error values (e_1 to e_3 in the figure) describing

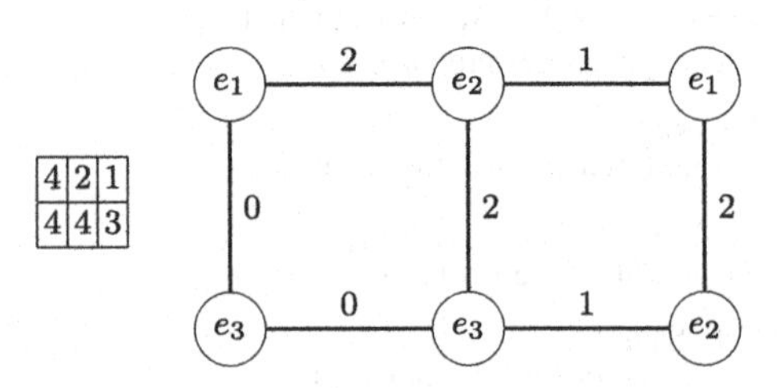

Fig. 1 Grid representation (*left*) with corresponding graph/lattice (*right*) featuring "slope" labels on edges

the expected distortion at that vertex as obtained from a "training set" (derived from "before and after" grid data). Identified VULS will describe local geometries each with a particular associated error pattern. This knowledge can then be used to predict errors in "unseen" grids so that some form of mitigating error correction can be applied.

The rest of this paper is organised as follows. In Sect. 2, we define the basic concepts of VULS together with an illustrative example. The REVULSM algorithm is then described in detail in Sect. 3. An experimental analysis of the approach is presented in Sect. 4 and Sect. 5 summarises the work and the main findings, and presents some conclusions.

2 The Problem Formulation

This section presents a formal definition of the concept of a VULS. Assume a connected labelled graph G comprised of a set of n vertices V, such that $V = \{v_1, v_2, \ldots, v_n\}$; and a set of m edges E, such that $E = \{e_1, e_2, \ldots, e_m\}$. The vertices are labelled according to a set of p vertex labels $L_V = \{l_{v_1}, l_{v_2}, \ldots, l_{v_p}\}$. The edges are labelled according to a set of q edge labels $L_E = \{l_{e_1}, l_{e_2}, \ldots, l_{e_q}\}$. A graph G can thus be conceptualised as comprising k one-edge subgraphs: $G = \{P_1, P_2, \ldots, P_k\}$, where P_i is a pair of vertices linked by an edge, thus $P_i = \langle v_a, v_b \rangle$ (where $v_a, v_b \in V$). The size of a graph G ($|G|$) can thus be defined in terms of its one edge sub-graphs, we refer to 1-edge subgraphs, 2-edge subgraphs and so on up to k-edge subgraphs. For undirected graphs, the edge $\langle v_a, v_b \rangle$ is equivalent to $\langle v_b, v_a \rangle$ (in this paper we assume undirected subgraphs). We use the notation $P_i.v_a$ and $P_i.v_b$ to indicate the vertices v_a and v_b associated with a particular vertex pair P_i, and the notation $P_i.v_a.label$ and $P_i.v_b.label$ to indicate the labels associated with $P_i.v_a$ and $P_i.v_b$ respectively. We indicate the sets of labels which might be associated with $P_i.v_a$ and $P_i.v_b$ using the notation $L_{P_i.v_a}$ and $L_{P_i.v_b}$ ($L_{P_i.v_a}, L_{P_i.v_b} \in L_V$). We indicate the edge label associated with P_i using the notaion $P_i.label$ ($P_i.label \in L_E$). We can use this notation with respect to any subgraph G_{sub} of G ($G_{sub} \subseteq G$).

For training purposes the graphs of interest are required to be labelled. However, we can also conceive of edge only labelled graphs and subgraphs. Given some edge only labelled subgraph ($G_{subedgelab}$) of some fully labelled graph G ($G_{subedgelab} \subseteq G$) comprised of k edges, there may be many different vertex labelings that can be associated with such a subgraph according to the nature of G. We thus define a function, $getVertexLabels$, that returns the potential list of labels S that can be assigned to the vertices in $G_{subedgelab}$ according to G:

$$getVertexLabels(G_{subedgelab}) \rightarrow S$$

where $G_{subedgelab} = \{P_1, P_2, \ldots, P_k\}$ and $S = [[L_{P_1.v_a}, L_{P_1.v_b}], [L_{P_2.v_a}, L_{P_2.v_b}], \ldots, [L_{P_k.v_a}, L_{P_k.v_b}]]$ (recall that $L_{P_i.v_a}$ and $L_{P_i.v_b}$ are the sets of potential vertex labels for vertex v_a and v_b associated with a one-edge subgraph P_i). Thus each element in S comprises two sub-sets of labels associated respectively with the start and end vertex for each edge in $G_{subedgelab}$; there is a one to one correspondence between each element (pair of label sets) in S with each element in $G_{subedgelab}$, hence they are both of the same size k (recall that k is the number of edges). We also assume that some canonical labelling is adopted.

According to the above, the formal definition of the concept of a VULS is as follows. Given: (i) a k-edge edge labelled subgraph $G_{subedgelab} = \{P_1, P_2, \ldots, P_k\}$ ($G_{subedgelab} \subseteq G$), (ii) a list of labels that may be associated with the vertices in $G_{subedgelab}$, $S = [[L_{P_1.v_a}, L_{P_1.v_b}], [L_{P_2.v_a}, L_{P_2.v_b}], \ldots, [L_{P_k.v_a}, L_{P_k.v_b}]]$, and (iii) the proviso that $G_{subedgelab}$ is connected. If $\forall [L_i, L_j] \in S, |L_i| = 1, |L_j| = 1$ then $G_{subedgelab}$ is a k-edge VULS with respect to G.

So as to provide for a full and complete comprehension of the concept of VULS an example lattice is presented in Fig. 2. The VULS that exist in this lattice are itemized in Figs. 3, 4 and 5. If we consider one-edge subgraphs first, there are two possibilities: (i) graphs featuring edge x, and (ii) graphs featuring edge y. The list of possible vertices S associated with the first, obtained using the $getVertexLabels$ function, is $[[\{a\}, \{a\}]]$, while the list associated with the second is $[[\{a, b\}, \{b\}]]$ (this can be verified by inspection of Fig. 2). Considering edge x first, $\forall [L_i, L_j] \in S, |L_i| = 1$ and $|L_j| = 1$, so this is a VULS; however, considering edge y, $\forall [L_i, L_j] \in S$, $|L_i| \neq 1$ and $|L_j| = 1$ hence this is not a VULS. We now consider the two edge subgraphs by extending the one edge subgraphs. We can not enumerate all two edge subgraphs here due to space limitations but the two edge VULS are shown in Fig. 4. Taking the first VULS in Fig. 4, $\{P_1, P_2\}$, as an example, here $P_1.v_a = a$, $P_1.v_b = a$, $P_2.v_a = a$ and $P_2.v_b = a$, furthermore the edge labels associated with P_1 and P_2 are $P_1.label = x$ and $P_2.label = x$ respectively. In this case $S = [[a, a], [a, a]]$ thus $\forall [L_i, L_j] \in S, |L_i| = 1, |L_j| = 1$ therefore this is a two-edge VULS with respect to G.

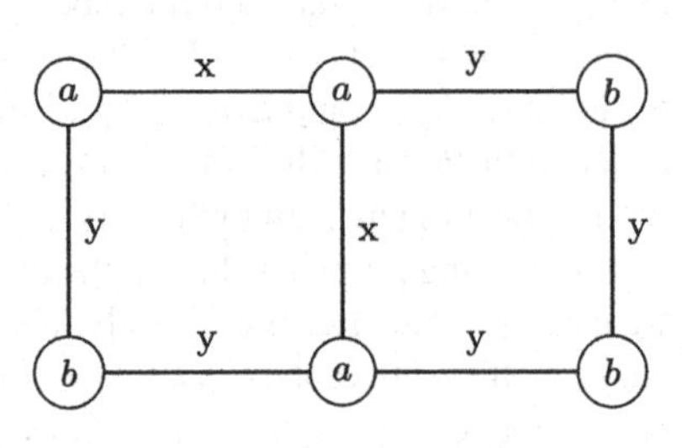

Fig. 2 Undirected example lattice

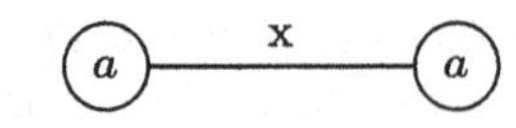

Fig. 3 One edge VULS generated from lattice in Fig. 2

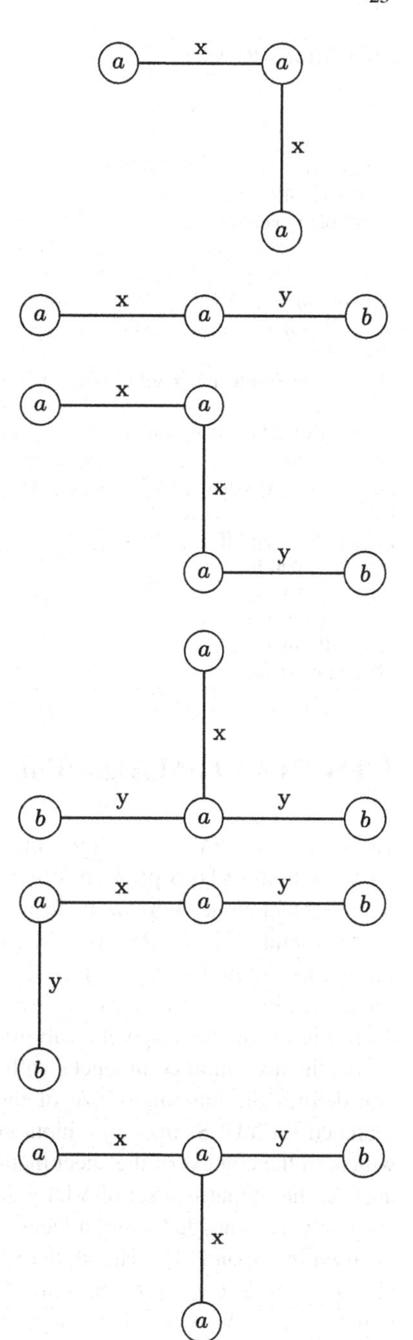

Fig. 4 Two edge VULS generated from lattice in Fig. 2

Fig. 5 Three edge VULS generated from lattice in Fig. 2

Algorithm 1 REVULSM

```
1: Input:
2: G_input = Input graph
3: Max=Max subgraph size
4: Output:
5: R = Set of VULS
6: Global variables:
7: G = set of subgraphs (VULS candidates) in G_input

8: procedure REVULSM(G_input, Max)
9:    R = ∅
10:   G = ∅
11:   G = Subgraph_Mining(G_input) (Algorithm 2)
12:   k=1
13:   while (k < Max) do
14:      for all G_sub ∈ G_k (where G_k is the set of k-edge subgraphs in G) do
15:         if IdentifyVULS(G_sub, G_k) == true (Algorithm 3) then
16:            R = R ∪ G_sub
17:         end if
18:      end for
19:      k++
20:   end while
21:   Return R
22: end procedure
```

3 The REVULSM Algorithm

The proposed REVULSM algorithm is defined in this section. The algorithm is founded on the VULS properties presented above and makes use of a graph representation technique "borrowed" from gSpan.

The pseudo code for REVULSM is presented in Algorithms 1, 2 and 3. Algorithm 1 presents the high level control structure, while Algorithm 2 presents the detail for generating all subgraphs (VULS candidates), and Algorithm 3 the detail for determining whether a specific sub-graph is a VULS or not. Considering Algorithm 1 first, the algorithm commences with an input graph G_{input} and a parameter Max that defines the maximum size of the VULS. If we do not limit the size of the searched-for VULSs the entire input graph may ultimately be identified as a VULS which, in the context of the sheet metal forming target application, will not be very useful. The output is a set of VULS R (the set R may include overlaps). Note that all graphs are encoded using a Depth First Search (DFS) lexicographical ordering (as used in gSpan [14]). The global variable G (line 7 in Algorithm 1) is the set of all subgraphs in G_{input}. At the start of the REVULSM procedure, the sets G and R will be empty. We proceed in a depth first manner to generate all subgraphs (VULS candidates) G by calling algorithm 2 (line 11). Then we identify VULS from all subgraphs G starting from one-edge subgraphs ($k = 1$), then two edge sub-graphs ($k = 2$), and so on. We continue in this manner until $k = Max$ (line 13–20). On each iteration algorithm 3 is called (line 15) to determine whether G_{sub} is a VULS

Algorithm 2 Subgraph_Mining

```
1: Input:
2: G_input = Input graph
3: Output:
4: G = set of subgraphs in G_input
5: Global variables:
6: G_temp=set of subgraphs generated so far

7: procedure Subgraph_Mining(G_input)
8:     G = ∅
9:     G_temp = ∅
10:    G_1=the set of one-edge subgraphs in G_input
11:    sort G_1 in DFS lexicographic order
12:    for each edge e ∈ G_1 do
13:        G_temp = Subgraph(e,1,Max)
14:        G = G ∪ G_temp
15:        G_input = (G_input − e) (remove e from G_input)
16:    end for
17:    Return G
18: end procedure

19: procedure Subgraph(e, size, Max)
20:    if size > Max then
21:        return ∅
22:    end if
23:    generate all e's potential extension subgraphs c in G_input with one edge growth by right
         most extension
24:    for each c do
25:        if c is minimal DFSCode then
26:            G_temp = G_temp ∪ c
27:            Subgraph(G_temp,size+1,Max)
28:        end if
29:    end for
30:    Return G_temp
31: end procedure
```

or not with respect to the k-edge subgraphs G_k. If it is VULS, it will be added to the set R.

Algorithm 2 comprises two procedures. The first, $Subgraph\_Mining(G_{input})$, is similar to that found in gSpan. We are iteratively finding all subgraphs, up to a size of Max. We commence (line 10–11) by sorting all the one-edge subgraphs, contained in input graph G_{input}, into DFS lexicographic order and storing them in G_1. Then (lines 12–16), for each one edge subgraph e in G_1 we call the $Subgraph$ procedure (line 13), which finds all super graphs for each one edge graph e up to size Max in a DFS manner, and stores the result in G_{temp}; which is then added to G (line 14). Finally, we remove e from G_{input} (line 15) to avoid generating again any duplicate subgraphs containing e.

The $Subgraph(e, size, Max)$ procedure generates all the super graphs of the given one edge subgraph e by growing e by adding edges using the right most

Algorithm 3 IdentifyVULS

```
1: Input:
2: g = a single k-edge subgraph (potential VULS)
3: G_k = a set of k-edge subgraphs to be compared with g
4: Output:
5: true if g is a VULS, false otherwise

6: procedure IdentifyVULS(g, G_k)
7:     isVULS = true
8:     S = the list of potential vertex labels that may be assigned to g
9:     for all [L_i, L_j] ∈ S do
10:        if either |L_i| ≠ 1 or |L_j| ≠ 1 then
11:            isVULS = false
12:            break
13:        end if
14:    end for
15:    return isVULS
16: end procedure
```

extension principle. For each potential subgraph c, if c is described by a minimal DFSCode (line 25) the process is repeat (in a DFS style) so as to generate all the super graphs of e (line 27). The process continues in this recursive manner until the number of edges in the super graphs to be generated ($size$) is greater than Max, or no more graphs can be generated.

Algorithm 3 presents the pseudo code for identifying whether a given sub-graph g is a VULS or not with respect to the current set of k-edge sub-graphs G_k from which g has been removed. The algorithm returns $true$ if g is a VULS and $false$ otherwise. The process commences (line 8) by generating the potential list of vertex labels S that can be matched to g according to the content of G_k (see previous section for detail). The list S is then processed and tested. If there exists a vertex pair whose possible labelling is not unique (has more than one possible labelling that can be associated with it) g is not a VULS and the procedure returns $false$, otherwise g is a VULS and the procedure returns $true$.

4 Experiments and Performance Study

This section describes the performance study that was conducted to analyse the generation and application of the concept of VULS. The reported experiments were all applied to a real application of sheet metal forming, more specifically the application of AISF [5, 11] to the fabrication of flat-topped pyramid shapes manufactured out of sheet steel. This shape was chosen as it is frequently used as a benchmark shape for conducting experiments in the context of AISF (although not necessarily with respect to error prediction). Nine graphs were generated from this data using three different grid sizes and different numbers of edge and vertex labels; in addition a range of values were used for the Max parameter. The rest of this sub-section is organised as follows. The performance measures used with respect to the evaluation are itemised in Sect. 4.1, more detail concerning the data sets used for the evaluation is given in Sect. 4.2, and the obtained results are presented and discussed in Sect. 4.3.

4.1 Experimental Performance Measurement

Four performance measures were used to analyse the effectiveness of the proposed REVULSM: (i) run time (seconds), (ii) number of VULS identified, (iii) discovery rate and (iv) coverage rate. The last two merit some further explanation. The discovery rate is the ratio of VULS discovered with respect to the total number of subgraphs of size less than Max (Eq. 1). The coverage rate is the ratio of the number of vertices covered by the detected VULS compared to the total number of vertices in the input graph (Eq. 2); with respect to the sheet steel forming example application high coverage rates are desirable.

$$discovery\ rate\ (\%) = \frac{number\ of\ VULS}{number\ of\ subgraphs} \tag{1}$$

$$coverage\ rate\ (\%) = \frac{number\ of\ vertices\ covered\ by\ VULS}{number\ of\ vertices\ in\ input\ graph} \tag{2}$$

4.2 Data Sets

The data sets used for the evaluation consisted of before and after "coordinate clouds"; the first generated by a CAD system, the second using an optical measuring technique. These were transformed into grid representations, referenced using a X-Y-Z coordinate system, such that the before grid could be correlated with the after grid and error measurements obtained. A fragment of the before grid data, with associated error values (mm), is presented in Table 1. The before grid data was then translated into a graph such that each grid square was represented by a vertex linked to each of its neighbouring squares by an edge. Each vertex was labelled with an error value while the edges were labelled according to the difference in Z of the two end vertices (the "slope" connecting them). Furthermore, the vertex and edge labels were discretised so that they were represented by nominal values (otherwise every edge pair was likely to be unique). This was then the input into the REVULSM algorithm.

As noted above, from the raw data, different sized grid representations, and consequently graph representations, may be generated. For experimental purposes three grid formats were used 6×6, 10×10 and 21×21. We can also assign different numbers of edge labels to the vertices and edges, for the evaluation reported here values of two and three were used in three different combinations. In total nine different

Table 1 Format of raw input data

x	y	z	Error
0.000	0.000	0.000	0.118
1.000	0.000	0.000	0.469
2.000	0.000	0.000	0.469
3.000	0.000	0.000	0.472
0.000	1.000	0.000	0.471
1.000	1.000	−1.402	0.088
2.000	1.000	−4.502	1.308
3.000	1.000	−4.676	1.907
...	...	...	...

Table 2 Summary of AISF graph sets

Graph set	# Vertices	# Edge labels	# Vertex labels	Graph set	# Vertices	# Edge labels	# Vertex labels
AISF1	36	3	2	AISF6	100	2	3
AISF2	36	2	2	AISF7	441	3	2
AISF3	36	2	3	AISF8	441	2	2
AISF4	100	3	2	AISF9	441	2	3
AISF5	100	2	2				

graph data sets were generated, numbered AISF1 to AISF9. AISF1 to AISF3 were generated using a 6×6 grid, while AISF4 to AISF6 were generated using a 10×10 grid, and AISF7 to AISF9 were generated using a 21×21 grid. Some statistics concerning these graph sets are presented in Table 2.

4.3 Experimental Results and Analysis

For experimental purposes REVULSM was implemented in the JAVA programming language. All experiments were conducted using a 2.7 GHz Intel Core i5 with 4 GB 1333 MHz DDR3 memory, running OS X 10.8.1 (12B19). The results obtained are presented in Figs. 6, 7, 8, 9, 10, 11, 12, 13, 14, 15, 16, 17. Figures 6, 7, 8 give the run time comparisons with respect to the nine graph sets. Figures 9, 10, 11 give the number of discovered VULS in each case. Figures 12, 13, 14 present a comparison of the recorded discovery rates with respect to the nine graph sets considered. Finally Figs. 15, 16, 17 give a comparison of the coverage rates.

From Figs. 6, 7, 8 it can be seen, as might be expected, that as the value of the Max parameter increases the run time also increases because more subgraphs and hence more VULS are generated. The same observation is true with respect to the size of the graph; the more vertices the greater the required runtime.

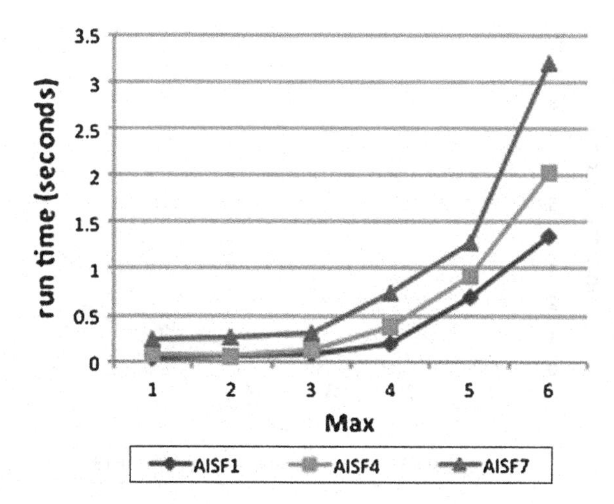

Fig. 6 Run time comparison using 3 edge and 2 vertex labels (AISF1, AISF4 and AISF7)

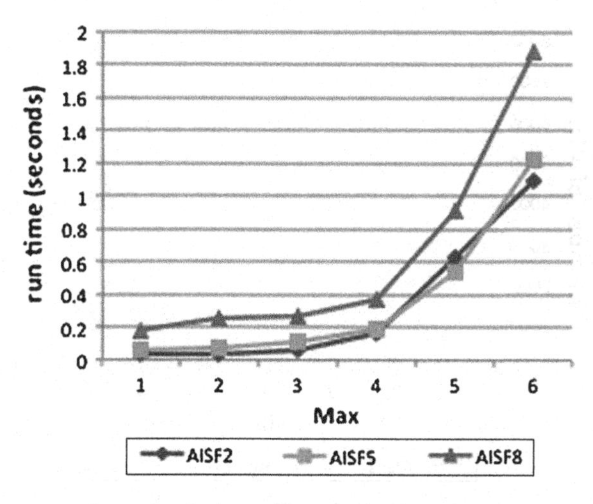

Fig. 7 Run time comparison using 2 edge and 2 vertex labels (AISF2, AISF5 and AISF8)

From Figs. 9 (6×6 grid), 10 (10×10 grid) and 11 (21×21 grid) it can be observed that as Max increases the number of VULS will also increase, again this is as might be expected. Comparing AISF1, 4 and 7 with AISF2, 5 and 8 respectively, it can be seen that as the number of edge labels increases while the number of vertex labels is kept constant the number of VULS also increases (AISF1 and AISF2 have the same number of vertex labels; as do AISF4 and AISF5, and AISF7 and AISF8). This is because the likelihood of VULS existing increases as the input graph becomes more diverse. However, comparing AISF2, 5 and 8 with AISF3, 6 and 9 respectively it can be seen that as the number of vertex labels increases, while the number of edge labels is kept constant, the number of VULS generated will decrease (AISF2 and

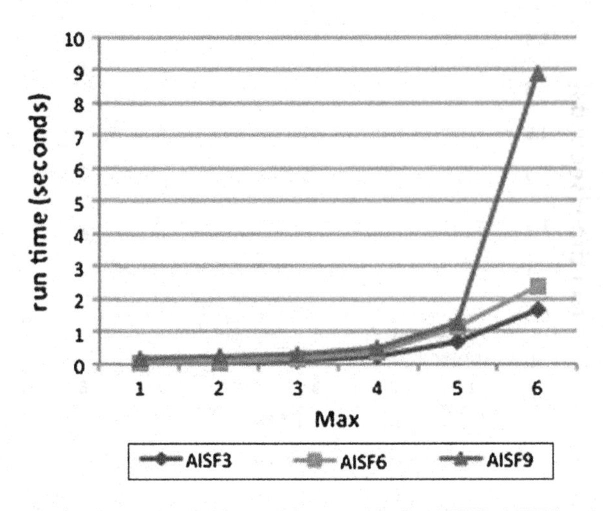

Fig. 8 Run time comparison using 2 edge and 3 vertex labels (AISF3, AISF6 and AISF9)

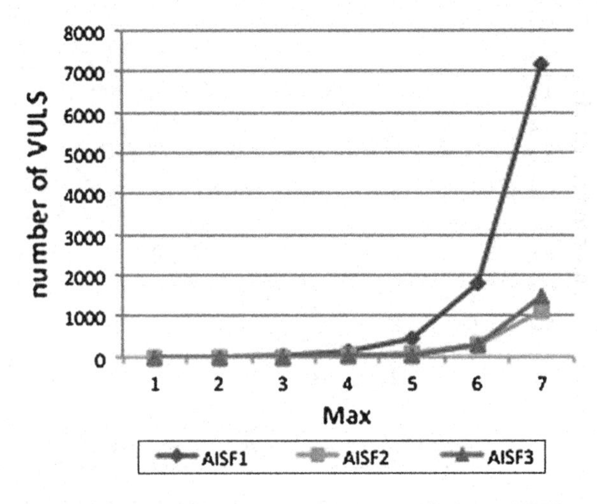

Fig. 9 Comparison of number of VULS generated (AISF1, AISF2 and AISF3)

AISF3 have the same number of edge labels; as do AISF5 and AISF6, and AISF8 and AISF9); because, given a high number of vertex labels, the likelihood of VULS existing decreases.

Figures 12 (6×6 grid), 13 (10×10 grid) and 14 (21×21 grid) show the recorded discovery rate values. Comparing AISF1, 4 and 7 with AISF2, 5 and 8 respectively; when the number of edge labels increases, while the number of vertex labels is kept constant, the discovery rate increases. This is because regardless of the number of edge labels a graph has (all other elements being kept constant) the number of subgraphs contained in the graph will not change, while (as indicated by the experiments

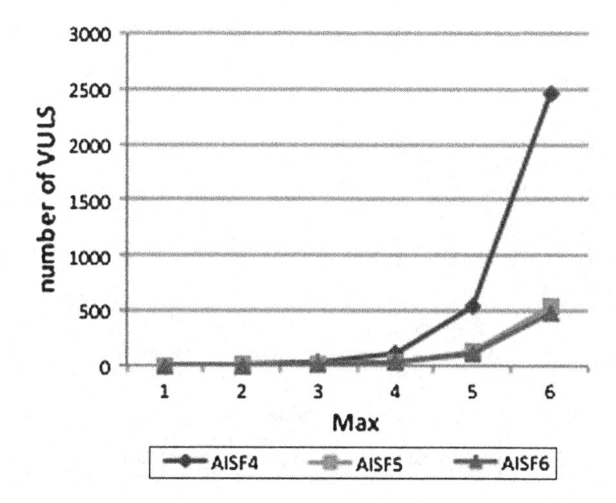

Fig. 10 Comparison of number of VULS generated (AISF4, AISF5 and AISF6)

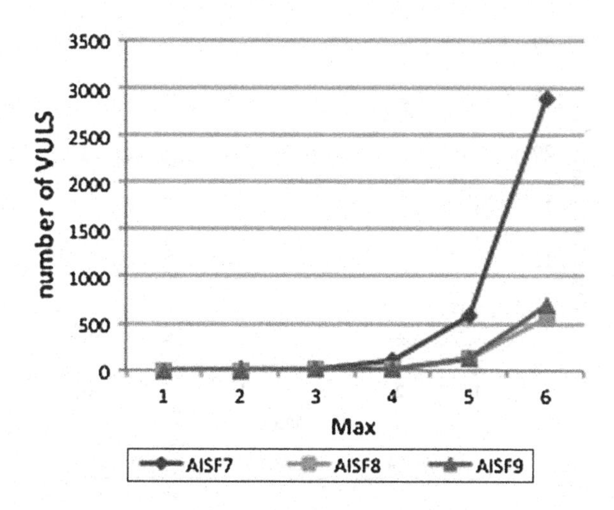

Fig. 11 Comparison of number of VULS generated (AISF7, AISF8 and AISF9)

reported in Figs. 9, 10, 11) the number of identified VULS increases as the number of edge labels increases. Conversely, comparing AISF2, 5 and 8 with AISF3, 6 and 9 respectively, when the number of vertex labels increases while the number of edge labels is kept constant, the discovery rate will decrease because (as already noted) the number of VULS generated decreases as the number of vertex labels increases. It can also be noted that as the Max value increases, the discovery rate does not always increase, as shown in the case of AISF4, 5 and 6. This is because as the Max value increases, the number of VULS goes up as does the number of subgraphs, but they may not both increase at the same rate.

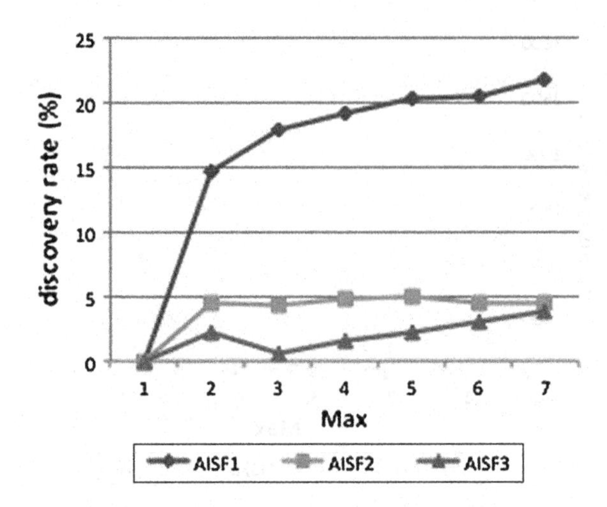

Fig. 12 Comparison of discovery rate (AISF1, AISF2 and AISF3)

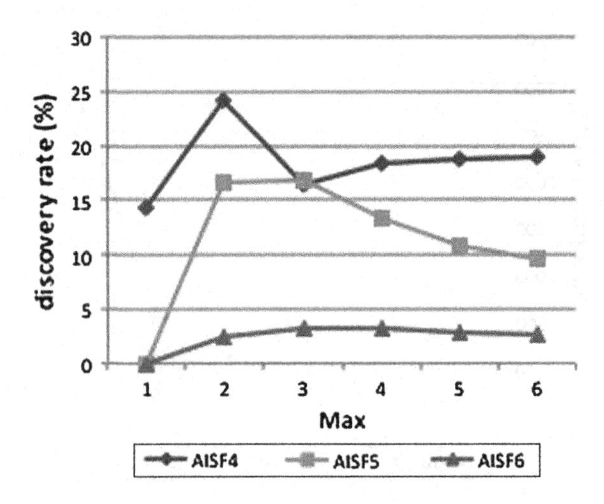

Fig. 13 Comparison of discovery rate (AISF4, AISF5 and AISF6)

Figures 15, 16 and 17 show the coverage rate. From the figures it can be observed that as Max increases the coverage rate also increases. This is to be expected, however it is interesting to note that the coverage rate in some cases reaches 100% (when $Max = 6$ with respect to AISF1, ASF2 and AISF4). One hundred percent coverage is desirable in the context of the sheet metal forming application so that unique patterns associated with particular error distributions (vertex labels) can be identified for all geometries. Comparing AISF1, 4 and 7 with AISF2, 5 and 8 respectively, the more edge labels a graph has the more VULS will be generated (see above); as a result more vertices will be covered by VULS and hence the coverage rate will go up. On the other hand, comparing AISF2, 5 and 8 with AISF3, 6 and 9 respectively,

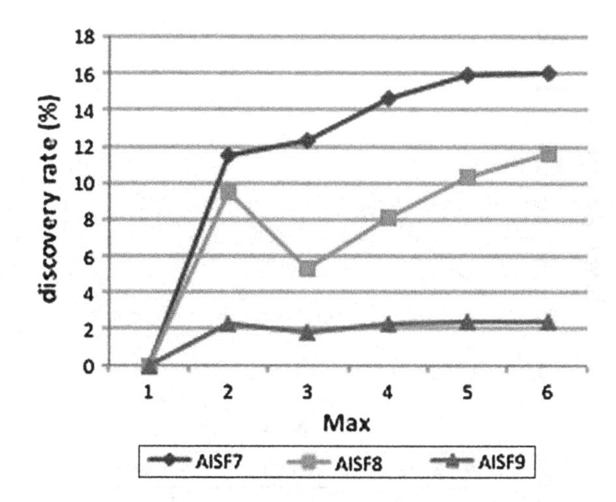

Fig. 14 Comparison of discovery rate (AISF7, AISF8 and AISF9)

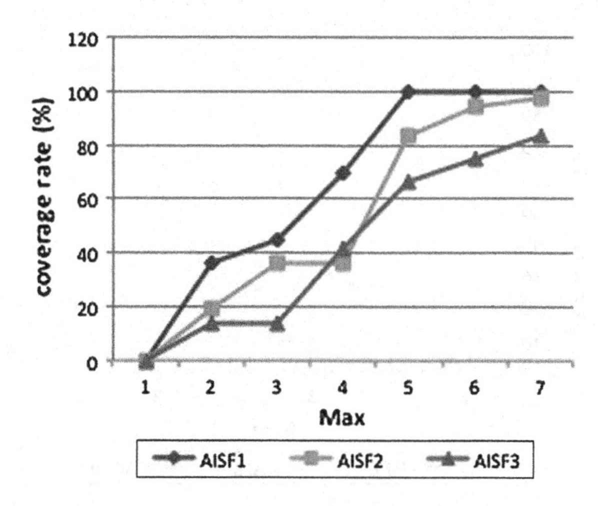

Fig. 15 Comparison of coverage rate (AISF1, AISF2 and AISF3)

the more vertex labels a graph has the less VULS will be generated (see above); as a result fewer vertices will be covered by VULS vertices and hence the coverage rate will go down.

5 Conclusions and Further Study

In this paper we have proposed the mining of VULS and presented the REVULSM algorithm. The reported experimental results demonstrated that the VULS idea is sound and that REVULSM can effectively identify VULS in real data. Having

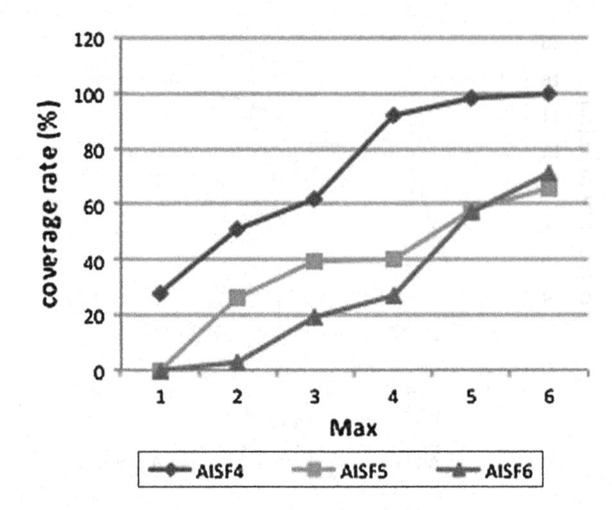

Fig. 16 Comparison of coverage rate (AISF4, AISF5 and AISF6)

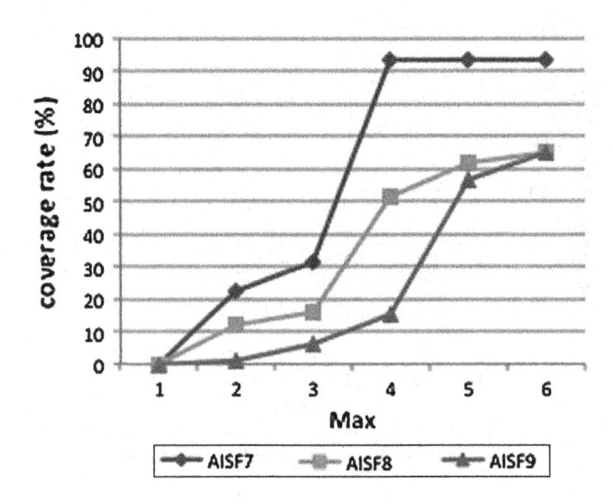

Fig. 17 Comparison of coverage rate (AISF7, AISF8 and AISF9)

established a "proof on concept" there are many interesting research problems related to VULSM that can now be pursued. For instance, currently, when the Max parameter is high REVULSM will run out of memory, although for the purpose of error prediction in sheet metal forming it can be argued that there is no requirement for larger VULS, it may be of interest to investigate methods whereby the efficiency of REVULSM can be improved. Finally, at present, REVULSM finds all VULS up to a predefined size, it is conjectured that efficiency gains can be made if only minimal VULS are found.

Acknowledgments The research leading to the results presented in this paper has received fund- ing from the European Union Seventh Framework Programme (FP7/2007-2013) under grant agreement number 266208.

References

1. Agrawal, R., Srikant, R.: fast algorithms for mining association rules. In: Proceedings of the 20th International Conference on Very Large Data Bases(VLDB '94), pp. 487–499 (1994).
2. Agrawal, R., Srikant, R.: Mining sequential patterns. In: Proceedings of the Eleventh International Conference on Data Engineering(ICDE '95), pp. 3–14 (1995).
3. Asai, T., Abe, K., Kawasoe, S., Sakamoto, H., Arikawa, S.: Efficient substructure discovery from large semi-structured data. In. In Proc. 2002 SIAM Int. Conf. Data Mining, pp. 158–174 (2002).
4. Cafuta, G., Mole, N., tok, B.: An enhanced displacement adjustment method: Springback and thinning compensation. Materials and Design 40, 476–487 (2012).
5. El-Salhi, S., Coenen, F., Dixon, C., Khan, M.S.: Identification of correlations between 3d surfaces using data mining techniques: Predicting springback in sheet metal forming. In: Research and Development in Intelligent Systems XXIX, pp. 391–404 (2012).
6. Firat, M., Kaftanoglu, B., Eser, O.: Sheet metal forming analyses with an emphasis on the springback deformation. Journal of Materials Processing Technology 196(1–3), 135–148 (2008).
7. Han, J., Cheng, H., Xin, D., Yan, X.: Frequent pattern mining: Current status and future directions. Data Mining and Knowledge Discovery 15(1), 55–86 (2007).
8. Huan, J., Wang, W., Prins, J., Yang, J.: SPIN: mining maximal frequent subgraphs from graph databases. In: Proceedings of the 10th ACM SIGKDD International Conference on Knowledge Discovery and Data Mining, pp. 581–586 (2004).
9. Inokuchi, A., Washio, T., Motoda, H.: An apriori-based algorithm for mining frequent substructures from graph data. In. In Principles of, Data Mining and Knowledge Discovery, pp. 13–23 (2000).
10. Jeswiet, J., Micari, F., Hirt, G., Bramley, A., andJ. Allwood, J.D.: Asymmetric single point incremental forming of sheet metal. CIRP Annals Manufacturing Technology 54(2), 88–114 (2005).
11. Khan, M.S., Coenen, F., Dixon, C., El-Salhi, S.: Finding correlations between 3-d surfaces: A study in asymmetric incremental sheet forming. Machine Learning and Data Mining in Pattern Recognition Lecture Notes in Computer Science 7376, 366–379 (2012).
12. Liu, W., Liang, Z., Huang, T., Chen, Y., Lian, J.: Process optimal ccontrol of sheet metal forming springback based on evolutionary strategy. In. In Intelligent Control and Automation, 2008. WCICA 2008. 7th World Congress, pp. 7940–7945 (June 2008).
13. Nasrollahi, V., Arezoo, B.: Prediction of springback in sheet metal components with holes on the bending area, using experiments, finite element and neural networks. Materials and Design 36, 331–336 (2012).
14. Yan, X., Han, J.: gSpan: Graph-based substructure pattern mining. In: Proceedings of the 2002 International Conference on Data Mining, pp. 721–724 (2002).
15. Yan, X., Han, J.: Close Graph: mining closed frequent graph patterns. In: Proceedings of the 9th ACM SIGKDD International Conference on Knowledge Discovery and Data Mining, pp. 286–295 (2003).
16. Zhu, F., Yan, X., Han, J., Yu, P.S.: gPrune: a constraint pushing framework for graph pattern mining. In: Proceedings of 2007 Pacific-Asia Conference on Knowledge Discovery and Data Mining (PAKDD'07), pp. 388–400 (2007).

Hierarchical Single Label Classification: An Alternative Approach

Esra'a Alshdaifat, Frans Coenen and Keith Dures

Abstract In this paper an approach to multi-class (as opposed to multi-label) classification is proposed. The idea is that a more effective classification can be produced if a coarse-grain classification (directed at groups of classes) is first conducted followed by increasingly more fine-grain classifications. A framework is proposed whereby this scheme can be realised in the form of a classification hierarchy. The main challenge is how best to create class groupings with respect to the labels nearer the root of the hierarchy. Three different techniques, based on the concepts of clustering and splitting, are proposed. Experimental results show that the proposed mechanism can improve classification performance in terms of average accuracy and average AUC in the context of some data sets.

1 Introduction

Classification is concerned with the creation of a global model to be used for predicting the class labels of new data. Classification can be viewed as a three-step process: (i) generation of the classifier using appropriately formatted "training" data, (ii) testing of the effectiveness of the generated classifier using test data and (iii) application of the classifier. The first two steps are sometimes combined for experimental purposes. There exist many possible models for classifier generation; the classifier can, for example, be expressed in terms of rules, decision trees or mathematical formula.

E. Alshdaifat (✉) · F. Coenen · K. Dures
Department of Computer Science, University of Liverpool, Ashton Building,
Ashton Street, Liverpool L69 3BX, United Kingdom
e-mail: esraa@liv.ac.uk

F. Coenen
e-mail: coenen@liv.ac.uk

K. Dures
e-mail: dures@liv.ac.uk

M. Bramer and M. Petridis (eds.), *Research and Development in Intelligent Systems XXX*, 39
DOI: 10.1007/978-3-319-02621-3_3,

The performance of classifiers can also vary greatly according to the nature of the input data, how the input data is preprocessed, the adopted generation mechanism and number of classes in the data set. To date no one classification model has been identified that is, in all cases, superior to all others in terms of classification effectiveness [9]. In an attempt to improve classifier performance ensemble models have been proposed. An ensemble model is a composite model comprised of a number of classifiers. There is evidence to suggest that ensembles are more accurate than their individual component classifiers [11].

Classification can be categorised according to: (i) the size of the set of classes C from which class labels may be drawn, if $|C| = 2$ we have a simple "yes-no" (binary) classifier, if $|C| > 2$ we have a multi-class classifier, and (ii) the number of class labels that may be assigned to a record (single-label classification versus multi-label classification), for most applications we typically wish to assign a single label to each record. The work described in this paper is directed at multi-class single-label classification, especially where $|C|$ is large. Multi-class classification is challenging for two reasons. The first is that the training data typically used to generate the desired classifier often features fewer examples of each class than in the case of binary training data. The second reason is that it is often difficult to identify a suitable subset of features that can effectively serve to distinguish between large numbers of classes (more than two class labels). The "Letter Recognition" data set [1], frequently used as a benchmark dataset with respect to the evaluation of machine learning algorithms, is an example of a data set with a large number of class labels; the data set describes black-and-white rectangular pixel displays each representing one of the twenty six capital letters available in the English alphabet.

Three most straightforward approaches for solving multi-class classification problems include: (i) using stand-alone classification algorithm such as decision tree classifiers [14], and Bayesian classification [12], (ii) building a sequence of binary classifiers [17], and (iii) adopting an ensemble approach such as bagging [3] or boosting [10]. The solution proposed on this paper is to adopt an ensemble approach founded on the idea of arranging the classifiers into a hierarchical form. Class hierarchy models are usually directed at multi-label classification where we have multiple-class sets (as opposed to single-label classification); each level in the hierarchy is directed at a specific class set. In our case the proposal is that the hierarchy be directed at single-label classification. Classifiers at the leaves of our hierarchy feature binary classifiers, while classifiers at nodes further up the hierarchy feature classifiers directed at groupings of class labels. The research challenge is how best to organise (group) the class labels so as to produce a hierarchy that generates an effective classification. A number of alternative techniques are proposed in this paper, whereby this may be achieved, founded on ideas concerned with clustering and splitting mechanisms to divide the class labels across nodes.

The rest of this paper is organised as follows. Section 2 gives a review of related work on multi-class classification. Section 3 describes the proposed hierarchical classification approach. Section 4 presents an evaluation of the proposed hierarchical classification approach as applied to a range of different data sets. Section 5 summarizes the work and indicates some future research directions.

2 Literature Review

In this section we review some of the previous work related to the work described in this paper. The section is organised as follows: we first (Sect. 2.1) consider classification algorithms that can be directly used to address multi-class classification problems; then we go on to (Sect. 2.2) consider ensemble mechanisms in the context of the multi-class classification problem.

2.1 Using Stand-Alone Classification Algorithms to Solve Multi-Class Classification Problems

Some classification algorithms are specifically designed to address binary classification, for example support vector machines [18]. However, such algorithms can be adapted to handle multi-class classification by building sequence of binary classifiers. Other classification algorithms can directly handle any number of class labels; examples include: decision tree classifiers [14], Classification Association Rule Mining (CARM) [7] and Bayesian classification [12]. Among these decision trees algorithms are of interest with respect to the work described in this paper because it can be argued that our hierarchies have some similarity with the concept of decision trees. Decision trees have a number of advantages with respect to some other classification techniques: (i) they can be constructed relatively quickly, (ii) they are easy to understand (and modify) and (iii) the tree can be expressed as a set of "decision rules" (which is of benefit with respect to some applications). Decision trees are constructed by inducing a split in the training data according to the values associated with the available attributes. The splitting is frequently undertaken according to "Information Gain" [14], "Gini Gain" [5], or "Gain Ratio" [14]. Each leaf node of a decision tree holds a class label. A new example is classified by following a path from the root node to a leaf node, the class held at the identified leaf node is then considered to be the class label for the example [11]. Amongst the most frequently quoted decision tree generation algorithms are: ID3 [14], C4.5 [15] and CART [5].

2.2 Using Ensemble Classifiers to Solve Multi-Class Classification Problems

The simplest form of ensemble can be argued to comprise a suite of binary classifiers trained using n different binary training sets, where n is the total number of class labels to be considered, each one trained to distinguish the examples in a single class from the examples in all remaining classes [17]. This scheme is often referred to as the One-Versus-All (OVA) [16] scheme. The one-versus-all scheme is conceptually simple, and has been independently proposed by numerous researchers [16].

A variation of the one-versus-all scheme is All-Versus-All (AVA) [17] where a classifier is constructed for every possible pair of classes. Given n classes this will result in $(n(n-1)/2)$ classifiers. To classify a new example, each classifier "votes", and the new example is assigned the class with the maximum number of votes. AVA tends to be superior to OVA [11]. The problem with this type of classification is that binary classifiers are typically sensitive to errors; if a classifier produces an erroneous result it can adversely affect the final vote count. In order to improve on this type of binary classification; it is possible to introduce "error-correcting output codes", which change the definition of the class a single classifier has to learn [8].

A more sophisticated form of ensemble is one where the classifiers are arranged in concurrent (parallel) [3] or sequential (serial) [10] form. In the concurrent ensemble methodology, the original dataset is partitioned into several subsets from which multiple classifiers are induced concurrently. The simplest approach, often referred to as "Bagging" (the name was obtained from the phrase "Bootstrap Aggregation") [13], is where "sampling with replacement" is used [3]. A well-known bagging algorithm is the "Random Forest" algorithm [4], which combines the output from a collection of decision trees.

In the sequential case there is interaction between the different classifiers (the outcome of one feeds into the next). Thus it is possible to take advantage of knowledge generated in a previous iteration to guide the learning in the next iterations. One well studied form of sequential ensemble classification is known as "Boosting", where a sequence of weak classifiers are "chained" together to produce a single composite strong classifier in order to achieve a higher combined accuracy than that which would have been obtained if the weak classifiers were used independently. A well-known boosting algorithm is Adaboost [10] .

Regardless of how an ensemble system might be configured, an important issue is how results are combined to enhance classification accuracy. The simplest approach is to use some kind of voting system [2]. Voting algorithms can be divided into two types: those that adaptively change the distribution of the training set based on the performance of previous classifiers (as in boosting methods) and those that do not (as in Bagging).

3 The Hierarchical Single-Label Classification Framework

As noted in the introduction to this paper the proposed hierarchical classification mechanism is a form of ensemble classifier. Each node in our hierarchy holds a classifier. Classifiers at the leaves conduct fine-grained (binary) classifications while the classifiers at non-leaf nodes further up the hierarchy conduct coarse-grained classification directed at categorising records using groups of labels. An example hierarchy is presented in Fig. 1. At the root we classify into two groups of class labels $\{a, b, c, d\}$ and $\{e, f, g\}$. At the next level we split into smaller groups, and so on till we reach classifiers that can associate single class labels with records. Note that Fig. 1 is just an example of the proposed hierarchical model; non-leaf child nodes

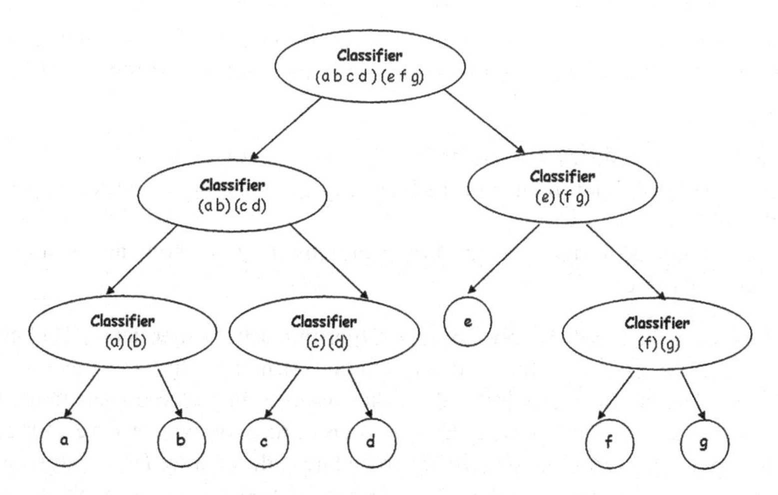

Fig. 1 Classification hierarchy example

may end up with overlapping classifications due to the result returned from the used clustering algorithms.

The challenge of hierarchical single-label classification, as conceived in this paper, is how best to distribute groupings of class labels at non-leaf nodes in the class hierarchy. A number of different mechanisms are considered based on clustering and splitting criteria: (i) k-means clustering, (ii) hierarchical clustering, and (iii) data splitting. The motivation for using ensemble classifiers arranged in a hierarchical form so as to improve the accuracy of the classification were thus: (i) the established observation that ensemble methods tend to improve classification performance [11], and (ii) dealing with smaller subsets of class labels at each node might produce better results. The remainder of this section is organised as follows. Section 3.1 explains the generation of the hierarchical model, while Sect. 3.2 illustrates its operation.

3.1 Generating the Hierarchical Model

In the proposed model classifiers nearer the root of the hierarchy conduct coarse-grain classification with respect to subsets of the available set of classes. Classifiers at the leaves of the hierarchy conduct fine-grain (binary) classification. To create the hierarchy a classifier needs to be generated for each node in the hierarchy using an appropriately configured training set. The process is illustrated in Fig. 1 where the root classifier classifies the data set into two sets of class labels $\{a, b, c, d\}$ and $\{e, f, g\}$, the level two classifiers are then directed at further subsets and so on. The sets of class labels (the label groupings) are identified by repeatedly dividing the data using one of the proposed clustering or splitting approaches.

The proposed hierarchy generation algorithm is presented in Algorithm 1. The algorithm assumes a data structure, called *hierarchy*, comprised of the following fields:

1. Classifier: A classifier (root and body nodes only, set to null otherwise).
2. Left: Reference to left branch of the hierarchy (root and body nodes only, set to null otherwise).
3. Right: Reference to right branch of the hierarchy (root and body nodes only, set to null otherwise).

Considering the algorithm presented in Algorithm 1 in further detail. The algorithm in this case assumes a binary tree (hierarchy) structure (this is not necessarily always the case but the assumption allows for a more simplified explanation with respect to this paper). The *Generate_Hierarchy* procedure is recursive. On each recursion the input to the *Generate_Hierarchy* procedure is the data set D (initially this is the entire training set). If the number of classes represented in D is one (all the records in D are of the same class) a leaf hierarchy node will be created (labeled with the appropriate class label). If the number of classes featured in D is two, a binary classifier is constructed (to distinguish between the two classes). The most sophisticated part of the *Generate_Hierarchy* procedure is the following option, which applies if the number of classes featured in D is more than two. In this case the records in D are divided into two groups $D1$ and $D2$ each with a meta-class label, $K1$ and $K2$, associated with it. A Classifier is then constructed to discriminate between $K1$ and $K2$. The *Generate_ Hierarchy* procedure is then called again, once with $D1$ (representing the left branch of the hierarchy) and once with $D2$ (representing the right branch of the hierarchy).

Two types of classifiers were considered with respect to the nodes within our hierarchy: (i) straight forward single "stand-alone" classifiers (Fig. 2a), and (ii) bagging ensembles (Fig. 2b). With respect to the first approach a simple decision tree classifier was generated for each node in the hierarchy. With respect to bagging, the data set D associated with each node was randomly divided into N disjoint partitions

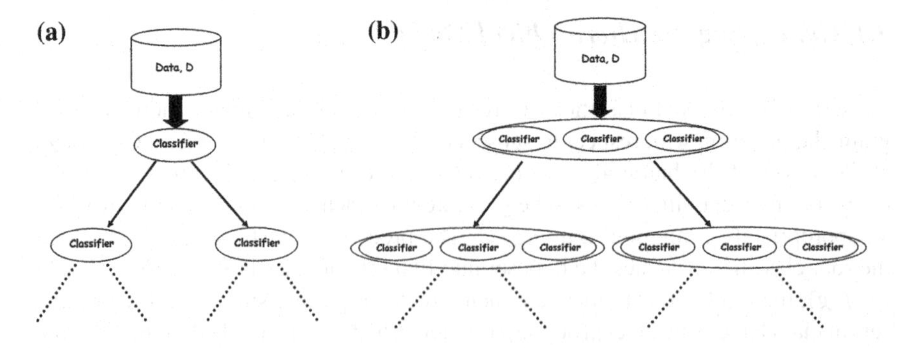

Fig. 2 Hierarchical classification, **a** using a single classifier at each node and **b** using a Bagging ensemble at each node

Algorithm: *Generate_Hierarchy*
Input:

1. Data partition, D, which is a set of training tuples and their associated class labels.
2. Classification method, a procedure for building *classifier* at a given hierarchy node.
3. Clustering (or splitting) method, a procedure for splitting the classes between nodes.

Output: A hierarchical classifiers.

Method:

```
create a hierarchy node N;
if number of classes featured in D is one class, C, then
    return N as a leaf node labeled with C,
    ( assign null value to hierarchy left, and right);
else
    if number of classes featured in D is two classes then
        create a hierarchy classifier (using classification method), to distinguish between the
        two (real) classes (binary classifier);
        assign null value to hierarchy left, and right;
    else
        cluster D into two clusters K1 and K2 (using clustering techniques);
        recast labels in D so that they correspond to K1 and K2;
        create a hierarchy classifier (using classification method), to distinguish between K1
        and K2;
        D1 = records in D containing class labels in K1;
        hierarchy right= Generate_Hierarchy (D1);
        D2 = records in D containing class labels in K2;
        hierarchy left = Generate_Hierarchy (D2);
    end
end
return N;
```

Algorithm 1: Ensemble hierarchy generation algorithm

and a classifier generated for each (in the evaluation reported in Sect. 4, $N = 3$ was used).

As already noted, with respect to dividing D during the hierarchy generation process, three different techniques were considered: (i) k-means, (ii) hierarchical clustering, and (iii) data splitting. Of these k-means is the most well-known and commonly used partitioning method where the input is divided into k partitions (in our model $k = 2$ was used because of the binary nature of our hierarchy used for experimental purposes). Hierarchical clustering creates a hierarchical decomposition of the given data. In the context of the work described in this paper a divisive hierarchical clustering approach (top-down) was used because this fits well with respect to our vision of hierarchical ensemble classification. This process commences with all records in one cluster, in each successive iteration, a cluster is split into smaller clusters until a "best" cluster configuration is arrived at (measured using cluster cohesion and separation measures). Data splitting comprises a simple "cut" of the data into two groups so that each contains a disjoint subset of the entire set of class labels.

3.2 Classification Using Hierarchical Model

Section 3.1 explained the generation of the desired hierarchical model. After the model has been generated, the intention is to use it to classify new unseen data records. In this section the process whereby this is achieved is explained. Recall from the above that during the generation process sets of class labels are grouped. For simplicity, and in acknowledgement of the binary nature of our example hierarchies, we refer to these groups as the $left$ and $right$ groups. The procedure for using the hierarchy to classify a record, R, is summarized in Algorithm 2. The $Classifying\_Records\_Using\_Hierarchy$ procedure is recursive. On each recursion the algorithm is called with two parameters: (i) R, the record to be classified, and (ii) a pointer to the current node location in the hierarchy (at start this will be the root node). How the process proceeds then depends on the nature of the class label returned by the classifier at the current node in the hierarchy. If the returned class belongs to one of either the $left$ or $right$ groups $Classifying\_Records\_Using\_Hierarchy$ will be called again with the parameters: (i) either the left or right child node as appropriate, and of course (ii) the record R. If the returned class label is a specific class label (as opposed to some grouping of labels) this class label will be returned as the label to be associated with the given record and the algorithm terminated.

Algorithm: *Classifying_Records_Using_Hierarchy*
Input:

1. New unseen record, R.
2. Hierarchy node, N. (pointer to the hierarchy)

Output: The predicted class label of the input record R

Method:
use node's *classifier* to classify R;
if the returned class label is member of original (real) class labels **then**
    return class label;
else
    if the returned class label is member of right class labels group **then**
        Classifying_Records_Using_Hierarchy (R, node N right branch);
    else
        Classifying_Records_Using_Hierarchy (R, node N left branch);
    end
end

Algorithm 2: Ensemble hierarchy classification algorithm

4 Experimentation and Evaluation

In this section we present an overview of the adopted experimental set up and the evaluation results obtained. The experiments used 14 different data sets (with different numbers of class labels) taken from UCI data repository [1], which were processed using the LUCS-KDD-DN software [6]. Ten-fold Cross Validation (TCV) was used throughout. The evaluation measures used were average accuracy and average AUC (Area Under the receiver operating Curve).

The following individual experiments were conducted and the results recorded:

1. Decision Trees (DT): Stand-alone classification of the data using a single classifier (no hierarchy), the objective being to establish a bench mark.
2. Bagging like strategy (Bagging): Classification of the data using a bagging ensemble classifier comprised of three classifiers (no hierarchy), the objective was to establish a second benchmark.
3. K-means and Decision tree (K-means&DT): The proposed approach using k-means ($k = 2$) to group data with decision tree classifiers at each node.
4. Data splitting and Decision tree (DS&DT): The proposed approach using data splitting to group data with decision tree classifiers at each node.
5. Hierarchical clustering and Decision tree (HC&DT): The proposed approach using divisive hierarchical clustering to group data with decision tree classifiers at each node.
6. K-means and Bagging (K-means&B): The proposed approach using k-means ($k = 2$) to group data with a bagging ensemble classifier at each node.
7. Data splitting and Bagging (DS&B): The proposed approach using data splitting to group data with a bagging ensemble classifier at each node.
8. Hierarchical clustering and Bagging (HC&B): The proposed approach using divisive hierarchical clustering to group data with a bagging ensemble classifier at each node.

With respect to the bagging methods three decision tree classifiers were generated with respect to each node.

The results in terms of average accuracy are presented in Fig. 3 in the form of a collection of Fourteen histograms, one for each data set considered. The best method (from the above list) is indicated to the top right of each histogram (in some cases two best methods were identified). From the figure it can be observed that the proposed hierarchical techniques can significantly improve the classification accuracy with respect to eight out of the Fourteen data sets considered. In two out of these eight cases (where hierarchical classification worked well) bagging produced the best result; in the remaining cases the decision tree classifiers produced the best result.

Figure 4 shows the results in terms of the average AUC of the eight techniques considered in the evaluation (again the best result with respect to each dataset is indicated). From the figure it can be observed that the proposed hierarchical techniques can enhance the classification AUC with respect to eight of the Fourteen data sets considered. In the remaining six cases out of the Fourteen the stand-alone decision tree classifier produced the best AUC result (although in one case the same AUC result was the same as that produced using our hierarchical approach and in another case the same as that produced using bagging).

It is interesting to note that the proposed hierarchical techniques work well for datasets with high numbers of class labels (18 in the case of the Chess KRvK dataset and 15 in the case of the Soybean dataset). The reason for this is that a high number

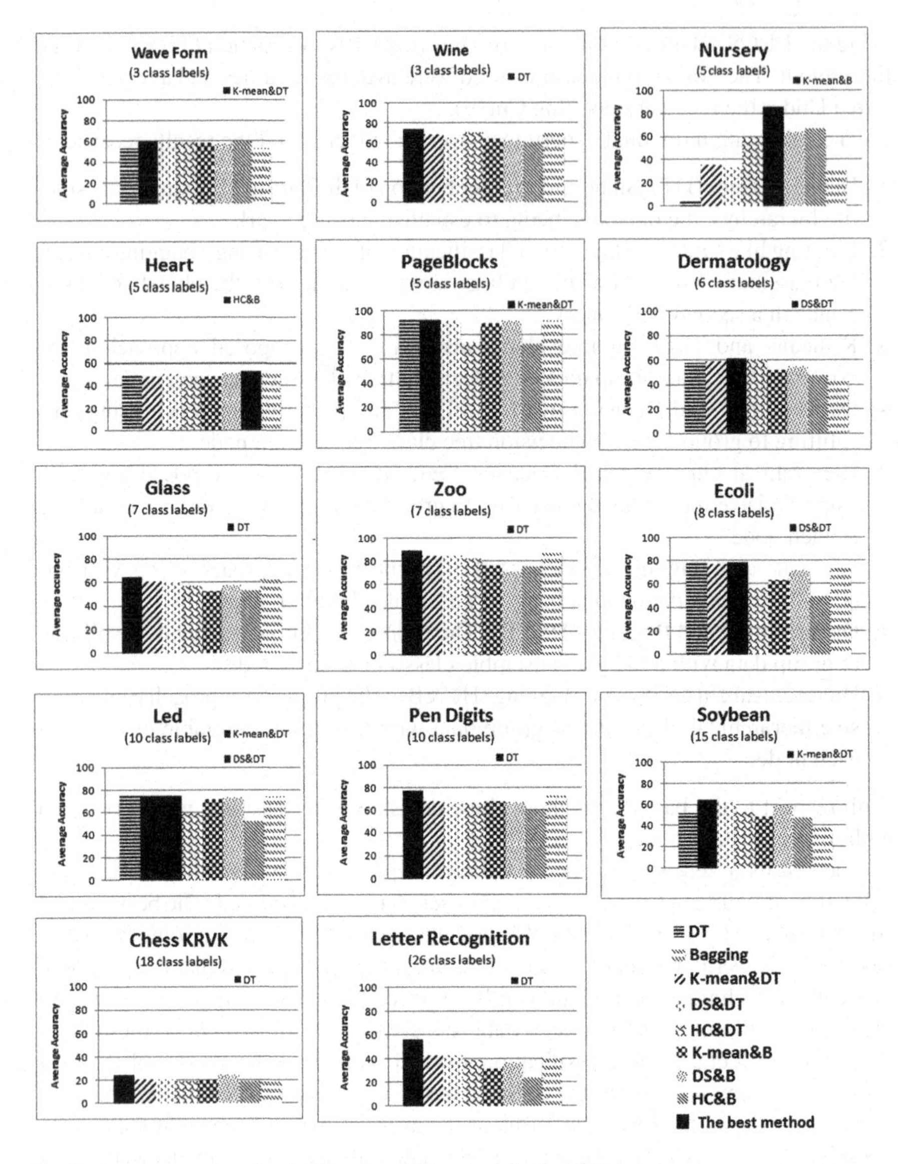

Fig. 3 Average accuracy for the Fourteen different evaluation data sets

of class labels allow for the generation of more sophisticated hierarchies, while a low number of classes does not. Thus data sets with high numbers of class labels are deemed to be much more compatible with the proposed approach.

The raw data on which the histograms presented in Figs. 3 and 4 were based is presented in Tabular form in Table 1 (the best result with respect to each dataset is highlighted).

Table 1 Average accuracy and AUC for the Fourteen different evaluation data sets

Data set	Classes	DT		Bagging		K-mean&DT		DS&DT		HC&DT		K-mean&Bagging		DS&B		HC&B	
		ACC	AUC	ACC	AUC	ACC	AUC	ACC	AUC	ACC	AUC	ACC	AUC	ACC	AUC	ACC	AUC
WaveForm	3	53.72	0.54	53.36	0.53	59.68	0.60	58.56	0.59	58.98	0.59	58.08	0.58	57.48	0.58	59.44	0.59
Wine	3	73.86	0.73	69.91	0.69	68.75	0.67	65.22	0.64	70.54	0.68	64.05	0.64	62.45	0.60	60.09	0.59
Nursery	5	5.15	0.03	32.71	0.16	35.69	0.19	33.51	0.27	59.70	0.32	86.20	0.46	64.55	0.32	67.32	0.36
Heart	5	48.80	0.28	51.15	0.28	47.97	0.31	49.89	0.32	47.48	0.26	47.41	0.24	51.42	0.28	52.86	0.25
PageBlocks	5	92.55	0.49	92.23	0.47	92.56	0.49	92.40	0.48	74.28	0.36	89.77	0.20	91.56	0.28	72.88	0.20
Dermatology	6	57.53	0.57	43.95	0.39	58.90	0.55	61.37	0.60	60.25	0.54	51.90	0.43	54.54	0.49	47.30	0.42
Glass	7	64.50	0.40	62.27	0.36	60.44	0.39	60.91	0.38	56.95	0.34	52.10	0.28	57.43	0.28	52.90	0.25
Zoo	7	89.00	0.53	87.27	0.53	85.00	0.50	85.00	0.50	82.09	0.51	76.27	0.44	71.36	0.41	75.45	0.43
Ecoli	8	78.07	0.34	73.61	0.31	76.90	0.33	78.72	0.35	56.29	0.28	63.59	0.23	71.63	0.23	49.44	0.22
Led	10	74.72	0.74	74.06	0.74	75.00	0.75	75.00	0.75	59.94	0.59	71.75	0.72	73.22	0.73	51.75	0.51
PenDigits	10	76.84	0.77	72.64	0.72	67.30	0.67	66.78	0.67	66.30	0.66	67.22	0.67	66.74	0.67	59.44	0.59
Soybean	15	52.12	0.52	37.18	0.30	64.62	0.65	63.02	0.56	52.67	0.47	48.74	0.38	59.06	0.49	48.01	0.39
ChessKRVK	18	24.05	0.13	18.20	0.09	20.45	0.13	20.50	0.15	18.80	0.13	19.85	0.10	23.95	0.11	18.95	0.10
LetRecog	26	56.05	0.56	41.82	0.42	42.99	0.43	42.76	0.43	38.59	0.39	31.12	0.31	36.92	0.37	24.26	0.24
Mean		60.50	0.47	57.88	0.43	61.16	0.48	60.97	0.48	57.35	0.44	59.15	0.41	60.17	0.42	52.86	0.37
Standerd Deviation		24.00	0.22	21.64	0.20	19.45	0.19	19.58	0.17	15.53	0.17	19.43	0.19	16.31	0.18	15.92	0.16

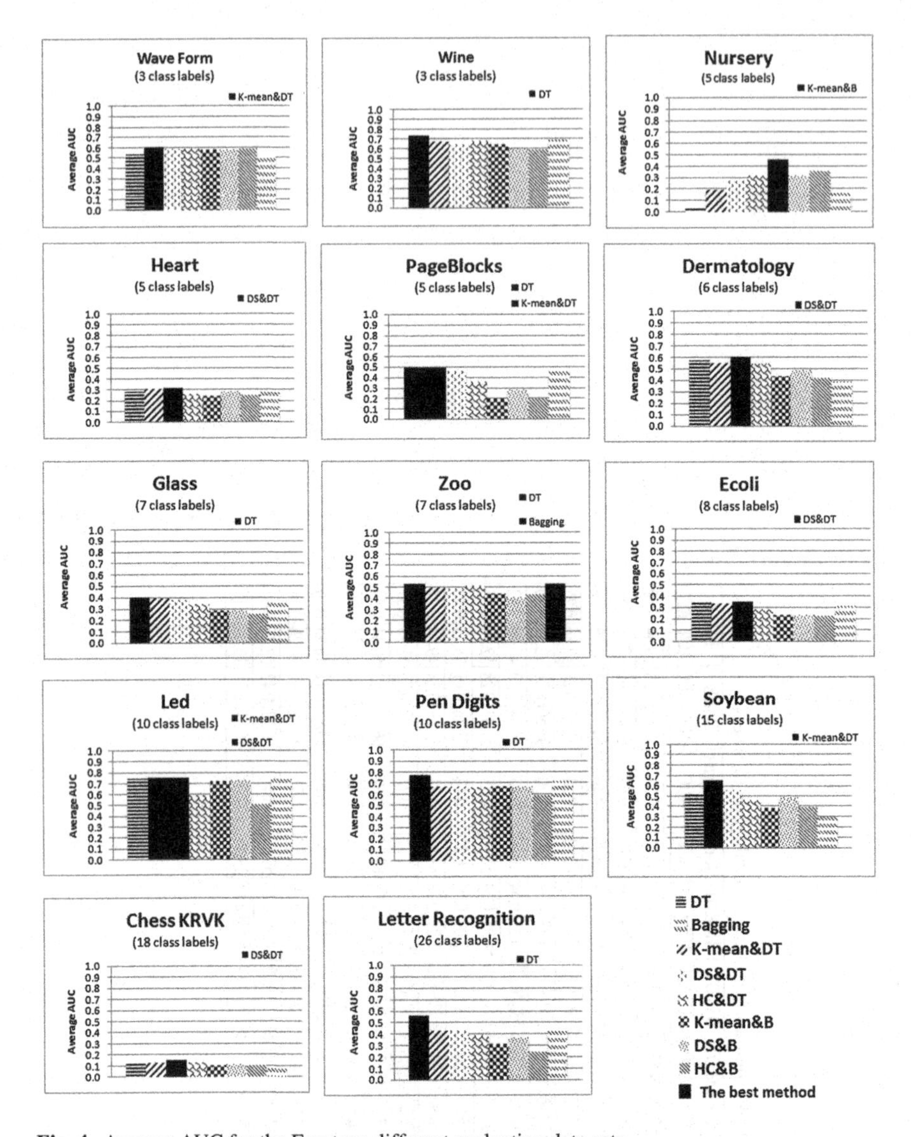

Fig. 4 Average AUC for the Fourteen different evaluation data sets

5 Conclusions

A classification technique to achieve multi-class classification using the concept of a classification hierarchy has been described. The idea is to conduct the classification starting in a coarse-grain manner, where records are allocated to groups, proceeding to a fine-grain manner until class labels can be assigned. To generate such a hierarchical classifier three different grouping techniques were considered: (i) k-mean,

(ii) hierarchical clustering and (iii) data splitting. We also considered the use of two different styles of classifier at each node: (i) decision tree classifiers and (ii) bagging ensemble classifiers.

From the evaluation presented in this paper it was demonstrated that the proposed hierarchical classification model could be successfully used to classify data in a more effective manner than when stand-alone classifiers were used. Each of the proposed methods tended to improve the accuracy and/or AUC of the classification process with respect to some of the data sets considered. Best results were produced using k-mean and decision tree classifiers (K-mean&DT), and data splitting and decision tree classifiers (ST&DT). An issue with the proposed approach is that if a record is miss classified early on in the process (near the root of the hierarchy) there is no opportunity for recovery, regardless of the classifications proposed at lower level nodes and the final leaf nodes. There is therefore no opportunity whereby the proposed approach can attempt to correct its self. To address this issue a number of alternatives will be considered for future work. The first of these is to use classification algorithms, such as Association Rule Mining Classification (CARM) or Bayesian classification, where a confidence or probability value will result. Consequently more than one branch in the tree can may be followed as a consequence of the nature of this value, and a final decision made according to some accumulated confidence or probability value. The use of other tree structures, than binary tree structures, will also be considered; again to allow for the possibility of exploring more than one alternative at each node. One way of achieving this is to use a Directed Acyclic Graph (DAG) structure, instead of a tree hierarchy structure so as to include a greater number of possible class label combinations at each level.

References

1. Bache, K., Lichman, M.: UCI machine learning repository (2013). http://archive.ics.uci.edu/ml
2. Bauer, E., Kohavi, R.: An empirical comparison of voting classification algorithms: Bagging, boosting, and variants. Machine Learning pp. 105–139 (1999).
3. Breiman, L.: Bagging predictors. Machine Learning (2), 123–140 (1996).
4. Breiman, L.: Random forests. In. Machine Learning, pp. 5–32 (2001).
5. Breiman, L., Friedman, J., Olshen, R., Stone, C.: Classification and Regression Trees. Wadsworth and Brooks, Monterey, CA (1984).
6. Coenen, F.: The LUCS-KDD discretised/normalised arm and carm data library (2003). http://www.csc.liv.ac.uk/frans/KDD/Software/LUCS_KDD_DN
7. Coenen, F., Leng, P.: The effect of threshold values on association rule based classification accuracy. Journal of, Data and Knowledge Engineering (2), pp 345–360 (2007).
8. Dietterich, T.G., Bakiri, G.: Solving multiclass learning problems via error-correcting output codes. JAIR (1995).
9. Dunham, M.H.: Data Mining:Introductory and Advanced Topics. Prentice Hall (2003).
10. Freund, Y., Schapire, R., Abe, N.: A short introduction to boosting. Journal of Japanese Society for Artificial Intelligence (5), 771–80 (1999).
11. Jiawei, H., Micheline, K., Jian, P.: Data Mining: Concepts and Techniques. Morgan Kaufmann (2011).
12. Leonard, T., Hsu, J.S.: Bayesian Methods: An Analysis for Statisticians and Interdisciplinary Researchers. Cambridge University Press (2001).

13. Machov, K., Bark, F., Bednr, P.: A bagging method using decision trees in the role of base classifiers (2006).
14. Quinlan, J.R.: Induction of decision trees. Machine Learning (1), 81–106 (1986).
15. Quinlan, J.R.: C4.5: Programs for Machine Learning. Morgan Kaufmann (1993).
16. Rifkin, R.M., Klautau, A.: In defense of one-vs-all classification. Journal of Machine Learning Research pp. 101–141 (2004).
17. Tax, D.M.J., Duin, R.P.W.: Using two-class classifiers for multiclass classification. In: ICPR (2), pp. 124–127 (2002).
18. Vapnik, V.N.: The Nature of Statistical Learning Theory. Statistics for Engineering and Information Science. Springer (2000).

Classification Based on Homogeneous Logical Proportions

Ronei M. Moraes, Liliane S. Machado, Henri Prade and Gilles Richard

Abstract A classification method based on a measure of analogical dissimilarity has been proposed some years ago by Laurent Miclet and his colleagues, which was giving very good results. We restart this study on a slightly different basis. Instead of estimating analogical dissimilarity, we use the logical definition of an analogical proportion. We also consider other related logical proportions and their link with analogical proportion. The paper reports on an ongoing work and contributes to a comparative study of the logical proportions predictive accuracy on a set of standard benchmarks coming from UCI repository. Logical proportions constitute an interesting framework to deal with binary and/or nominal classification tasks without introducing any metrics or numerical weights.

1 Introduction

In problem solving, analogical reasoning [3, 5] is often viewed as a way to enhance reasoning processes by transferring conclusions observed in known situations to an only partially known situation. A particular instance of this reasoning style is derived from the notion of analogical proportion linking four situations or items a, b, c, d and stating that *a is to b as c is to d*, often denoted $a : b :: c : d$. This statement basically

L. S. Machado (✉) · R. M. Moraes
LabTEVE, Federal University of Paraiba, Cidade Universitaria s/n, Joao Pessoa, Brazil
e-mail: liliane@di.ufpb.br

R. M. Moraes
e-mail: ronei@de.ufpb.br

H. Prade · G. Richard
IRIT, University of Toulouse, 118 route de Narbonne, 31062 Toulouse Cedex 09, France
e-mail: prade@irit.fr

G. Richard
e-mail: richard@irit.fr

M. Bramer and M. Petridis (eds.), *Research and Development in Intelligent Systems XXX*,
DOI: 10.1007/978-3-319-02621-3_4,

expresses that a differs from b as c differs from d and vice-versa. In a classification context where a, b, c are in a training set, i.e., their class $cl(a)$, $cl(b)$, $cl(c)$ are known, and d is a new object whose class is unknown, this analogical proportion leads to a common sense inference rule which can be stated as:

if a is to b as c is to d, then cl(a) is to cl(b) as cl(c) is to cl(d)

In [2, 6, 13], this inference principle has been deeply investigated on diverse domains and even implemented, using a measure of analogical dissimilarity. More precisely, when dealing with classification problems as in [6, 12], the items a, b, c, d are generally represented as vectors of values. To fit with real life applications, the previous inference rule has to be (in some sense) weakened into: *if* an analogical proportion holds for a large number of properties between the four items, *then* an analogical proportion should also hold for the class of these items.

In the last past years, a logical model of analogical proportions has been proposed [7, 9, 10]. It has been shown that analogical proportions are a special case of so-called *logical proportions* [8]. Roughly speaking, a logical proportion between four terms a, b, c, d equates similarity or dissimilarity evaluations about the pair a, b with similarity or dissimilarity evaluations about the pair c, d. Among the set of existing logical proportions, four *homogeneous* ones (in the sense that they are equating similarities with similarities, or dissimilarities with dissimilarities), including the analogical proportion, enjoy remarkable properties and seem particularly attractive for completing missing values [10].

This paper suggests to use diverse homogeneous proportions to implement what could be called "logical proportion-based classifiers". We make use of analogical proportion, but also of the three other homogeneous proportions. The implementation provides good results and suggests that logical proportions offer a suitable framework for classification tasks.

The paper is organized as follows. After a short background on Boolean analogical proportion and three related proportions in Sect. 2, we briefly recall in Sect. 3 the work of [2] and we describe the new algorithm we suggest. In Sect. 4, we report experiments on datasets coming from the UCI. Our results are analyzed and compared to those obtained in [2]. We conclude in Sect. 5.

2 Logical Proportions: A Short Background

A *logical proportion* [9] is a particular type of Boolean expression $T(a, b, c, d)$ involving four variables a, b, c, d, (whose truth values belong to $\mathbb{B} = \{0, 1\}$). It is made of the conjunction of two distinct equivalences, involving a conjunction of variables a, b on one side, and a conjunction of variables c, d on the other side of $\equiv$, where each variable may be negated. Both $a \wedge \overline{b}$ and $\overline{a} \wedge b$ capture the idea of dissimilarity between a and b, while $a \wedge b$ and $\overline{a} \wedge \overline{b}$ capture the idea of similarity, positively and negatively. For instance, $(a\overline{b} \equiv c\overline{d}) \wedge (\overline{a}b \equiv \overline{c}d)$[1] is the expression

[1] $\overline{a}$ is a compact notation for the negation of a and $a\overline{b}$ is short for $a \wedge \overline{b}$, and so on.

of the *analogical proportion* [7]. As can be seen, analogical proportion uses only dissimilarities and could be informally read as *what is true for a and not for b is exactly what is true for c and not for d, and vice versa.* When a logical proportion does not mix similarities and dissimilarities in its definition, we call it *homogeneous*: For instance, analogical proportion is homogeneous.

More generally, it has been proved that there are 120 semantically distinct logical proportions that can be built [8]. Moreover, each logical proportion has exactly 6 lines leading to true (i.e., 1) in its truth table, the ten remaining lines lead to false (i.e., 0). Logical proportions are quite rare in the world of quaternary Boolean formulas, since there are 8080 quaternary operators that are true for 6 and only 6 lines of this truth table. Inspired from the well known numerical proportions, two properties seem essential for defining the logical proportions that could be considered as the best counterparts to numerical proportions:

- Numerical proportions remain valid when is exchanged (a, b) with (c, d). Similarly, logical proportions T should satisfy *symmetry* property:

$$T(a, b, c, d) \implies T(c, d, a, b) \tag{1}$$

- A valid numerical proportion does not depend on the representation of the numbers in a particular basis. In the same spirit, logical proportions should satisfy the *code independency* property:

$$T(a, b, c, d) \implies T(\overline{a}, \overline{b}, \overline{c}, \overline{d}) \tag{2}$$

  insuring that the proportion T holds whatever the coding of falsity with 0 or 1, or if we prefer, T holds whatever we consider a property or its negation for describing the objects.

Only 4 among the 120 proportions satisfy the two previous properties [10]. These four proportions also satisfy other properties associated with the idea of proportion such as central permutation

$$T(a, b, c, d) \implies T(a, c, b, d) \tag{3}$$

In Table 1, we provide the definitions of the 4 homogeneous logical proportions and in Table 2, we give their truth tables, restricted to the six valuations leading to truth value one.

These proportions, denoted by $A(a, b, c, d)$, $R(a, b, c, d)$, $P(a, b, c, d)$ and $I(a, b, c, d)$, are called *analogy*, *reverse analogy*, *paralogy* and *inverse paralogy*

Table 1 4 remarkable logical proportions: A, R, P, I

A	R	P	I
$(a\overline{b} \equiv c\overline{d}) \wedge (\overline{a}b \equiv \overline{c}d)$	$(a\overline{b} \equiv \overline{c}d) \wedge (\overline{a}b \equiv c\overline{d})$	$(ab \equiv cd) \wedge (\overline{a}\overline{b} \equiv \overline{c}\overline{d})$	$(ab \equiv \overline{c}\overline{d}) \wedge (\overline{a}\overline{b} \equiv cd)$

Table 2 A, R, P, I: Boolean truth tables

A	R	P	I
0000	0000	0000	0011
1111	1111	1111	1100
0011	0011	0110	0110
1100	1100	1001	1001
0101	0110	0101	0101
1010	1001	1010	1010

[9], and respectively express that a (resp. b) differs from b (resp. a) as c (resp. d) differs from d (resp. c), that a is to b as d is to c, that what a and b have in common, c and d have it also, and that what a and b have in common, both c and d miss it. The 4 homogeneous proportions are linked together:

$$A(a, b, c, d) \equiv R(a, b, d, c) \equiv P(a, d, c, b) \equiv I(a, \overline{b}, \overline{c}, d). \tag{4}$$

The semantical properties of these 4 proportions have been investigated in [10].

The idea of proportion is closely related to the idea of extrapolation, i.e., to guess/compute a new value on the ground of existing values. In other words, if for some reason, it is believed or known that a proportion holds between four binary items, three of them being known, then one may try to infer the value of the fourth one. The problem can be stated as follows. Given a logical proportion T and a 3-tuple (a, b, c), does there exist a Boolean value x such that $T(a, b, c, x) = 1$, and in that case, is this value unique? It is easy to see that there are always cases where the equation has no solution since the triple a, b, c may take $2^3 = 8$ values, while any proportion T is true only for 6 distinct 4-tuples. For instance, when we deal with analogy A, the equations $A(1, 0, 0, x)$ and $A(0, 1, 1, x)$ have no solution. And it has been proved, for instance that the analogical equation $A(a, b, c, x)$ is solvable iff $(a \equiv b) \vee (a \equiv c)$ holds. In that case, the unique solution is $x = a \equiv (b \equiv c)$. Similar results hold for the 3 remaining homogeneous proportions. As sketched in the introduction, this equation solving property can be the basis of a constructive classification rule over binary data by adopting the following inference rule: having 4 objects $\vec{a}, \vec{b}, \vec{c}, \vec{d}$ (three in the training set with classes $cl(\vec{a}), cl(\vec{b}), cl(\vec{c})$, the fourth being the object to be classified where $cl(\vec{d})$ is unknown:

$$\frac{\forall i \in [1, n],\ A(a_i, b_i, c_i, d_i)}{A(cl(\vec{a}), cl(\vec{b}), cl(\vec{c}), cl(\vec{d}))}$$

Then, if the equation $A(cl(\vec{a}), cl(\vec{b}), cl(\vec{c}), x)$ is solvable, we can allocate to $cl(\vec{d})$ the solution of this equation. Obviously, A could be replaced with R, P or I. Let us investigate an implementation of this principle.

3 Analogical Dissimilarity and Algorithms

Learning by analogy, as presented in [2] is a lazy learning technique which uses a measure of *analogical dissimilarity* between four objects. It estimates how far four situations are from being in analogical proportion. Roughly speaking, the analogical dissimilarity *ad* between 4 Boolean values is the minimum number of bits that have to be switched to get a proper analogy. For instance $\mathtt{ad}(1,0,1,0) = 0$, $\mathtt{ad}(1,0,1,1) = 1$ and $\mathtt{ad}(1,0,0,1) = 2$. Thus, $A(a,b,c,d)$ holds if and only if $\mathtt{ad}(a,b,c,d) = 0$. Moreover $\mathtt{ad}$ differentiates two types of cases where analogy does not hold, namely the eight cases with an odd number of 0 and an odd number of one among the 4 Boolean values, such as $\mathtt{ad}(0,0,0,1) = 1$ or $\mathtt{ad}(0,1,1,1) = 1$, and the two cases $\mathtt{ad}(0,1,1,0) = \mathtt{ad}(1,0,0,1) = 2$.

When, instead of having 4 Boolean values, we deal with 4 Boolean vectors in $\mathbb{B}^n$, we add the $\mathtt{ad}$ evaluations componentwise to get the analogical dissimilarity between 4 Boolean vectors, which leads to an integer belonging to the interval $[0, 2n]$. This number is a numerical measure of how far the four vectors are from building, componentwise, a complete analogy. It has been used in [2] to implement a classification algorithm where we have as input a training set S of classified items, a new item $\overrightarrow{d}$ to be classified, and an integer k. The algorithm proceeds as follows:

Step 1: Compute the analogical dissimilarity $\mathtt{ad}$ between $\overrightarrow{d}$ and all the triples in S^3 that produce a solution for the class of $\overrightarrow{d}$.

Step 2: Sort these n triples by the increasing value of $\mathtt{ad}$ wrt with $\overrightarrow{d}$.

Step 3: If the k-th triple has the integer value p for $\mathtt{ad}$, then let k' be the greatest integer such that the k'-th triple has the value p.

Step 4: Solve the k' analogical equations on the label of the class. Take the winner of the k' votes and allocate this winner as the class of $\overrightarrow{d}$.

This simple approach provides remarkable results and, in most of the cases, outperforms the best known algorithms (see [6]). Our approach is a bit different and we consider a simpler analogical dissimilarity that we denote ad^* which, on binary values a, b, c, d, is equal to 0 when $A(a,b,c,d)$ holds, and is 1 otherwise. If we consider $A(a,b,c,d)$ as a numerical value 0 or 1, the function:

$$\mathtt{ad}^*(a,b,c,d) = ((a-b-c+d)^2)^{1/2}, \text{ for } a,b,c,d \in \{0,1\}, \tag{5}$$

it is nothing more than $\mathtt{ad}^*(a,b,c,d) = 1 - A(a,b,c,d)$. Of course, other functions can be used and can provide the same results, as for instance: $\mathtt{ad}^{*1}(a,b,c,d) = |a-b-c+d|$.

It means that we do not differentiate the two types of cases where analogy does not hold. Considering Boolean vectors in $\mathbb{B}^n$, we add the $\mathtt{ad}^*$ evaluations componentwise, to get a final integer belonging to the interval $[0, n]$ (instead of $[0, 2n]$ in the previous case). Namely,

$$\mathrm{ad}^*(\vec{a}^k, \vec{b}^k, \vec{c}^k, \vec{d}) = \sum_{i=1,n} \mathrm{ad}^*(a_i^k, b_i^k, c_i^k, d_i). \tag{6}$$

Our algorithm is as follows, taking as input a training set S, the item to be classified $\vec{d}$, and a integer p, with $0 \leq p \leq n$, which means for each quadruple analyzed, we accept at most p features for which the considered proportion does not hold true.

Step 1: Compute $\mathrm{ad}^*(\vec{a}^k, \vec{b}^k, \vec{c}^k, \vec{d})$ between $\vec{d}$ and all triples $(\vec{a}^k, \vec{b}^k, \vec{c}^k) \in S^3$.

Step 2: If $\mathrm{ad}^*(\vec{a}^k, \vec{b}^k, \vec{c}^k, \vec{d}) \leq p$, solve the class analogical equation. If there is a solution, consider this solution as a vote for the predicted class of $\vec{d}$.

Step 3: Take the winner of the votes and allocate it as the class of $\vec{d}$.

In terms of worst case complexity, our algorithm is still cubic, due to step 1. But, it avoids the sorting step of the previous algorithm, leading to a lower complexity than the one of [2]. These options are implemented and the results are shown below.

4 First Experiments and Discussion

To ensure a fair comparison with the approach in [2], we use the same first four datasets with binary and nominal attributes, all coming from the University of California at Irvine repository [1]. We consider the following datasets.

- MONK 1,2 and 3 problems (denoted MO.1, MO.2 and MO.3). MONK3 problem has noise added.
- SPECT heart data (SP. for short).

To binarize, we use the standard technique, replacing a nominal attribute having n different values with n binary attributes. We summarize the datasets structure and our results in Table 3.

As can be seen, the results obtained are very similar to those obtained in [6], although we are using a counterpart ad^* of the logical view of the analogical proportion for estimating the analogical dissimilarity of four vectors, instead of the more discriminating measure introduced in [6]. Note that for very small values of p some instances $\vec{d}$ may not find a triple for predicting a class for them. This may lead to a low accuracy, which then increases when p is less requiring. For too large values of p then the accuracy generally decreases since the number of features for which an analogical proportion holds becomes too small and then we allow "poor quality" triples to vote.

Besides, instead of using $\mathrm{ad}^*(a, b, c, d) = 1 - A(a, b, c, d)$, one may replace A by reverse analogy R, or paralogy P, or inverse paralogy I. We have checked that the results remain the same if we use R, or P instead of A. This is in perfect agreement with the existence of permutations changing one proportion into another as highlight in Eq. (4).

Table 3 Results

Information	MO.1	MO.2	MO.3	SP.
nb. of nominal attributes	7	7	7	22
nb. of binary attributes	15	15	15	22
nb. of training instances	124	169	122	80
nb. of test instances	432	432	432	172
nb. of class	2	2	2	2
Our algorithm accuracy (best results)	97.22	99.77	97.22	57.22
Best value of p	1	0	2	4
Miclet *et al.* best result [6][a]	98	100	96	58
Decision table	97.22	66.21	97.22	71.12
PART	92.59	74.31	98.15	81.28
Multilayer perceptron	100.00	100.00	93.52	72.73
Logistic	71.06	61.57	97.22	66.31
BayesNet	71.06	61.11	97.22	75.94

[a] Results were provided by [6] as presented here

However, when considering $\mathtt{ad}^*(a, b, c, d) = 1 - I(a, b, c, d)$, first results indicate that the accuracy decreases substantially. This does not really come as a surprise, since looking at the truth table of I, we then allow the simultaneous presence of patterns such as $(1, 0, 1, 0)$ and $(1, 0, 0, 1)$, in complete disagreement with the measure of analogical dissimilarity $\mathtt{ad}$, even if I would apparently agree with the idea that objects that share (almost) nothing should belong to different classes. Ultimately, when a triple (a, b, c) belongs to S^3, it does not imply that the triple $(a, \overline{b}, \overline{c})$ belongs to S^3, and then, thanks again to Eq. 4, the set of voters for I may be different from the set of voters of A, R and P, for which the set is the same.

We made comparisons with some classifiers found in the literature and for these ones, we used the Weka package [4] with default values. It is possible to note the Multilayer Perceptron achieved the best accuracy for MO.1 and MO.2 databases, but its performance is lower for the MO.3 database. PART classifier provides best classification for MO.3 and SP. databases, but its performance is not so good for MO.2. Decision Table provides good classifications for MO.1 and MO.2 databases. Similar results were found by [6]. Logistic and BayesNet classifiers have good performance only for MO.3 database. So, for these four databases and this set of classifiers, only our algorithm and [6] achieved good results for three databases. It shows that our algorithm is a competitive approach for classification tasks of databases with binary and nominal features.

5 Concluding Remarks

This preliminary study shows that the logical view of analogical proportion leads to good results in classification. More experiments currently under progress should bring more light on the behaviour of the classification process with regard to p. Using the same type of algorithm and the extension to multi-valued logics as done in [11] will allow us to deal with numerical values and continuous attributes.

The algorithm, when compared with others found in the literature, provides competitive results for three of of the four benchmarks considered. New comparisons with other classifiers and other databases are under progress to better understand the applicability and the competitiveness of the approach.

Acknowledgments This work is supported by CNPq, processes 246939/2012-5, 246938/2012-9, 310561/2012-4, 310470/2012-9 and INCT-MACC (process 181813/2010-6).

References

1. Bache, K., Lichman, M.: UCI machine learning repository (2013). http://archive.ics.uci.edu/ml
2. Bayoudh, S., Miclet, L., Delhay, A.: Learning by analogy: A classification rule for binary and nominal data. Proc. Inter. Conf. on, Artificial Intelligence IJCAI07 pp. 678–683 (2007).
3. Gentner, D., Holyoak, K.J., Kokinov, B.N.: The Analogical Mind: Perspectives from Cognitive Science. Cognitive Science, and Philosophy. MIT Press, Cambridge, MA (2001).
4. Hall, M., Frank, E., Holmes, G., Pfahringer, B., Reutemann, P., Witten, I.H.: The weka data mining software: An update. SIGKDD Explorations **11**(1), 10–18 (2009).
5. Melis, E., Veloso, M.: Analogy in problem solving. In: Handbook of Practical Reasoning: Computational and Theoretical Aspects. Oxford Univ. Press (1998).
6. Miclet, L., Bayoudh, S., Delhay, A.: Analogical dissimilarity: definition, algorithms and two experiments in machine learning. JAIR, 32 pp. 793–824 (2008).
7. Miclet, L., Prade, H.: Handling analogical proportions in classical logic and fuzzy logics settings. In: Proc. 10th Eur. Conf. on Symbolic and Quantitative Approaches to Reasoning with Uncertainty (ECSQARU'09), Verona, pp. 638–650. Springer, LNCS 5590 (2009).
8. Prade, H., Richard, G.: Logical proportions—typology and roadmap. In: E. Hüllermeier, R. Kruse, F. Hoffmann (eds.) Computational Intelligence for Knowledge-Based Systems Design: Proc. 13th Inter. Conf. on Information Processing and Management of Uncertainty (IPMU'10), Dortmund, June 28 - July 2, *LNCS*, vol. 6178, pp. 757–767. Springer (2010).
9. Prade, H., Richard, G.: Reasoning with logical proportions. In: Proc. 12th Inter. Conf. on Principles of Knowledge Representation and Reasoning (KR'10), Toronto, Ontario, Canada, May 9–13, (F. Z. Lin, U. Sattler, M. Truszczynski, eds.), pp. 545–555. AAAI Press (2010).
10. Prade, H., Richard, G.: Homogeneous logical proportions: Their uniqueness and their role in similarity-based prediction. In: G. Brewka, T. Eiter, S.A. McIlraith (eds.) Proc. 13th Inter. Conf. on Principles of Knowledge Representation and Reasoning (KR'12), Roma, June 10–14, pp. 402–412. AAAI Press (2012).
11. Prade, H., Richard, G.: Analogical proportions and multiple-valued logics. In: L.C. van der Gaag (ed.) Proc. 12th Europ. Conf. on Symbolic and Quantitative Approaches to Reasoning with Uncertainty (ECSQARU'13), Utrecht, July 8–10, *LNCS*, vol. 7958, pp. 497–509. Springer (2013).
12. Prade, H., Richard, G., Yao, B.: Enforcing regularity by means of analogy-related proportions-a new approach to classification. International Journal of Computer Information Systems and Industrial Management Applications **4**, 648–658 (2012).
13. Stroppa, N., Yvon, F.: Analogical learning and formal proportions: Definitions and methodological issues. ENST Paris report (2005).

Knowledge Discovery and Data Mining II

Predicting Occupant Locations Using Association Rule Mining

Conor Ryan and Kenneth N. Brown

Abstract Heating, ventilation, air conditioning (HVAC) systems are significant consumers of energy, however building management systems do not typically operate them in accordance with occupant movements. Due to the delayed response of HVAC systems, prediction of occupant locations is necessary to maximize energy efficiency. In this paper we present two approaches to occupant location prediction based on association rule mining which allow prediction based on historical occupant movements and any available real time information, or based on recent occupant movements. We show how association rule mining can be adapted for occupant prediction and evaluate both approaches against existing approaches on two sets of real occupants.

1 Introduction

Office buildings are significant consumers of energy: buildings typically account for up to 40 % of the energy use in industrialised countries [1], and of that, over 70 % is consumed in the operation of the building through HVAC and lighting. A large portion of this is consumed under static control regimes, in which heating, cooling and lighting are applied according to fixed schedules, specified when the buildings were designed, regardless of how the buildings are actually used. To improve energy efficiency, the building management system should operate the HVAC systems in response to the actual behaviour patterns of the occupants. However, heating and cooling systems have a delayed response, so to satisfy the needs of the occupants,

C. Ryan (✉) · K. N. Brown
Cork Constraint Computation Centre, Department of Computer Science,
University College Cork, Cork, Ireland
e-mail: cryan@4c.ucc.ie

K. N. Brown
e-mail: k.brown@cs.ucc.ie

M. Bramer and M. Petridis (eds.), *Research and Development in Intelligent Systems XXX*,
DOI: 10.1007/978-3-319-02621-3_5,

the management system must predict the occupant behaviour. The prediction system should be accurate at both bulk and individual levels: the total number of occupants of a building or a zone determine the total load on the HVAC system, while knowing the presence and identity of an occupant of an individual office allows us to avoid waste through unnecessary heating or cooling without discomforting the individual.

We believe that in most office buildings, the behaviour of occupants tends to be regular. An occupant's behaviour may relate to the time of day, the day of the week or the time of year. Their behaviour on a given day may also depend on their location earlier on that day or on their most recent sequence of movements. We require a system which is able to recognize these time and feature based patterns across different levels of granularity from observed data. Further, many office users now use electronic calendars to manage their schedules, and information in these calendars may support or override the regular behaviour. The reliability of the calendar data will depend on the individual maintaining it, so the prediction system needs to be able to learn occupant-specific patterns from the calendars.

We propose the use of association rule mining for learning individual occupant behaviour patterns. We wish to find patterns of any kind which can be used to predict occupant movements, for which association rule mining is ideal as it is designed to find any useful patterns in a dataset. We use the Apriori algorithm [2], and show how the algorithm can be extended to represent time series, incorporating calendar entries. We then propose a number of transformations of the learning mechanism, pruning itemsets and rules to focus in on useful rules, and extending the generation of itemsets in areas where useful patterns will be found. Finally we describe a further modification of this approach which incorporates time-independent sequences. We evaluate the performance on two sets of actual occupant data, and show up to 76 and 86 % accuracy on each set respectively.

The remainder of this paper is organized as follows: Sect. 2 provides an overview of association rules and the existing work on location prediction. Sections 3 and 4 detail the modifications we make to Apriori to make timeslot-specific and timeslot-independent predictions respectively. In Sect. 5 we outline the datasets we use for evaluation and the other approaches we evaluate against and present our results. We conclude the paper in Sect. 6.

2 Related Work

Existing methods for predicting occupant locations include bayesian networks [3], neural networks [4], state predictors [5], hidden markov models [6], context predictors [7], eigenbehaviours [8].

The Bayesian network approach presented in [3] predicts the occupant's next location based on the sequence of their previous locations and the current time of day and day of the week. Based on the current room and the day/time, it also predicts the duration of the occupant's stay in the current room. This results in separate predictions for the occupant's next location and for the time they will move.

The neural network approach uses a binary codification of the location sequences as input to a neural network. In [4] both local and global predictors are considered. A local predictor is a network which is trained on and predicts a particular occupant, and thus deals only with codified location sequences. The global predictor takes all occupants' location sequences, along with associated occupant codes, as training data, and can make predictions for any occupant.

The state predictor approach in [5] uses a two-level context predictor with two-state predictors. This method selects a two-state predictor based on the occupant's sequence of previous locations. Each state within the selected predictor is a prediction; the current state is used as the prediction, and the state may then change depending on whether the prediction was accurate. Being a two-state predictor, each possible location has two corresponding states, so a maximum of two incorrect predictions for any given sequence is necessary to change future predictions, resulting in fast retraining if an occupant changes their behaviour. The second level of this predictor can alternatively store the frequencies of the possible next locations for each sequence. This makes it equivalent to a markov model approach.

These approaches all predict the occupant's next location, and with the exception of the Bayesian network, only use the occupant's recent locations. Our application requires longer term predictions and we believe there may be more general associations between the occupants' locations at different times which allow for such predictions. Association rule mining is intended to discover general patterns in data and so we propose to investigate whether association rule mining can be used to predict occupant locations.

Association rule mining was introduced in [2] as an unsupervised approach to finding patterns in large datasets. The original application was discovering patterns in datasets of transactions, where each transaction was a market basket, i.e. a set of purchased items. In that application items were literals, simple strings which are either present or absent in a transaction; however the algorithm can be applied without modification to sets of attribute/value pairs. We chose Apriori as it is the most basic association rule mining algorithm and thus simplest to modify.

Let U be a universe of items. A dataset D is a set of instances $\{I_1 \ldots I_n\}$, where each instance is a set of items from U. An itemset X is a subset of U. The frequency of X, $freq(X)$, is the number of instances I in D for which $X \subseteq I$, while the support is $supp(x) = freq(X)/|D|$. An association rule is an implication of the form $X \Rightarrow Y$ where X and Y are itemsets such that $X \cap Y = \emptyset$. This rule states that each instance which contains X tends to contain Y. The support of the rule is $supp(X \cup Y)$. The confidence of the rule is how often it is correct as a fraction of how often it applies $conf(X \Rightarrow Y) = supp(X \cup Y)/supp(X)$.

The purpose of an association rule mining algorithm is to find the set of rules which are above user-specific thresholds of confidence and support. The first step is to find all itemsets which are 'frequent' according to the support threshold. Association rules are then generated from these itemsets, and any rules which fall below the user-specified minimum confidence are discarded. Confidence is used to measure the reliability of a rule in terms of how often it is correct according to the training data. Finding the frequent itemsets is the more difficult step, as the desired itemsets must be found among the $2^{|U|} - 1$ itemsets which can be generated.

Apriori uses breadth first search to find all frequent itemsets. First all itemsets of size 1 are enumerated. Itemsets whose support falls below the support threshold (infrequent itemsets) are removed, as any superset of an infrequent itemset will also be infrequent. Candidate itemsets of size 2 are then generated by combining all frequent itemsets of size 1, and infrequent itemsets of size 2 are removed. This process continues, finding frequent itemsets of size n by generating candidates from the itemsets of size n-1 and removing infrequent itemsets, until an n where no frequent itemsets exist is reached.

Once the frequent itemsets have been found, for each frequent itemset X all rules of the form $Y \Rightarrow X - Y$ where $Y \subset X$ and $Y \neq \emptyset$ are generated, and those which do not obey the confidence threshold are discarded.

3 Adapting Association Rule Mining for Occupant Prediction

The first task in applying association rule mining is to determine the format of the dataset. We define an instance to be a single day for a single occupant, recording for each time slot the location of the occupant. It also includes a set of scheduled locations, specifying where the occupant's calendar stated they would be. Finally, each instance records which occupant and day of the week it applies to. Thus the set of attributes in our dataset is $A = \{d, o, l_i \ldots l_j, s_i \ldots s_j\}$, where d is the day, o is the occupant, is the occupant's location at time slot n, and s_n is the location the occupant l_n was scheduled to be in at time n. Our objective then is to find rules which predict the value of an attribute in $\{l_i \ldots l_j\}$ based on the other attributes. In order to be able to compare confidences meaningfully, we restrict our attention to rules which predict single attributes.

Although this format is all that is needed to run Apriori, it is unlikely to produce usable results. The items in our dataset have semantics which are critical for the eventual application, but Apriori by default treats them all as equivalent. The location attributes $\{l_i \ldots l_j\}$ represent an ordered list of time/location pairs which it is our objective to predict. However, Apriori has no concept of the importance of or ordering over these items, so it will produce rules which run counter to the order, i.e. rules which use later locations to predict earlier locations, and which make useless predictions, e.g. predicting timetable entries.

A further important attribute distinction is that $\{l_i \ldots l_j\}$ and $\{s_i \ldots s_j\}$ are actual location data, whereas d and o are data labeling the location data, i.e. meta-data. Due to this their values are in a sense fixed. For example, in an instance which describes occupant A's movements on a Monday, d and o are fixed at Monday and A respectively, whereas all the other attributes can, in principle, take any value in their domain. This affects the meaning of the support metric as the maximum support for any itemset which includes d or o will be less than 1. Since support is used to determine which itemsets are considered frequent, patterns which occur frequently for certain days and/or agents will be rated as less frequent due to the inclusion of other days and agents in the dataset.

A problem with regard to the content of the data is that the many common patterns tend to be the least interesting, while we require low frequency patterns to be found in order to make predictions in unusual circumstances. Consider for example an occupant who has a 90 % chance of being in their office in any timeslot from 9 am to 5 pm. In this case, any pattern of the form "in at N implies in at M" where N and M are between 9 and 5 will have support of at least 80 %, thus all such patterns will be found. But there is no real correlation there; all these patterns could be summarized simply as "the occupant is likely to be in". At the extreme opposite end, we have days when the occupant does not turn up at all, due to illness or other reasons—a very obvious pattern which would be represented by rules such as "out at 9, 10, 11 implies out at 12". Such rules could have confidence close to 100 % if the occupant tends to be in in the morning, but if absences are rare the itemset behind the rule will have such low support it won't even be a candidate. Since enumerating every itemset is not feasible, we wish to eliminate the common uninteresting ones and focus on the less common but interesting ones.

3.1 Candidate/Rule Pruning

As mentioned above, standard Apriori has no concept of the relationships between the items in an instance which exist in occupancy data. Due to this it will by default generate some useless rules. The important features are that the location attributes $\{l_i \dots l_j\}$ represent an ordered list and that they are the only attributes we wish to predict. As an itemset which does not contain any of these attributes cannot produce a rule which predicts any of them, we eliminate itemsets which do not contain some subset of $\{l_i \dots l_j\}$ during candidate elimination.

With regard to rule generation, we only wish to predict the future based on the past (i.e. rules which obey the ordering of $\{l_i \dots l_j\}$)), and we only wish to predict a single location at a time in order to allow meaningful comparison of the rules at rule selection time. Thus our rule generation is as follows: for every itemset $\{l_i \dots l_j, x_i \dots x_j\}\}$, where l is a location item and x is any other type of item, l_j is the consequent and all other items are the antecedent.

3.2 Support Modification

In 2.1 we provided the typical definition of support, the proportion of the instances which contain the itemset/rule. To deal with the reduction in support for itemsets which contain metadata items, we redefine support as $supp(X) = freq(X)/max(freq(X))$. For market basket items, which can in principle occur in every instance, this is the same definition. In the case of our metadata attribute/value pairs however, this definition results in a different value which is normalized such that the maximum value of $supp(X)$ is always 1 for comparison to other support values.

Using this modified support threshold in Apriori allows it to find itemsets when have a lower support due to their metadata attributes. However this greatly increases the area of the itemset lattice which is explored for any given support threshold. Thus, in order to conserve memory, we mine each possible combination in a separate pass. For every combination of metadata attributes/values C, we initialize Apriori with all itemsets of size $|C|+1$ which are a superset of C, instead of standard 1-itemsets. This allows the generation of every itemset which contains that metadata combination in a separate pass.

3.3 Windowing

Some important patterns have such low support that trying to find them by simply lowering the support threshold would result in a combinatorial explosion. Instead we will use the structure of the data to target them specifically. An example of such a pattern is a full day of absence: a very obvious pattern, but one which occurs so infrequently that it won't be learned. As our location attributes form an ordered list we can define subsets of them which are consecutive, temporal windows over the location data. By mining these subsets individually, we can reduce the size of the space of itemsets while still discovering the itemsets which describe consecutive elements of the low support patterns.

We define a window as: where $Win(n, m) =< d, o, l_{n...n+m}, S_{i...j} >$ where i and j denote the first and last timeslots, and n and m denote the beginning and length of the window respectively. In the windowing phase, we search within every window of the chosen length. This approach ignores patterns which span times which do not fit within a window. We choose to focus on patterns which occur in consecutive time slots as predicting occupant locations based on their most recent movements has been shown to work by the other approaches discussed in Sect. 2.

For distinct patterns windowing is sufficient to find rules which will make the correct predictions should the pattern recur. Taking the example of an occupant who is absent all day, within each window we will learn that consecutive hours of absence imply absence in the next hour. Taken in combination, these rules will state that at any hour of the day, consecutive absence implies continued absence, although we are still not learning sequences in the same sense as the approaches in Sect. 2, as the individual rules are still tied to specific time slots. These rules are added to the rules mined from the complete instances.

3.4 Rule Selection

Once the rules are generated we need a mechanism to choose a rule to make a prediction. When a prediction is required, values for any subset of the possible attributes can be supplied as an itemset V. A target for the prediction l_t is also given. We search the generated rules for all rules $X \Rightarrow Y$ where $X \subseteq V$ and $Y = \{l_t\}$. From these we select the rule with the highest confidence as the prediction.

4 Ordered Association Rules

In order to be able to predict occupant locations the timeslot-specific approach above requires that the occupant's behaviour correlates with the time of day. Our evaluation shows that on occupants for whom this does not hold the approach performs poorly. However, existing approaches which find patterns that don't relate to specific times are able to make accurate predictions on such occupants. We now describe a modification to Apriori which allows it to find time-independent sequences of locations using the order of the timeslots. As with the other timeslot-independent approaches, this approach relies on the occupant's most recent locations, and cannot make predictions beyond the next timeslot.

4.1 Ordered Itemsets

In order to represent time-independent sequences we define a new set of attributes $\{q_0 \dots q_j\}$ which represent an ordered list of consecutive locations. These attributes are similar to the timeslot attributes $\{l_i \dots l_j\}$, but rather than being a list of time/location pairs, $\{q_0 \dots q_j\}$ is a list of ordering/location pairs. Thus the first attribute is always '0', as it is the first element in the list. As we deal only with consecutive sequences, for any list of length $j+1$, all elements $0 \dots j$ must be present.

Ordered itemsets are itemsets of the form: $\{q_0 \dots q_j\}$. Each ordered itemset is essentially the set of itemsets $\{\{l_k \dots l_{j+k}\} : n \leq k \leq m - j\}$, where n and m are the minimum and maximum timeslots in the dataset respectively, represented as a single list. An ordered itemset may be instantiated to a time specific itemset by choosing a starting timeslot k and adding k to every attribute in the list, turning the itemset $\{q_0 \dots q_j\}$ into the itemset $\{l_k \dots l_{j+k}\}$ for the chosen timeslot k.

For example, take the ordered itemset {0=>O, 1=>A, 2=>O}, which signifies that an occupant is in their office, leaves for an hour, and then returns. If we set k to be 12:00, this itemset becomes {12=>O, 13=>A, 14=>O}, which states that an occupant is in their office at 12:00, leaves at 13:00, and returns at 14:00.

The individual timeslot-specific itemsets which the ordered itemset represents could be found separately by our original approach, however it would require each of them to occur separately with sufficiently high support, and for rules to be generated for each variation it would similarly require each to separately have sufficiently high confidence. Searching for the sequence of movements over all timeslots provides two advantages. First, that a pattern which recurs will be supported even if it recurs at different times, resulting in low support for each individual instance of the pattern, allowing us to find a pattern we otherwise wouldn't. Second, that when we generate an ordered rule from an ordered itemset, it can apply in cases where the pattern had low or even no previous support.

4.2 Confidence and Support

Confidence and support are defined in the same way for ordered itemsets as for timeslot-specific itemsets. However, *freq(X)* for an ordered itemset *X* counts the number of occurrences of the sequence in each transaction, i.e. $freq(\{q_0 \dots q_j\}) = \sum_{k=n}^{m-j} freq(\{l_k \dots l_{j+k}\})$. This results in values greater than 1 for support, however the anti-monotonicity property still holds, and so using a support threshold to eliminate candidates is still valid for this definition of support, although the threshold no longer represents the fraction of the dataset in which the itemset occurs. We considered alternative definitions for support for ordered itemsets, however they failed to correctly represent the relative frequency of the itemsets of different sizes and/or broke the anti-monotonicity property Apriori relies on.

Confidence for ordered rules is still the fraction of the times that it applies that it is correct, however as with support it is now considered over all occurrences of the time-independent sequence that the rule is based on.

4.3 Candidate Generation

To generate itemsets we use a modified form of Apriori's candidate generation. Since the first attribute of any ordered itemset must be '0' and the attributes must be consecutive, any two itemsets of the same length will have the same attributes. This means that we cannot generate candidates by combining itemsets of the same length, as the only case where a longer itemset would be generated would be when the itemsets have different values for the same attributes, which will result in a support of zero.

Instead, for every possible pair of ordered itemsets of length *j*, we increment the attributes of one of the itemsets by one, shifting the sequence to the right by one timeslot, and then combine them. Thus we combine two itemsets $\{q_0 \dots q_j\}$ and $\{r_0 \dots r_j\}$ if $\forall n : 1 \leq n \leq j-1 : q_{n+1} = r_n$, essentially if the latter sequence can provides one item to be appended onto the former sequence. Aside from this modification, candidate generation proceeds as previously described.

4.4 Rule Selection

Rule selection proceeds in the same manner as in the timeslot-dependent approach, except that ordered rules are instantiated to check their applicability. Given the prediction target l_t and the attributes/values to predict on V, for each ordered rule $\{q_0 \ldots q_{j-1}\} \Rightarrow \{q_j\}$, we set $j = t$ to get the rule $\{l_{t-j} \ldots l_{t-1}\} \Rightarrow \{l_t\}$, before checking whether $\{l_{t-j} \ldots l_{t-1}\} \subseteq V$. As before, the applicable rule with the highest confidence is chosen. The ordered and timeslot-specific rules can be combined into a single ruleset and used simultaneously, however simply combining them provides no advantage over selecting and using only the more appropriate ruleset for the dataset.

5 Experimental Evaluation

To test our approach we use two datasets: data recorded by occupants of the 4C lab in UCC, and data from the Augsburg Indoor Location Tracking Benchmarks [9]. We also evaluate three methods which were used in [6] and [10], a HMM, an Elman net and a frequency predictor. The HMM and Elman net were evaluated using the respective tools in MATLAB, while we implemented the frequency predictor ourselves based on the frequency analysis context predictor in [5].

To gather data to test our approach, six occupants of the 4C lab in University College Cork including the authors manually recorded their movements over a period of 5–15 months using google calendar. Each occupant recorded their location by room code if within a campus building, or marked themselves as 'away' if off campus. The data was recorded from 8am to 6pm with half-hour granularity, with any occupancy of significantly shorter duration than 30 min. filtered out. The occupants also recorded their timetables for the time period, which recorded the locations they were scheduled to be in in the same format as the record of their actual movements. 20 locations were frequented by the occupants including the 'away' location. The test set for this evaluation was the most recent 2 months of data for each occupant, while the training set was all the preceding data each occupant had recorded, which covered between 3 and 13 months.

The Augsburg dataset contains data on four occupants for 1–2 weeks in summer and 1–2 months in fall. The format of the dataset is a series of timestamped locations for each occupant. As the data is not broken down into timeslots, we only compare the sequence based approaches on this version of the dataset. In order to be able to apply our timeslot-dependent approach to the data, we converted it to the same timeslot format as our gathered data. An occupant's location in each timeslot is the location in which they spent the majority of that timeslot. Following this conversion there are seven locations frequented by the occupants including 'away'. We compare all approaches on this version of the dataset.

5.1 Experiments

We generate time-dependent rules from each training set using a minimum support and confidence of 0.2 and 0.5 respectively. During windowing we use a window size of 3 slots and a minimum support of 0.05. We generate time-independent rules using no support or confidence threshold in order to maximize the coverage of the resulting ruleset. Instead of the support threshold we limit the length of generated rules to 2 items as shorter sequences have proven to be more reliable predictors of occupants' next locations.

Following are the configurations used for the comparison approaches. In the frequency predictor we use a maximum order of 2, again due to shorter sequences being more reliable predictors. For the hidden markov model we use 7–8 hidden states depending on the dataset as this maximized accuracy. The elman net uses the MATLAB default settings for layer delays and hidden layer size as no other combinations of values tested produced higher overall accuracy.

A feature of the frequency predictor is that it continues to train as it predicts; in our evaluation we allow all the sequence-based predictors to retrain with the days they have already predicted included in their training set, in order to maximize their accuracy. Timeslot Apriori predicts with only the initial training run, as the time taken to train makes retraining after every predicted day unfeasible.

We test all approaches on their accuracy in predicting the occupant's exact location in every time slot. As the sequence-based approaches use only recent occupant movements they can only predict for the next timeslot. Timeslot Apriori is tested on its ability to predict Next-Slot and Next-Day, and with or without timetable data available. The former determines whether $l_i \dots l_{n-1}$ are available, where n is the time slot being predicted, 'Next Slot' if this information is available, and 'Next Day' if not. The latter determines whether the values of $s_i \dots s_j$ are available when predicting, and is marked 'no Timetable' if they are not.

5.2 Results

Table 1 shows the accuracy of Timeslot Apriori making different types of predictions on the UCC dataset. The highest accuracy is achieved on Next-Slot predictions, which confirms that recent occupant movements, on which all the sequence-based approaches rely, are the most reliable predictor of an occupant's next location. For Next-Slot predictions, the timetable only helps marginally. Next-Day predictions are significantly less accurate as they must be made based solely on the occupants'

Table 1 Timeslot apriori accuracy by prediction type on UCC dataset

Next-slot (%)	Next-day (%)	Next-slot (No TT) (%)	Next-day (No TT) (%)
86	75	85	71

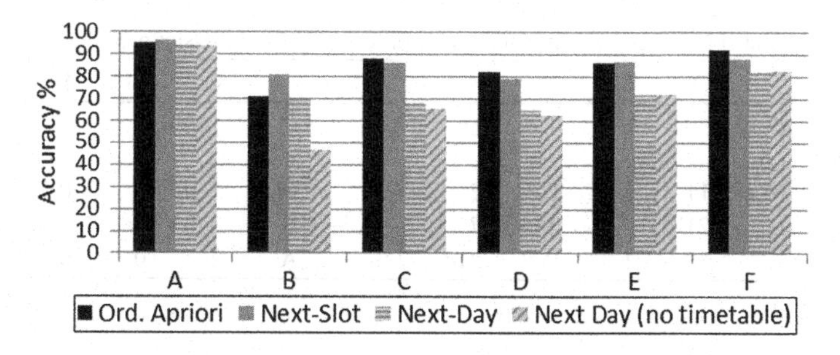

Fig. 1 Accuracy across all occupants on the UCC dataset

historical data without any knowledge of their movements during the day, however for these predictions the timetable does make a difference to the accuracy.

Figure 1 shows these results broken down by occupant, and includes Ordered Apriori making Next-Slot predictions. Occupant A's movements are very homogenous, so they are easily predicted by all prediction types. Occupant B has the most varied movements and the most scheduled events of all the occupants. This makes them harder to predict than the other occupants, however the addition of timetable data makes the largest difference on this occupant, especially on next-day predictions where it allows over 20 % higher accuracy. For the other occupants, whose movements follow general patterns with minimal scheduled events, their predictability is primarily contingent on the availability of real-time location data, resulting in one accuracy level for Next-Slot and a slightly lower one for Next-Day. However, Timeslot and Ordered Apriori do vary slightly on Next-Slot predictions, indicating that they do not make exactly the same predictions.

Figure 2 breaks the prediction accuracy down by timeslot, showing how the different prediction types fare at different times of day. In general we can see that they match in the morning, evening and around lunch, when the occupants are most predictable. Outside those times, Next-Day predictions drop down as the occupants' activities at those times are more variable, and real-time information is required to

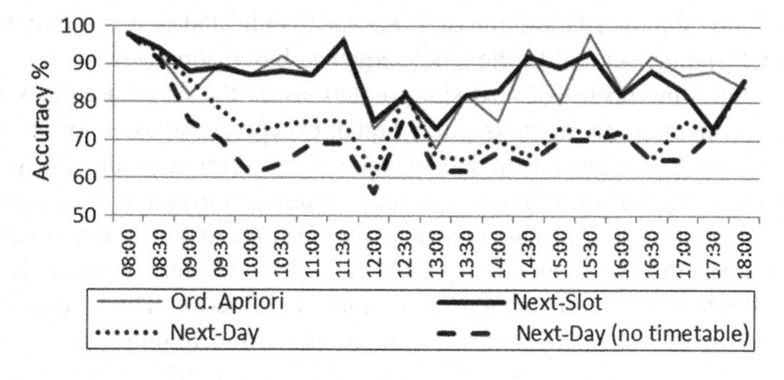

Fig. 2 Accuracy across the day on all occupants in the UCC dataset

Table 2 Occupant B and E confusion matrices for timeslot apriori next-slot predictions

		O	M	B	A		O	M	B	A
Actual Location	O	304	0	1	66	O	148	1	3	65
	M	11	19	0	2	M	1	6	0	0
	B	21	0	145	7	B	3	0	3	0
	A	37	1	0	214	A	29	0	0	581

Table 3 All approaches accuracy for next-slot on UCC dataset

HMM (%)	Elman net (%)	Frequency predictor (%)	Timeslot apriori (%)	Ord. apriori (%)
77	86	87	86	86

Table 4 All approaches accuracy for next-slot on Augsburg dataset

HMM (%)	Elman net (%)	Frequency predictor (%)	Timeslot apriori (%)	Ord. apriori (%)
74	76	77	39	76

maintain the accuracy level. Due to occupant B, Next-Day predictions are improved with the use of timetable data, particularly in the morning.

Table 2 shows the confusion matrices for occupants B and E, classifying the location either as in their own office (O), in a specific group meeting (M), any other room in the building (B) or away (A). The matrices show that the primary source of errors is reversing Office and Away, being uncertain whether the occupant will be in. For the group meeting and events elsewhere in the building, both occupants tend to be either predicted correctly or predicted to be in their office; recognizing that the occupant is in, but not that they will be leaving their office.

Table 3 shows that the sequence predictors generally match Timeslot Apriori for making Next-Slot predictions on the UCC dataset. The HMM performs worse because retraining it on a dataset this size was unfeasible, and so it was only trained on the initial training set, unlike the other sequence-based approaches.

Table 4 shows the results of next-slot predictions on the Augsburg dataset. The results for all methods are approximately equal, except for timeslot apriori, which performs poorly. The occupants in the Augsburg dataset follow predictable patterns in their movements, however they follow these patterns at irregular times. Timeslot Apriori cannot learn patterns independently from the time at which they occur; if an occupant repeats the same sequence of movements in a different timeslot, Timeslot Apriori will attempt to learn separate rules for each timeslot. This is exacerbated by the fact that the Augsburg dataset contains very little training data; Timeslot Apriori could potentially learn every possible instantiation of the occupants' patterns separately, but there aren't enough examples present to do so. The other approaches

Table 5 Comparison of sequence learners on original Augsburg dataset

Occupant (%)	HMM (%)	Elman net (%)	Frequency predictor (%)	Ord. apriori (%)
A	62	57	60	59
B	55	58	58	58
C	47	46	47	47
D	54	50	56	56
All	55	53	55	55

are successful as they are able to learn the sequences independent of the time at which they occur. Ordered Apriori is similarly time-independent and thus matches the other sequence-based approaches. As timeslot apriori performs poorly on this dataset, we do not attempt next-day predictions.

Table 5 shows the accuracy of the sequence learners on the unmodified Augsburg dataset. As this version of the dataset does not have timeslots, predictions are of the occupant's next location only with no concept of time. While the average accuracy across all four occupants is approximately the same for three of the approaches, there are minor variations in accuracy on each individual opponent. These results show that ordered apriori is also able to match existing methods on pure sequence data. The results are lower than the corresponding results in [10] as we include predictions when the occupant is leaving their own office and cases where a prediction was not made.

6 Conclusions and Future Work

In this paper we presented two approaches for applying association rule mining to the problem of predicting future occupant locations. We implemented our approaches using modifications of a standard association rule mining algorithm and presented experimental results which show that our modifications can predict actual occupant movements with a high degree of accuracy.

Compared to standard approaches, our timeslot-dependent approach has some advantages and disadvantages. Our aim with this approach is to predict for any time slot using whatever information is available, whether it be the occupant's recent movements on the same day or simply their historical patterns, allowing it to use a wider variety of data to make a wider variety of predictions. This is successful on the UCC dataset, however it performs poorly on the Augsburg set even for next-slot predictions due to the time-independent nature of their movements.

Our sequence based approach matches the capabilities of the existing approaches, giving it the same accuracy as those approaches, although the same limitations in terms of the information which is used and what predictions can be made. Using the rulesets generated by both approaches, we are essentially able to predict with whichever approach is more suitable for any given prediction. Thus we can match the accuracy of existing methods for the predictions they can make, while being able to make a wider array of predictions.

Since timeslot apriori and ordered apriori do not make exactly the same predictions for next-slot prediction, it is possible that intelligent selection of the rules from both sets that we could improve accuracy over what either approach achieves alone. We intend to investigate this possibility as part of our future work.

The approaches evaluated in this paper only consider patterns in the occupant's specific location. There may be patterns which support more general predictions, such as that an occupant will be out of their office without predicting exactly where they will be. There is existing work [11] on extending Apriori to mine association rules with taxonomies. As part of our future work we intend to similarly extend our approach, allowing us to make more generalized predictions.

The eventual goal is to integrate this approach with occupant localization systems such as [12], and predictive control systems such as [13]. Using occupant localization data, a system based on our approach could provide the predictions necessary for more energy efficient building control.

Acknowledgments This work is funded by Science Foundation Ireland under the grants ITOBO 07.SRC.I1170 and Insight 12/RC/2289

References

1. World Business Council for Sustainable Development, Energy Efficiency in Buildings: Facts and Trends - Full Report, (2008).
2. Agrawal, R., Srikant, R.: Fast Algorithms for Mining Association Rules in Large Databases. Proceedings of the 20th International Conference on Very Large Databases, 487–499, (1994).
3. Petzold, J., Pietzowski, A., Bagci, F., Trumler, W. and Ungerer, T.: Prediction of Indoor Movements Using Bayesian Networks. LoCA 2005, LNCS 3479, 211–222, (2005).
4. Vintan, L., Gellert, A., Petzold, J., and Ungerer, T.: Person Movement Prediction Using Neural Networks. Technical Report, Institute of Computer Science, University of Augsburg, (2004).
5. Petzold, J., Bagci, F., Trumler, W., Ungerer, T., and Vintan, L.: Global State Context Prediction Techniques Applied to a Smart Office Building. The Communication Networks and Distributed Systems Modeling and Simulation Conference, San Diego, CA, USA, (2004).
6. Gellert, A., Vintan, L.: Person Movement Prediction Using Hidden Markov Models. Studies in Informatics and, Control, Vol. 15, No.1 (2006).
7. Voigtmann, C., Sian Lun Lau, David, K.: A Collaborative Context Prediction Technique. Vehicular Technology Conference, (2011).
8. Eagle, N., Pentland, A. S.: Eigenbehaviors: Identifying structure in routine. Behavioral Ecology and Sociobiology, Vol. 63, No. 7, 1057–1066, (2009).
9. Petzold, J.: Augsburg Indoor Location Tracking Benchmarks. Context Database, Institute of Pervasive Computing, University of Linz, Austria. http://www.soft.uni-linz.ac.at/Research/ContextDatabase/index.php, (2005).
10. Petzold, J., Bagci, F., Trumler, W., and Ungerer, T.: Comparison of Different Methods for Next Location Prediction. EuroPar 2006, LNCS 4128, 909–918, (2006).
11. Srikant, R., and Agrawal, R.: Mining Generalized Association Rules. Proceedings of the 21st International Conference on Very Large Databases, 407–419, (1995).

12. Najib, W., Klepal, M., Wibowo, S.B.: MapUme: Scalable middleware for location aware computing applications. International Conf. on Indoor Positioning and Indoor Navigation (IPIN), (2011).
13. Mady, A. E.D., Provan, G., Ryan, C., Brown, K. N.: Stochastic Model Predictive Controller for the Integration of Building Use and Temperature Regulation. AAAI (2011).

Contextual Sentiment Analysis in Social Media Using High-Coverage Lexicon

Aminu Muhammad, Nirmalie Wiratunga, Robert Lothian and Richard Glassey

Abstract Automatically generated sentiment lexicons offer sentiment information for a large number of terms and often at a more granular level than manually generated ones. While such rich information has the potential of enhancing sentiment analysis, it also presents the challenge of finding the best possible strategy to utilising the information. In SentiWordNet, negation terms and lexical valence shifters (i.e. intensifier and diminisher terms) are associated with sentiment scores. Therefore, such terms could either be treated as sentiment-bearing using the scores offered by the lexicon, or as sentiment modifiers that influence the scores assigned to adjacent terms. In this paper, we investigate the suitability of both these approaches applied to sentiment classification. Further, we explore the role of non-lexical modifiers common to social media and introduce a sentiment score aggregation strategy named SMARTSA. Evaluation on three social media datasets show that the strategy is effective and outperform the baseline of using aggregate-and-average approach.

1 Introduction

Sentiment analysis involves the extraction and assessment of subjectivity and opinion in text. Its main goal is to effectively determine, firstly, whether a piece of natural language text (word/phrase, sentence or document) expresses opinion [23], secondly,

A. Muhammad (✉) · N. Wiratunga · R. Lothian · R. Glassey
IDEAS Research Institute, Robert Gordon University, Aberdeen, Scotland
e-mail: a.b.muhammad1@rgu.ac.uk

N. Wiratunga
e-mail: n.wiratunga@rgu.ac.uk

R. Lothian
e-mail: r.m.lothian@rgu.ac.uk

R. Glassey
e-mail: r.j.glassey@rgu.ac.uk

M. Bramer and M. Petridis (eds.), *Research and Development in Intelligent Systems XXX*, 79
DOI: 10.1007/978-3-319-02621-3_6, © Springer International Publishing Switzerland 2013

the sentiment orientation or polarity (i.e. positive or negative) of opinionated text [23, 34], and thirdly, the strength of such polarity [32, 33]. Sentiment analysis has been predominantly performed in the domain of movies and general product reviews [24, 34]. However there has been a growing interest in other user-generated content on the web (e.g. in blogs and discussion forums) referred to as the social media content [15]. This content is fundamentally different from review related data in that whereas reviews usually have a specific subject (i.e. the product/service being reviewed), social media content tends to be diverse in topics. Similarly, social media is more likely to contain non-lexical terms/content such as non-standard spellings, punctuations and capitalization; sequence of repeating characters and emoticons that may affect the expression of sentiment.

Sentiment analysis can be organised into machine learning (ML) or lexicon-based (LB). With ML, an algorithm is trained with sentiment labelled data and the learnt model is used to classify new documents. This method requires the initial labelled data typically generated through labour-intensive human annotation. The LB method, on the other hand, involves the extraction and aggregation of terms' sentiment scores offered by a lexicon (i.e prior polarities) to make sentiment prediction. An important component of an LB method for sentiment analysis is to account for the polarity with which a sentiment-bearing term appears in text (i.e. contextual polarity) as this can be different from its prior polarity. For example, the prior polarity of the term 'good' is positive but when the term is immediately preceded by the modifier 'not' in text, its (contextual) polarity becomes negative. Several strategies have been proposed to account for contextual polarity, crucial amongst which is to account for the effect of modifiers also called valence shifters (i.e. negation e.g. 'not' and 'never', intensification e.g. 'very' and 'highly', and diminishing e.g. 'slightly' and 'a-bit'). This has been performed, typically, when no sentiment information is offered by the lexicon in use (e.g. in [31] and [33]). With high-coverage lexicons (e.g. SentiWordNet [6]), sentiment information is made available for a large number of terms including valence shifters. Furthermore, as SentiWordNet is a general knowledge lexicon, it does not include sentiment information about non-lexical features that influence sentiment expression in social media. Here we focus on the following research questions in relation to using the lexicon for sentiment analysis:

- Does treating negation terms as sentiment modifiers rather than sentiment-bearing terms improve sentiment analysis?
- Does treating intensifiers/diminishers as sentiment modifiers rather than sentiment-bearing terms improve sentiment analysis?
- Does accounting for non-lexical features improve sentiment analysis?

In this paper, we compare sentiment classification accuracy with valence shifters treated as sentiment-bearing terms against when they are decoupled from their sentiment scores and treated as score modifiers of other terms. Although there are many examples in which valence shifters are accounted for as sentiment modifiers [2, 16, 27], this is the first time they are compared against their sentiment-bearing properties. As such our work gives insight into their behaviour both as sentiment-bearing terms and as sentiment modifiers. Furthermore, we introduce and evaluate performance of strategies to account for non-lexical modifiers in social media.

The remainder of this paper is organised as follows. Section 2 describes related work. Sentiment analysis using SentiWordNet is presented in Sect. 3 while the proposed algorithm (SMARTSA) is introduced in Sect. 4. Experimental results and discussions appear in Sect. 5, followed by conclusions in Sect. 6.

2 Related Work

Research in sentiment analysis can be organised into two broad methods: ML and LB. Typical text representation for ML is an unordered list of terms that appear in documents (i.e. bag-of-words). A binary representation based on term presence or absence attained up to 87.2 % accuracy on movie review dataset [23]. The addition of phrases that are used to express sentiment (i.e. appraisal groups) as features in the binary representation resulted in further improvement of 90.6 % [35] while best result of 96.9 % was achieved using term-frequency/inverse-document-frequency (tf-idf) weighting [25]. Feature selection mechanisms have also been proposed for ML. For instance, Rahman et al. [19] developed an approach for selecting bi-gram features and Riloff et al. [28] proposed feature space reduction based on subsumption hierarchy. However, despite the high performance of ML sentiment classifiers, they suffer poor domain adaptability because they often rely on domain specific features from their training data. Further, as human-labelled data is required, the method tends to be time-consuming and costly where such data is not readily available such as in social media. Also, with the dynamic nature of social media, language evolves rapidly which may render any previous learning less useful.

The LB method does not necessarily require a text representation or learning step prior to the analysis and, as such, leaves room for any linguistic rule to be applied at the analysis time. Broadly, sentiment lexicons are either manually, or automatically generated. With manually generated lexicons such as General Inquirer [30] and Opinion Lexicon [14], sentiment polarity values are assigned purely by humans and typically have limited coverage. As for the automatically generated lexicons, two semi-unsupervised methods are common, *corpus-based* and *dictionary-based*. Both methods begin with a small set of seed terms. For example, a positive seed set such as 'good', 'nice' and 'excellent' and a negative seed set could contain terms such as 'bad', 'awful' and 'horrible'. The methods leverage on language resources and exploit relationships between terms to expand the sets. The two methods differ in that corpus-based uses collection of documents while the dictionary-based uses machine-readable dictionaries as the lexical resource.

Prior work using LB method include [11], in which 657 and 679 adjectives were manually annotated as positive and negative seed sets respectively. Thereafter, the sets were expanded to conjoining adjectives in a document collection based on the connectives 'and' and 'but' where 'and' indicates similar and 'but' indicates contrasting polarities between the conjoining adjectives. This work was limited to lexicon creation and was not extended to sentiment classification. In [34], a phrasal lexicon was generated from reviews which was later used for sentiment classification based on majority vote of terms from the lexicon. Sentiment score modification based on term co-occurrence with valence shifters have been proposed. In [16, 27, 31], polarities of terms that are under the influence of negators (e.g. 'not', and 'never'), are inverted or shifted (i.e. inverted and reduced by a certain weight) and that of those terms that are under the influence of intensifiers (e.g. 'very' and 'really') and diminishers (e.g. 'slightly' and 'a-little') are increased and decreased respectively. Similarly, approaches for combining valence shifters with adjectives were proposed [2]. A key challenge for accounting for valence shifters in sentiment analysis is determining the terms modified by such valence shifters when they occur in text (i.e. their scope). This is even more challenging with social media because of the informal and non-standard use of language as a result of which standard parsers often give the wrong results [21]. An alternative approach is to assign the scope of modifiers to adjacent terms in text [13, 29].

Discourse analysis has also been employed to adjust prior polarity of terms. For instance, in [12] term weighting was integrated into prior polarities based on discourse structures in which the terms occur. Their approach shows that even simple variation of term weights from the beginning to the end of text (in ascending order of importance) could improve a positive/negative sentiment classification. Further improvement was reported when a more sophisticated weighting approach based on Rhetorical Structure Theory [17] was introduced, although, also when a correction value, learned from training data was applied. Similarly, using discourse connectives (e.g 'but', 'and' and 'although') to modify terms prior polarity have been shown to improve sentiment classification [18].

The domain of social media presents additional challenges to sentiment analysis as users often employ non-standard but creative means to express sentiment. In more recent work, lexicon-based sentiment analysis was extended to incorporate term prior polarity modification based on non-lexical modifiers in addition to lexical contextual valence shifters [21, 32, 33]. Such non-lexical modifiers include term elongation by repeating character (e.g. 'haaappppyy' in place 'happy'), capitalization of terms (e.g. 'GOOD' in place 'good') and internet slang (e.g. 'lol', 'rotf'). In these works, manually generated lexicons were used as the source of prior polarity scores. These lexicons require manual effort to be generated or updated and, for this reason, tend to be of limited coverage and contain limited information about the covered terms.

SentiWordNet is an automatically generated lexicon which associates entries in WordNet [8] with a positive, negative and objective scores. The lexicon covers a large number of terms at word sense level. This is important since sentiment can change between different parts-of-speech (PoSs) of the same term or different word-senses of the same PoS. The lexicon was used for sentiment analysis of movie reviews [20], it was extended with score modification based on discourse analysis in [12], valence shifters in [1, 26] and WordNet semantic relations in [5]. In all these works, either the sentiment-bearing or the modification property of valence shifters was assumed. This could result in applying the wrong strategy for handling the terms. We explore score modification based on valence shifters further, with particular emphasis on social media content.

3 Sentiment Analysis Using SentiWordNet

SentiWordNet [7] was generated by using WordNet dictionary. Each synset (i.e. a group of synonymous terms on a particular meaning) in WordNet is associated with three numerical scores indicating the degree of association of the synset with positive, negative and objective text. In generating the lexicon, seed (positive and negative) synsets were expanded by exploiting *synonymy* and *antonymy* relations in WordNet, whereby synonymy preserves while antonymy reverses the polarity with a given synset. Since there is no direct synonym relation between synsets in WordNet, the relations: see_also, similar_to, pertains_to, derived_from and attribute were used to represent synonymy relation while direct antonym relation was used for the antonymy. Glosses (i.e. textual definitions) of the expanded sets of synsets along with that of another set assumed to be composed of objective synsets were used to train eight ternary classifiers. The classifiers are used to classify every synset and the proportion of classification for each class (positive, negative and objective) were deemed to be the scores for the synset. In an enhanced version of the lexicon (i.e. SentiWordNet 3.0 [6]), the scores were optimised by a random walk using the PageRank [3] approach. This starts with manually selected synsets and then propagates sentiment polarity (positive or negative) to a target synset by assessing the synsets that connect to the target synset through the appearance of their terms in the gloss of the target synset. SentiWordNet can be seen to have a tree structure as shown in Fig. 1. The root node of the tree is a term whose child nodes are the four basic PoS tags in WordNet (i.e. noun, verb, adjective and adverb). Each PoS can have multiple word senses as child nodes. Sentiment scores illustrated by a point within the triangular space in the diagram are attached to word-senses. Subjectivity increases (while objectivity decreases) from lower to upper, and positivity increases (while negativity decreases) from right to the left part of the triangle.

Typical sentiment classification of text using SentiWordNet begins with text preprocessing operations. First, input text is broken into unit tokens (tokenization) and each token is assigned a lemma (i.e. corresponding dictionary entry) and PoS. Although scores are associated with word-senses, disambiguation is usually not performed as it does not seem to yield better results than using either the average score across all senses of a term-PoS or the score attached to the most frequent sense of the term (e.g. in [4, 20, 26]. Using PoS and lemma, terms' positive and negative scores are extracted from the lexicon and summed. Net positive and negative scores are normalized by the total number of sentiment terms found in the text to counter the effect of varying text sizes. The sentiment class of the input text is deemed to be positive if the net positive score exceeds the net negative score and negative otherwise. We refer to this approach as BASE where the sentiment score for a post p is determined using the following equation:

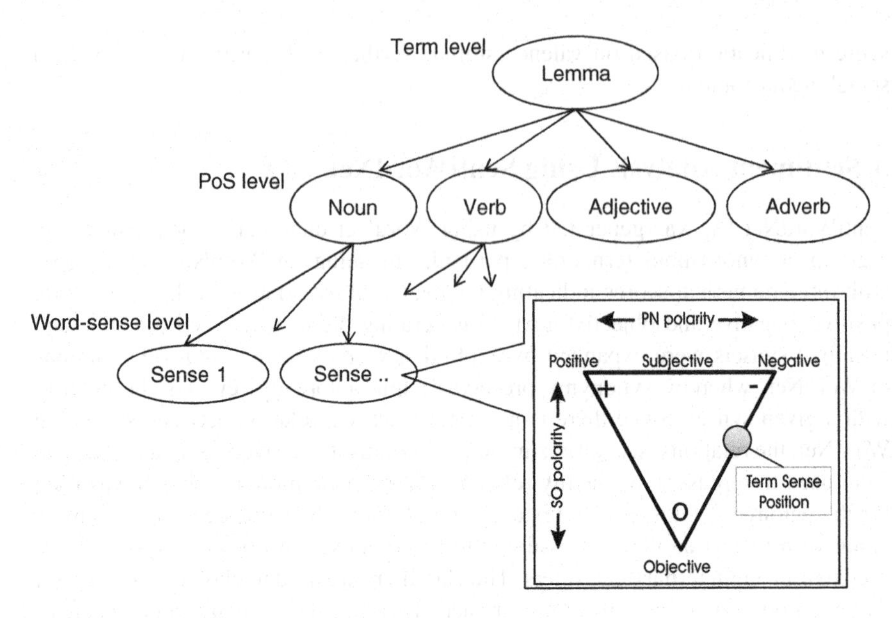

Fig. 1 Diagram showing structure of SentiWordNet

$$score(p)_{dim} = \frac{\sum_{w_j \in p} score(w_j)_{dim}}{|n|} \tag{1}$$

Where $score(p)_{dim}$ is the score of the post p in the sentiment dimension of dim (dim is either positive or negative) while $score(w_j)_{dim}$ is the sentiment score of the term w_j in p extracted from SentiWordNet, and $|n|$ is the number of sentiment-bearing terms in p.

4 SMARTSA System

The system (see Algorithm 1) is composed of four main components as shown in Fig. 2. These are text pre-processing, prior polarities extraction, contextual score modification and normalization components. Input text is passed to the pre-processing component from which PoS and lemma are determined. Prior polarities extracted from lexicons are modified, normalized and aggregated to make positive/negative sentiment classification.

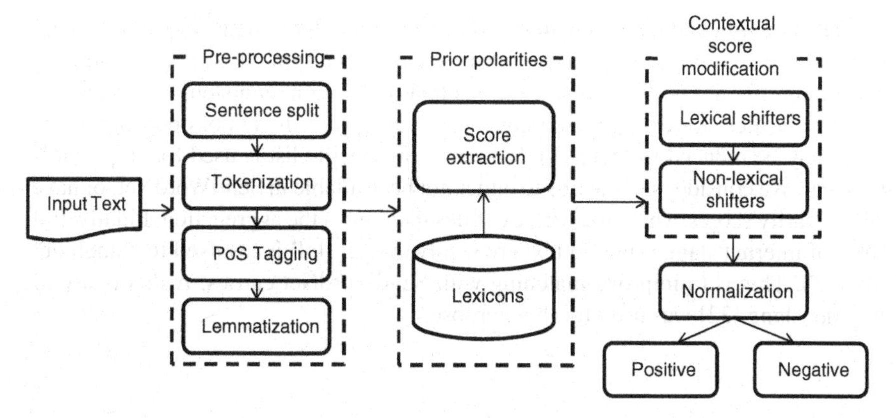

Fig. 2 Diagram showing the components of the proposed system (SMARTSA)

Algorithm 1 : SMARTSA

1: **INPUT:** d, Resources $\{SWN, Intensifiers, Diminishers, Emoticons, Slang\}$
2: **OUTPUT:** $class$ ▷ document sentiment class
3: Initialise: $posScore, negScore$
4: $fullForms(d)$ ▷ extract full forms of internet slang
5: $Annotate(d)$ ▷ apply PoS, Lemma, Sentiment, IsCapitals and HasRepeatLetter annotations
6: **for all** $w_i \in d$ **do**
7: $ModifyScores(w_i, neighbourhood)$ ▷ modify sentiment annotations
8: **if** $w_i.pos > 0$ **then**
9: $posScore \leftarrow posScore + w_i.pos$
10: $nPos \leftarrow nPos + 1$ ▷ increment number of positive terms
11: **end if**
12: **if** $w_i.neg > 0$ **then**
13: $negScore \leftarrow negScore + w_i.neg$
14: $nNeg \leftarrow nNeg + 1$ ▷ increment number of negative terms
15: **end if**
16: **end for**
17: **if** $posScore/nPos > negScore/nNeg$ **then return** $positive$
18: **else return** $negative$
19: **end if**

4.1 Pre-Processing

In this component, standard text pre-processing operations are applied in order to transform input text to equivalent SentiWordNet entries. These operations include splitting the text into its constituent sentences and tokens. Thereafter, each token is annotated with their respective PoS and lemma. We use Stanford CoreNLP[1] pipeline

[1] http://nlp.stanford.edu/software/corenlp.shtml

for sentence split and lemmatization. However, we use TweetNLP[2] [9] for tokenization and PoS tagging because it recognises social media symbols such as emoticons. Stemming was not performed since terms that have a common stem may have different polarities (for example, 'defense' and 'defensive', though of opposing polarities, may be reduced to 'defens'). Similarly, no specific list is used for stop-words removal. We consider such words to either not be available in SentiWordNet, or have zero polarity scores. This, in effect, excludes them from the aggregation. Finally, full form of internet slang found in text are resolved (e.g. 'lol' is resolved to "laugh out loudly"). This is to improve matching with SentiWordNet entries. A dictionary of internet slang [32] was used for this purpose.

4.2 Prior Polarities Extraction

Given PoS and lemma, sentiment scores are extracted from lexicons. Here, the main lexicon is SentiWordNet. However since the lexicon does not include sentiment-bearing symbols used in social media, we use an additional lexicon for emoticons in which positive emoticons (e.g. :-), :), :o), :]) are assigned positive score of $\frac{3}{5}$ and negative score of 0 while negative emoticons (e.g. :(, :-(, :[ ) are assigned positive score of 0 and negative score of $\frac{3}{5}$ [32].

4.3 Contextual Polarity Modification

We propose the following prior polarity modification to account for the effect of modifiers on sentiment-bearing terms.

4.3.1 Lexical Shifters

When a lexical shifter (i.e. intensifier or diminisher) is detected within the neighbourhood of a sentiment-bearing term, defined as a 5-token (before and after) window; the dominant prior polarity (i.e. maximum of positive and negative scores) of the target term is increased (in the case of intensifier) or decreased (in the case of diminisher) relative to the strength of the intensifier/diminisher. Sentiment scores of the intensifiers/diminishers are not utilised in the aggregation. We used the list provided in [32] where intensifiers/diminishers are assigned weight of 1 or 2, which we convert to 50 and 100 % respectively. Figure 3 shows the classification of an example text with intensifier classified using BASE and SMARTSA. The text, which is clearly negative is incorrectly classified as positive using sentiment scores of the intensifier (i.e.

---

[2] http://www.ark.cs.cmu.edu/TweetNLP

```
Text:              It's    really    awful

Positive Score :   0.0     0.438     0.25
Negative Score:    0.0     0.065     0.542

BASE:
    Positive:    ( 0.0 + 0.438  +  0.25 ) / 3   | Class =
    Negative:    ( 0.0 + 0.065  +  0.542 ) / 3  | Positive

SMARTSA:
    Positive:    ( 0.0 +   0.0  +  0.25 ) / 1                        | Class =
    Negative:    ( 0.0 +   0.0  +  0.542 x ( 100% + 50% ) ) / 1      | Negative
```

Fig. 3 Example of BASE and SMARTSA approaches to sentiment scoring

BASE approach). However, it is correctly classified using the intensification strength of the intensifier (i.e. SMARTSA approach).

As for negation terms, we examined the performance of two strategies namely, *invert* and *shift* in relation to BASE. In the invert, positive and negative scores (from SentiWordNet) of the target term are interchanged while for the shift, the approach proposed by [21] is used, whereby the score of the target word is reversed and reduced by a weight of $\frac{1}{5}$. In both strategies, scores associated with negation words (as listed in [32]) were not utilised. The result shows performance degradation using either of the methods (see Sect. 5.1). Therefore, in SMARTSA negation terms are treated as sentiment-bearing.

4.3.2 Non-Lexical Shifters

In SMARTSA, we implement the following strategies to account for non-lexical modifiers and features in social media text:

Sequence of Repeating Letters: When a sequence(s) of three or more letters is detected, the target term is identified by first reducing the number of the letter to a maximum of two and check with dictionary. If the intermediate word is not found, the repeating letters are further reduced to one letter, one sequence at a time. If the term is identified, its dominant polarity (maximum of positive and negative scores) is increased with the intensification weight of the word 'very'. For example, 'happpyyy' is resolved to 'very happy'.

Multiple Exclamation/Question Mark: The occurrence of three or more consecutive exclamation or question marks or a mixture of both is treated as sentiment intensification of dominant polarities of the terms that are in the neighbourhood (a 5-token window before and after the term) using the intensification weight of the word 'very'. For example, 'happy!!!!' is resolved to 'very happy'.

Capitalisation: In the case of capitalisation of terms, the target term is intensified with the intensification weight equivalent to that of the word 'very'. The term is intensified only if the whole sentence in which it appears is not capitalised. This is because in such cases the capitalisation may not be for emphasis but writing style. For example, if 'BAD' occurs in a sentence whose other terms are not capitalised, it is treated as equivalent to 'very bad'.

4.4 Normalization

Unlike typical sentiment lexicons, SentiWordNet offers sentiment polarity for terms in both dimensions of positive and negative. Therefore, simple normalization of net document score by dividing by the number of terms in a dimension is inadequate as it can result in over inflation (or over penalization) of one dimension over the other. In SMARTSA, we introduce a normalization approach which, in addition to addressing the effect of varying text sizes, is aware of the possibility of having the same term contributing scores to both dimensions. In this approach, the net positive (or negative) score is divided by the number of terms that have non-zero score for the positive (or negative) polarity dimension.

After the afore-mentioned score modification strategies, sentiment is aggregated as follows:

$$score(p)_{dim} = \frac{\sum_{w_j \in p} mod\_score(w_j)_{dim}}{|n_{dim}|} \tag{2}$$

Where $score(p)_{dim}$ is the score of the post p in the sentiment dimension of dim (dim is either positive or negative) while $mod\_score(w_j)_{dim}$ is the modified sentiment score of the term w_j in p extracted from SentiWordNet, and $|n_{dim}|$ is the number of terms in p that have non-zero score for dimension dim.

5 Experimental Evaluation

We conduct a comparative study to evaluate the proposed algorithm (SMARTSA). The aim of the study is, first, to observe the modification and sentiment-bearing effects of lexical valence shifters. Second, the effectiveness of non-lexical, social media focused prior polarity modifications introduced in this paper. Accordingly we conduct comparative study of the following:

1. ZeroR: A classifier that always predicts the majority class.
2. BASE: A standard algorithm for sentiment classification using SentiWordNet (see Sect. 3).
3. SMARTSA +invert: Extending SMARTSA with accounting for negation using invert approach (see Sect. 4)

4. SMARTSA +shift: Extending SMARTSA with accounting for negation using shift approach (see Sect. 4)
5. SMARTSA-I/D: Excluding sentiment modification based on intensifiers/ diminishers from SMARTSA (i.e. treating such terms as sentiment bearing using their lexicon scores)
6. SMARTSA-NS: Excluding accounting for non-lexical shifters from SMARTSA
7. SMARTSA: Social media oriented classifier introduced in this paper (see Algorithm 1).

We compare the systems against three human annotated social media data sets: Digg, MySpace [22][3] and Twitter [10].[4] For Digg and MySpace, we followed [21] and convert their original 5-point scale annotation to a positive/negative annotation.

Result is presented as average value of F1 score for both (positive and negative) categories to quantify classification quality. *Precision* (Pr) and *recall* (R) are also reported for completeness.

5.1 Results and Discussion

Table 1 shows some statistics extracted from the datasets. Twitter is fairly balanced in number of positive and negative comments. Digg has negative number of comments higher than positive, while MySpace has positive number of comments higher than negative. Proportion of negation terms in negative comments is higher than in positive comments across all the three data sets. Proportion of intensifiers and diminishers do not show clear inclination towards any of the classes. The use of emoticons is prevalent in Twitter more than in Digg and MySpace perhaps due to

Table 1 Datasets statistics

Characteristics	Data set					
	Digg		MySpace		Twitter	
	Positive	Negative	Positive	Negative	Positive	Negative
#Com	107	221	400	105	182	177
#Neg	20	163	98	49	15	40
#Int & Dim	57	226	192	66	179	125
#Emoticon	7	0	49	8	30	24
#Cap (mostly abbreviation)	35	168	814	125	309	275
#Rep letter	1	5	39	5	8	14
#Rep ! and/or ?	12	12	235	31	28	17

The table shows, for each dataset, the number of: comments (#Com), negators (#Neg), Intensifiers & Diminishers (# Int & Dim), emoticons (#Emoticon), word capitalizations (#Cap), repeating letters (#Rep letter) and repeating exclamation and/or question marks (#Rep ! and/or ?)

[3] The dataset are obtained from the cyberemotions project: http://www.cyberemotions.eu

[4] Available from http://www.sentiment140.com.

the fact that Twitter comment is restricted to 140 characters. Likewise, the statistics reveals that emoticons are associated with positive comments more than negative. The number of capitalizations observed from the datasets is mostly for abbreviation rather than emphasis and is fairly evenly distributed between the two classes. While proportion of repeating letter is fairly the same across the two classes, proportion of repeating exclamation/question mark in positive comments is higher than in negative comments.

Sentiment classification result is presented in Table 2. It shows SMARTSA to consistently outperform the other classifiers/settings on all the three data sets. However, the difference is statically significant (at 95 % confidence level) only in MySpace which is composed mostly of positive comments. This indicates that the implemented strategies are especially useful in detecting positive sentiment. Both SMARTSA-NS and SMARTSA-I/D outperformed BASE in all cases with SMARTSA-NS performing best. This shows that the contribution of the intensifier/diminisher analysis aspect of SMARTSA is more than that of the non-lexical shifters analysis. However, This can be attributed to the fact that even in social media, lexical modifiers are used more often

Table 2 Performance results (%) of the classifiers on Digg, MySpace and Twitter data sets

Classifier	Positive			Negative			Avg F1
	Pr	R	F1	Pr	R	F1	
Digg							
ZeroR	0.0	0.0	0.0	67.4	100	80.5	40.3
BASE	56.7	71.0	63.1	84.0	73.8	78.6	70.8
SMARTSA +invert	55.3	71.1	62.2	83.4	71.9	77.2	69.7
SMARTSA +shift	56.5	71.7	63.2	83.9	71.3	77.1	70.2
SMARTSA-I/D	57.4	72.9	64.2	84.9	73.6	78.9	71.6
SMARTSA-NS	61.3	71.0	65.8	84.8	78.3	81.4	73.6
SMARTSA	62.5	74.8	68.1	86.5	78.3	82.3	75.1
MySpace							
ZeroR	79.2	100	88.4	0.0	0.0	0.0	44.2
BASE	90.5	57.2	70.0	26.0	71.4	38.1	54.1
SMARTSA +invert	86.8	58.0	69.5	28.0	70.8	40.1	54.8
SMARTSA +shift	88.2	60.1	71.5	27.3	68.9	39.1	55.3
SMARTSA-I/D	90.9	**75.0**	**82.2**	**42.9**	71.4	**53.6**	**67.9**
SMARTSA-NS	91.2	**74.6**	**82.1**	**42.9**	72.4	**53.9**	**68.0**
SMARTSA	91.5	**78.3**	**84.4**	**46.6**	72.4	**56.7**	**70.5**
Twitter							
ZeroR	50.7	100	67.3	0.0	0.0	0.0	33.7
BASE	68.4	70.3	69.4	68.6	66.7	67.6	68.5
SMARTSA +invert	64.6	68.2	66.4	64.0	64.1	64.0	65.2
SMARTSA +shift	65.7	68.4	67.0	65.2	63.8	64.5	65.7
SMARTSA-I/D	69.0	69.9	69.4	68.6	67.8	68.9	68.8
SMARTSA-NS	71.1	70.3	70.7	69.8	70.6	70.2	70.5
SMARTSA	76.0	69.8	72.8	71.4	**77.4**	74.3	73.5

Bold-face indicates statistically significant difference in relation to BASE at 95 % confidence level than non-lexical shifters (see statistics in Table 1). For negation, the BASE approach performed better than both SMARTSA +invert and SMARTSA +shift. This shows that sentiment analysis is better when negation terms are treated as sentiment-bearing rather than sentiment influencers of other terms. A possible explanation to this is that negation terms are not as much modifiers of sentiment terms than they are associated with negative text. For example, "I'm not for the movie" is clearly negative even though the negation term 'not' does not seem to modify any sentiment-bearing term.

6 Conclusion

In this paper, we addressed the problem of sentiment analysis in the social media using a lexicon that has high term coverage (SentiWordNet). Our results show that, although intensifiers and diminshers are associated with sentiment scores in the lexicon, performance improves when such terms are treated as modifiers rather than sentiment-bearing. However, sentiment analysis is better when negation terms are treated as sentiment bearing terms. Furthermore, the results show that accounting for non standard sentiment modifiers that are often used in the social media domain improves sentiment analysis using SentiWordNet.

References

1. Agrawal, S., Siddiqui, T.: Using syntactic and contextual information for sentiment polarity analysis. In: Proceedings of the 2nd International Conference on Interaction Sciences: Information Technology, Culture and Human, ICIS '09, pp. 620–623. ACM, New York, NY, USA (2009)
2. Benamara, F., Cesarano, C., Picariello, A., Reforgiato, D., Subrahmanian, V.: Sentiment analysis: Adjectives and adverbs are better than adjectives alone. In: Proceeding of ICWSM. (2007)
3. Brin, S., Page, L.: The anatomy of a large-scale hypertextual web search engine. In: Seventh International World-Wide Web Conference (WWW 1998) (1998)
4. Denecke, K.: Using sentiwordnet for multilingual sentiment analysis. In: ICDE Workshop (2008)
5. Devitt, A., Ahmad, K.: Sentiment polarity identification in financial news: A cohesion-based approach. In: Proceedings of the 45th Annual Meeting of the Association of Computational Linguistics, pp. 984–991. Prague, Czech Republic (2007)
6. Esuli, A., Baccianella, S., Sebastiani, F.: Sentiwordnet 3.0: An enhanced lexical resource for sentiment analysis and opinion mining. In: Proceedings of the Seventh conference on International, Language Resources and Evaluation (LREC10) (2010)
7. Esuli, A., Sebastiani, F.: Determining term subjectivity and term orientation for opinion mining. In: Proceedings of EACL-06, 11th Conference of the European Chapter of the Association for Computational Linguistics, Trento, IT. (2006)
8. Fellbaum, C.: WordNet: An Electronic Lexical Database. MIT Press (1998)
9. Gimpel, K., Schneider, N., O'Connor, B., Das, D., Mills, D., Eisenstein, J., Heilman, M., Yogatama, D., Flanigan, J., Smith, N.A.: Part-of-speech tagging for twitter: annotation, features, and experiments. In: Proceedings of the 49th Annual Meeting of the Association for

Computational Linguistics: Human Language Technologies: short papers - Volume 2, HLT '11, pp. 42–47. Association for Computational Linguistics, Stroudsburg, PA, USA (2011)
10. Go, A., Bhayani, R., Huang, L.: Twitter sentiment classification using distant supervision. Processing pp. 1–6 (2009)
11. Hatzivassiloglou, V., McKeown, K.R.: Predicting the semantic orientation of adjectives. In: Proceedings of the 35th Annual Meeting of the ACL and the 8th Conference of the European Chapter of the ACL, pp. 174–181. New Brunswick, NJ (1997)
12. Heerschop, B., Goossen, F., Hogenboom, A., Frasincar, F., Kaymak, U., de Jong, F.: Polarity analysis of texts using discourse structure. In: Proceedings of the 20th ACM international conference on Information and knowledge management CIKM'11, pp. 1061–1070. ACM, Glasgow UK (2011)
13. Hogenboom, A., van Iterson, P., Heerschop, B., Frasincar, F., Kaymak, U.: Determining negation scope and strength in sentiment analysis. In: Proceedings of the IEEE International Conference on Systems, Man and Cybernetics, pp. 2589–2594 (2011)
14. Hu, M., Liu, B.: Mining and summarizing customer reviews. In: Proceedings of the tenth ACM SIGKDD international conference on Knowledge discovery and data mining, pp. 168–177 (2004)
15. Kaplan, A.M., Haenlein, M.: Users of the world, unite! the challenges and opportunities of social media. Business Horizons **53**(1), 59–68 (2010)
16. Kennedy, A., Inkpen, D.: Sentiment classification of movie reviews using contextual valence shifters. Computational Intelligence **22**, 2006 (2006)
17. Mann, W., Thompson, S.: Rhetorical structure theory: Toward a functional theory of text organization. Text **8**(3), 243–281 (1998)
18. Mukherjee, S., Bhattacharyya, P.: Sentiment analysis in twitter with lightweight discourse analysis. In: Proceedings of the 24th International Conference on Computational Linguistics (COLING 2012) (2012)
19. Mukras, R., Wiratunga, N., Lothian, R.: Selecting bi-tags for sentiment analysis of text. In: Proceedings of the Twenty-seventh SGAI International Conference on Innovative Techniques and Applications of Artificial Intelligence (2007)
20. Ohana, B., Tierney, B.: Sentiment classification of reviews using sentiwordnet. In: 9th IT&T Conference, Dublin, Ireland (2009)
21. Paltoglou, G., Thelwall, M.: Twitter, myspace, digg: Unsupervised sentiment analysis in social media. ACM Transactions on Intelligent Systems and Technology **3**(4) (2012)
22. Paltoglou, G., Thelwall, M., Buckley, K.: Online textual communications annotated with grades of emotion strength. In: Proceedings of 31st International Workshop of Emotion: Corpora for research emotion and affect (2010)
23. Pang, B., Lee, L.: Polarity dataset v2.0, 2004. online (2004). http://www.cs.cornell.edu/People/pabo/movie-review-data/.
24. Pang, B., Lee, L.: Seeing stars: Exploiting class relationships for sentiment categorization with respect to rating scales. In: Proceedings of ACL, pp. 115–124 (2005)
25. Pang, B., Lee, L.: Opinion mining and sentiment analysis. Foundations and Trends in Information Retrieval **2**(1), 1–135 (2008)
26. Pera, M., Qumsiyeh, R., Ng, Y.K.: An unsupervised sentiment classifier on summarized or full reviews. In: Proceedings of the 11th International Conference on Web Information, Systems Engineering, pp. 142–156 (2010)
27. Polanyi, L., Zaenen, A.: Contextual valence shifters, vol. 20. Springer, Dordrecht, The Netherlands (2004)
28. Riloff, E., Patwardhan, S., Wiebe, J.: Feature subsumption for opinion analysis. In: Proceedings of the 2006 Conference on Empirical Methods in Natural Language Processing (EMNLP-06) (2006)
29. Santos, R.L., He, B., Macdonald, C., Ounis, I.: Integrating proximity to subjective sentences for blog opinion retrieval. In: Proceedings of the 31st European Conference on Research on Advances in Information Retrieval (ECIR'09) (2009)

30. Stone, P.J., Dexter, D.C., Marshall, S.S., Daniel, O.M.: The General Inquirer: A Computer Approach to Content Analysis. MIT Press, Cambridge, MA (1966)
31. Taboada, M., Brooke, J., Tofiloski, M., Voll, K., Stede, M.: Lexicon-based methods for sentiment analysis. Computational Linguistics **37**, 267–307 (2011)
32. Thelwall, M., Buckley, K., Paltoglou, G.: Sentiment strength detection for the social web. Journal of the American Society for Information Science and Technology **63**(1), 163–173 (2012)
33. Thelwall, M., Buckley, K., Paltoglou, G., Cai, D., Kappas, A.: Sentiment strength detection in short informal text. Journal of the American Society for Information Science and Technology **61**(12), 2444–2558 (2010)
34. Turney, P.D.: Thumbs up or thumbs down? semantic orientation applied to unsupervised classification of reviews. In: Proceedings of the Annual Meeting of the Association for, Computational Linguistics, pp. 417–424 (2002)
35. Whitelaw, C., Garg, N., Argamon., S.: Using appraisal groups for sentiment analysis. In: 14th ACM International Conference on Information and Knowledge Management (CIKM 2005), pp. 625–631 (2005)

Profiling Spatial Collectives

Zena Wood

Abstract Much research has been undertaken in analysing an individual's behaviour based on their movement patterns, but the behaviour of the collectives that the individuals may participate in, remains largely under-researched. The movement of a collective has similarities to that of an individual but also distinct features that must be accounted for. This research focuses on the development of a method that allows the motion of a class of collectives, known as spatial collectives, to be analysed. This method, referred to as the Three Level Analysis (TLA) method, uses string matching to produce a profile for a spatial collective which gives a detailed analysis of its movement patterns; such profiles could then be used to identify the type of spatial collective. A computer program has been developed that allows the method to be applied to a spatiotemporal dataset.

1 Introduction

A *collective* is a group of individuals that are bound together by some principle of unity, referred to as the *source of their collectivity*, that makes it natural for us to consider them as collectively constituting a single entity. Such phenomena exist across multiple domains in both the natural world (e.g., flocks, herds, shoals) and the social world (e.g., crowds, orchestras, committees).

The movement patterns exhibited by collectives, deemed *collective motion*, could be considered one of the most important properties that we would wish to reason about. For example, consider the use of an intelligent traffic monitoring system that tracks the movement of vehicles within a city. A cluster of stationary vehicles could be interpreted as a traffic jam. Movement patterns could also be used to detect when an accident has occurred and, therefore, the need for officials at the scene or a

Z. Wood (✉)
University of Exeter, Exeter, UK
e-mail: Z.M.Wood@exeter.ac.uk

M. Bramer and M. Petridis (eds.), *Research and Development in Intelligent Systems XXX*,
DOI: 10.1007/978-3-319-02621-3_7,

suitable diversion route to be put in place. An understanding of how crowds move and where they are likely to gather could assist in urban planning [1] and in emergency management [11]. Despite the ubiquity of collectives in our everyday lives, it would appear that their behaviour has been under-researched; we do not currently possess the necessary tools that would allow us to reason about or predict their behaviour based on their movement patterns.

Much research has been undertaken into movement pattern analysis (e.g., [7, 15, 16]). Although some of this research includes an examination of groups of individuals (e.g., [3, 5, 11]), they often focus on the level of the individuals, omitting important information about the collective itself. Ultimately the aim of the research reported here is to develop a system that, using a combination of visual and automatic analysis, identifies whether or not a collective is present within a spatiotemporal data set and if so, the type of collective it is. This relies on our understanding of the different types of collective that exist, being able to identify them within a spatiotemporal dataset and developing a method for differentiating between the different types of collective. Research has been carried out in developing a taxonomy of collectives [19] and in methods to extract collectives from a spatiotemporal dataset [5, 6, 17]; however, there is little research distinguishing between the different types of collective based on an analysis of their movement patterns.

Analysing the behaviour of a collective can be much more complex than analysing that of an individual. Although they share some commonalities there exists important distinctions that must be accounted for. In particular, it must be clear how the motions of the individuals relate to each other but also how they relate to motion of the collective that they are participating in [15]. Collective motion ultimately arises from the aggregated motions of its individual members. However, there could be important qualitative differences between the motions of the individual members and that of the collective. Consider a crowd which slowly drifts west. The collective, when considered as a single unit, can be observed as moving in a westerly direction but the individuals as moving around randomly.

This paper introduces a method that allows collective motion to be analysed in detail taking account of the important features of collective motion. Referred to as Three-Level Analysis (TLA) method, the method examines three levels of a collective's motion: the movement of a collective when considered as a point, the evolution of the region occupied by the collective and the movements of the individuals. These three levels could be thought of as three distinct levels of spatial granularity. Each movement pattern that is examined is analysed in terms of the *segments* that it comprises to produce a *profile* of that movement pattern. A segment represents the chunk of motion between two time steps (Fig. 1a). The use of segments could be seen as similar to the use of episodes by [8, 13] or the use of primitives by [4, 7]. It is distinct from the use of episodes in [20] who use the term to represent 'a maximal chunk of homogeneous process at a given level of granularity'.

Represented as a string, a profile denotes the segments that a movement pattern comprises. The way in which the movement information is encoded allows a string similarity algorithm to be used to calculate the similarity between the motions of the individual members but also the motion of the collective and those of its members.

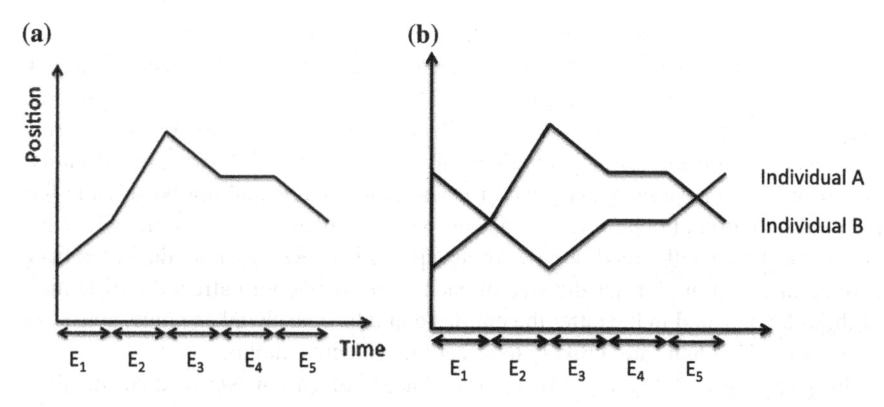

Fig. 1 Decomposing trajectories. **a** A trajectory comprising six time steps and five segments. **b** Two trajectories showing the importance of direction

All of the information that is calculated within the method is presented as a *consolidated profile*. Such a profile could be used to distinguish between different types of collectives; this could be considered an important feature of the proposed method.

Many types of collective exist, far too many to be considered here. The current version of the TLA method assumes that the members of a collective have already been identified and that the movement patterns of each member is known. Therefore, this paper concentrates on how the method could be applied to *spatial collectives*. A method to extract them from a spatiotemporal dataset has already been established [17] thus allowing the proposed method to be applied.

An overview of existing research is given (Sect. 2) followed by an explanation of spatial collectives (Sect. 3). Section 4 introduces the TLA method which is then applied to a sample spatiotemporal dataset (Sect. 5). The paper concludes with a review of the developed method and a discussion of how the method is likely to be expanded (Sect. 6).

2 Existing Research

Much of the existing research into collectives focuses on the levels of the individuals and there appears to be no research that considers the full range of collectives [15, 18]. The main purpose of the proposed method is to analyse collective motion in more detail and allow different types of collective to be identified from their profile. Although no existing frameworks have been identified for comparison some existing research could be considered that considers groups of spatial collectives and those that decompose movement patterns of individuals to form profiles.

Thériault et al. proposes a taxonomy that classifies sets of geographical entities that are located in a geographical space and share a thematic or functional relationship [16]. Like Mountain and Raper [13] and Thériault et al. [16] use spatial profiles to measure the location of sets of geographical entities at or over a given time period and spatiotemporal profiles to measure both location and time. The two profiles can be combined with measures of movement (e.g., speed) to allow the spatiotemporal behaviours of the entities to be compared. The taxonomy considers spatiotemporal properties that characterise how the sets of entities evolve within their environment. Two levels are considered: the individual entities and the set level. However, an entity is considered to have no shape, size or orientation. This would not be suitable for the representation of collectives. Consider two collectives: one whose members are closely packed together and one where the spaces between the individual members is quite large. In the former the size of each entity is likely to affect the dynamics of the collective and in the latter the entities could be considered as points since the size of the individuals are unlikely to affect the collective motion.

Instead of primitives, [8, 13] use the concept of an episode to conceptualise movement. Within [13] a piece of software, the Location Trends Extractor (LTE), is introduced. Developed as part of the HyperGeo EU project, the LTE was designed to allow a dataset to be analysed and trends extracted; a report was produced, referred to as a user profile, that could be analysed to establish a user's spatiotemporal behaviour. An analyst could visualise the types of information that could be extracted and establish a set of episodes. An episode is defined as a period of 'relatively homogeneous' activity [13] and are established when a sudden change occurs in one of the indices that has been calculated from the raw data (e.g., speed, direction, sinuosity). Since segmentation is data-driven, [8] note that this method is less arbitrary than looking at particular scales or periods.

3 Spatial Collectives

The TLA method focuses on analysing the collective motion exhibited by spatial collectives: groups of individuals that exhibit a form of spatial coherence at some point during their lifetime [17]. The spatially coherent behaviour may not be exhibited continuously and intermittent spatial coherence is possible. For example, consider a musical society such as a choir. The collective will come together regularly to rehearse or perform but between these occasions, although they are still considered to be a collective, they are unlikely to exhibit any spatial coherence. A choir's source of collectivity (i.e., the reason why they are considered a collective) is their collective intention to sing. This example also highlights that a source of a spatial collective's collectivity need not be spatial.

In order for a group of individuals to be identified as a possible spatial collective within a spatiotemporal dataset they must satisfy at least one of a set of spatial coherence criteria for a user-defined number of time steps. Three forms of spatial coherence are considered: common location, similar movement parameters, and formation; particular spatial collectives may satisfy one or a combination of these criteria. The common location criterion considers both an individual-based and location-based perspective. *Common location (individual-based)* identifies a group of individuals that has been observed as occupying, at a certain level of granularity, the same location (i.e., their positions fall within a sufficiently small region that, at the granularity under consideration, they can be considered as co-located) for a sufficient number of time steps. Distinctions can be made according to whether the common location is shared *intermittently* (e.g., an orchestra or reading group) or *continuously* (e.g., the trees within a forest) but also according to the identity of that location (i.e., is it always the same common location or does it vary). The *common location (location-based)* criterion identifies locations that have a sufficient number of individuals for a sufficient number of time steps. Groups of individuals that share similar values of movement parameters such as direction and speed are identified by the *similar movement* criterion; values are considered shared if they fall within a similarity threshold as given by the user. The *formation criterion* identifies groups of individuals who maintain their relative positions for a sufficient number of time steps allowing a relative view of space to be considered (i.e., the position of the individuals in relation to each other) instead of the absolute view of space as given by the common location criterion.

Although movement pattern analysis has been used to identify these groups within a spatiotemporal dataset, the extraction method does not really analyse the collective's motion and, therefore, little can be said about the type of spatial collective that has been identified. The Three Level Analysis (TLA) method is an example of a tool that would allow the motion to be analysed in greater detail thus allowing the different types of spatial collective to possibly be identified.

4 The Proposed System: The Three Level Analysis (TLA) Method

The Three-Level Analysis (TLA) method focuses on three aspects of collective motion: the motion of the collective when considered as a single entity, the region of space occupied by the collective and the motions of the individual members. For each of these aspects the relevant movement patterns are extracted and analysed to determine the composition of segments that they comprise. These segments are then encoded as a string to form a profile of that movement. The string encoding allows a string similarity algorithm to be used to calculate the similarity between the motions of the individuals relevant to each other and the motions of the individuals compared to that of the collective that they are members of. All information that has been calculated by the method is output as a consolidated profile; such a profile could then be used to distinguish the different types of spatial collective. The remainder of this section will detail each aspect of the method.

4.1 Extracting the Three Aspects of Motion

The method takes as its input the information that has been output from the method detailed in Sect. 3. Therefore, for each possible spatial collective that has been identified in a dataset it is known: which individuals were identified as being members of that collective during the given time period; which of the spatial coherence criteria the group satisfied (may be more than one); which time steps the group of individuals were identified as fulfilling the spatial coherence criteria (including the first and last time step); and, if identified by the common location criteria (individual-based), which common locations the individuals shared and at which time steps.

The first and last time steps at which the group of individuals were found to satisfy the coherence criteria mark the formation and termination of the spatial collective, denoted by T_{BEGIN} and T_{END} respectively and, therefore, the lifespan of the collective. All movement patterns considered by the TLA method will be bound by these two points in time.

The trajectories of all individuals identified as members of the collective within the relevant time period (i.e., between T_{BEGIN} and T_{END}) are extracted from the spatiotemporal dataset. A representative point (e.g., the centroid) is used to represent the overall movement of the collective. The position of the chosen point is calculated at each time step to produce a trajectory. The region occupied by the collective can be referred to as its *footprint*. Much research has been undertaken into establishing the most suitable footprint for a set of points [9]. The TLA method assumes that the algorithm to determine the footprint of a spatial collective has been specified and will use this to calculate the footprint of the collective at each time step; for illustrative purposes, this paper will use the convex hull to represent a spatial collective's footprint although as pointed out in [9] this does not necessarily give an accurate representation of the region occupied by the collectives.

Within [19], variable membership is noted as being an important property of collectives. However, the method used to identify the members of a spatial collective does not really allow for this to be taken into consideration. Although the individuals are identified as satisfying the spatial coherence criteria at certain time steps, it would be difficult to accurately identify whether they are members between these points if intermittent spatial coherence is deemed possible. Therefore, this method will assume an individual is a member of a collective at each and every time step between T_{BEGIN} and T_{END}. The overall report that is produced may establish that a particular type of overall profile indicates a spatial collective that exhibits intermittent spatial coherence. For example, large variations in the size of the footprint could indicate the presence of such a spatial collective.

4.2 Analysing the Extracted Motions

For each of the three aspects of collective motion the extracted movement pattern is analysed for the segments that it comprises. A set of segment types are defined that captures the salient qualitative features of the type of motion that is being considered. The motion of the collective is being considered as the trajectory of a representative point. Therefore the TLA method only considers two types of motion: point-to-point and the evolution of the footprint resulting in two sets of segment types being defined. It is important to note that the method presented here is meant to be application independent and includes the minimal set of segment types that are needed to analyse collective motion. It is possible that further segment types could be established (e.g., those relating to speed or a specific application) and the method extended to include them.

When considering the segment types for point-to-point motion the lack of movement should be considered. It is sometimes ignored but could give significant insight to the type of spatial collective. Contrast the collective motions of a protest march and an orchestra. Both groups are likely to be continuously spatially proximate for some period during their existence but the orchestra will remain at a location when coming together to perform or rehearse; a protest march is likely to be stationary for a much smaller duration. Therefore, the segment type stationary is defined to represent a period where an individual's displacement is below a given threshold (i.e., they have not changed their position between time steps). Ideally a displacement of 0 would indicate a stationary period but one must take into consideration the possibility of noise and errors within the dataset, especially if the data has been collected via GPS. Since different devices will have different levels of accuracy the user is asked to specify all thresholds prior to analysis.

If an entity is not stationary, the direction of motion could be examined. In reality an individual's direction could be constant or varied. However the method calculates the trajectory of an individual based on their position at each time step assuming linear motion; there is insufficient evidence to distinguish between curved and linear motion. This could be different if the method was analysing motion in real time but for now only linear motion will be considered.

A further type of distinction could be made according to the direction of motion. Although this requires imposing a reference frame on the dataset it will allow distinct motions to be characterised. Consider Fig. 1b. If the direction of motion is not considered, individuals *a* and *b* will be considered as having a trajectory comprising the same sequence of segments: three segments of linear motion, one stationary and finally linear motion. Considering direction shows that they are different and gives us a more detailed understanding of the motion. Individual *a* would be considered as having the segments: linear motion (south east), linear motion (south east), linear motion (north east), stationary, linear motion (north east). Individual *b* would have the segment combination: linear motion (north east), linear motion (north east), linear motion (south east), stationary, linear motion (south east). It should be noted that Fig. 1b depicts motion in one dimension; in reality two or three dimensions are more likely. Currently the method only distinguishes between North, North East, East, South East, South, South West, West and North West. A user may decide that further distinctions are necessary and the method extended but we currently only want to illustrate how the method could be used and present it in its most basic form.

The number of possible changes that a footprint can undergo, especially in relation to shape, is vast and it does not seem sensible to systematically define all possible segment types under these three categories. Therefore, when considering the segment types in relation to the evolution of the footprint, only the most salient features must be established. The most basic measure of change in shape would be size which could be measured by both area and perimeter. Each of these attributes could be considered as constant, increasing or decreasing leading to the segment types: constant area, decreasing area, increasing area, constant perimeter, decreasing perimeter and increasing perimeter. The method currently only considers two dimensional spaces but if expanded to three dimensions area could be replaced with volume.

For illustrative purposes the method currently calculates the convex hull for the footprint. However, if a suitable footprint algorithm was used it could be established whether a collective has split or merged resulting in three further possible segment types: splitting, merging and constant number of components. The shape of the footprint could also be considered (e.g., its circularity or elongation). The current version of this method omits these segment types.

4.3 Building the Profiles and Calculating Similarity

A profile is produced for each aspect of motion that is examined by the method, where a profile is a string of characters that encodes the combination of segments that the motion comprises. Each segment type is represented within the profile by a single character (e.g., stationary = S, linear motion north = N).

The analysis of the three aspects of collective motion is an important step in examining collective motion. However, if the method is to be comprehensive, mechanisms must be included that allow the motions of individuals to be compared to each other and that of the collective. Clearly for some collectives there may be no qualitative difference between the motion of the individuals and the motion of its members. For example, a jury during a court session will all be stationary. An analysis of the motion of the collective, when considered as a single entity, and the motion of the jurors will be qualitatively the same. However, for some collectives the two motions will be qualitatively distinct (e.g., the crowd that slowly drifts east whilst the individuals are randomly moving about). For both relationships a distinction could be made between coordinated and uncoordinated motion [15, 19]. However, this is not a clear distinction, there being a gradation from fully coordinated to fully uncoordinated. The question arises as to how this qualitative scale could be quantified.

The segment types that are used to build each profile represent qualitative differences in movement, and a comparison of these profiles should indicate whether the motions are qualitatively different or similar. By encoding the profiles as strings the method can compute how similar two motions are using a string similarity algorithm; the more similar the motions, the more coordinated they are considered to be.

String matching algorithms are used widely in data cleaning (e.g., identifying typographical and formatting errors) and database queries (e.g., identifying duplicated records). String similarity algorithms have also been used to compare movement trajectories (e.g., [10, 14]) but much of this research only compares two trajectories. Often it is the locations that are being compared and not the qualitative aspects of the motions.

Many different types of string matching algorithms have been developed including the Damerau-Levenshtein distance, Jaccard similarity algorithm and Dice similarity distance [12]. The simplest type of similarity algorithm will compare two strings. For example the Damerau-Levenshtein distance calculates the distance (i.e., similarity) between two strings as the minimum number of operations needed to turn one string into the second; the smaller the distance the more similar the two strings are thought to be. Possible operations include insertion, deletion or substitution of a character or transposition of two adjacent characters. Some algorithms (e.g., the Jaccard similarity algorithm) decompose a string into a set of tokens (e.g., words) with the similarity of two sets being used to evaluate the similarity of the strings. Tokens can be weighted according to their importance (e.g., how frequent a token appears in the text or how much 'information content' it is considered to carry).

The current implementation of the TLA method uses a simple string similarity algorithm based on the Damerau-Levenshtein distance. Traditionally this measure is used to compare two sequences. Within the TLA method, unless a collective only has two members, more than two motions must be compared. Therefore, the profile of each individual member is taken and compared to the remaining members' profiles. For each comparison the distance (i.e., measure of similarity) is calculated and stored. The final value is the average of the stored distances. When comparing the coordination between the collective and its individual members, the collective's profile is compared to the profile of each individual member and the distances averaged. Although computationally expensive, this method was adopted to understand if even a simple string similarity algorithm could be used for this purpose. A more thorough investigation may result in a more suitable algorithm being identified or a single algorithm being chosen. However, this paper aims to illustrate the benefits of using a string matching algorithm as part of the TLA method to identify the level of coordination within collective motion. Within the method each operation (i.e., insertion, deletion, substitution or transposition of segment types), is considered to be of equal importance and, therefore, no weights have been added.

4.4 The Consolidated Profile

A consolidated profile is output by the method presenting a summary of the analysis. For each spatial collective the consolidated profile would include: the profiles produced for each aspect of motion with a corresponding plot of that motion; a measure of similarity between the individuals' motions; and, a measure of similarity between the motion of the collective and that of its members.

5 Case Study: Application to AIS Data

To illustrate the proposed method a computer program has been developed in Matlab that allows the method to be applied to a sample spatiotemporal dataset. The chosen dataset records the movement of 480 ships that were fitted with an Automatic

Identification System (AIS) within the Solent over a twenty-four hour period. For each individual recording, the following information was collected: a ship's numerical identifier, its position, its status (e.g., underway) and its bearing; each record also had a date and time stamp. The dataset contains the position of each ship at five minute intervals. For some time steps the position of a ship is not known usually because it did not feature in the remainder of the observation period (i.e., it had stopped emitting its position or it was out of the range of the AIS receiver). At these points the ship's position was recorded as 'Nan'. The dataset had already been preprocessed and the spatial coherence criteria applied. This paper takes the results from this application to feed into the TLA method; a more detailed explanation of the dataset and the preprocessing can be found in [17]. Although many different spatial collectives were identified within this dataset (see [17]), this paper only analyses one of them identified using the common location (individual-based) criterion, to illustrate the method. Instead of displaying the consolidated profile, each aspect of this profile will be discussed in turn.

5.1 Extracting the Three Levels

Figure 2 shows the movement patterns that have been extracted for each of the three levels that are considered by the TLA method: the movement of the collective when represented by its centroid (Fig. 2a), the evolution of the footprint when represented by the convex hull (Fig. 2b), and the motion of the individual members (Fig. 2c).

5.2 Identification of Segments and Profiles

For each level within the TLA method the segments were identified and the profiles produced. Figure 3a and b gives a visualisation of the profiles for the footprint in terms of area and perimeter respectively. Figure 3c depicts the profile generated for the collective. For the footprint, a score of 0, 1 or -1 indicates constant, increased or decreased area or perimeter. For the collective motion: 0 stationary, 1 linear motion (North), 2 linear motion (North East), 3 linear motion (East), 4 linear motion (South East), 5 linear motion (South), 6 linear motion (South West), 7 linear motion (West) and 8 linear motion (North West).

5.3 Calculating Similarity Measures

Given the calculated profiles, a similarity measure of 29.99 % was calculated for the movement of the individual members in relation to each other. The movement of the collective was considered to have a similarity score of 33.54 % with the motions of

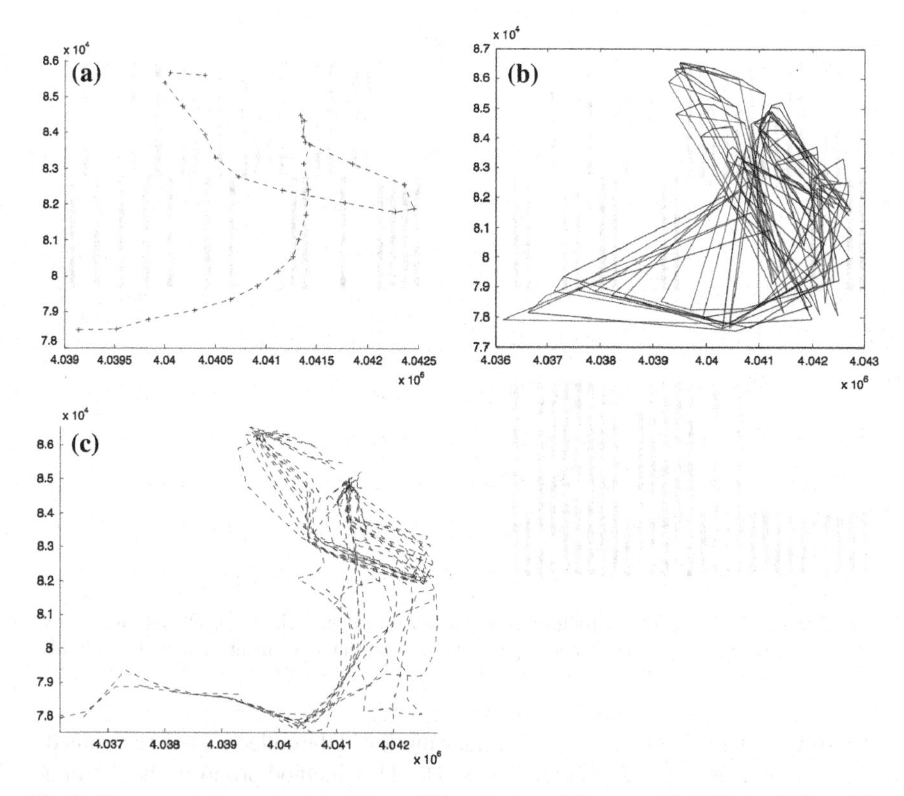

Fig. 2 The extracted movement patterns. **a** The trajectory of the centroid. **b** The evolution of the footprint. **c** The trajectories of the individual members

its individual members. These scores could be considered quite uncoordinated and reflects the lack of coordination between members towards the end of the collective motion.

6 Discussion

The TLA method has been developed to allow collective motion to be analysed in detail taking account of the important features of collective motion. This paper has shown that it could be applied to a sample spatiotemporal dataset; however, this example has raised important points for discussion and highlighted how the TLA method could be extended.

Currently the method only includes the minimal number of segment types that were thought necessary to show that collective motion could be examined in this way. The set of segment types could be expanded, especially in relation to the evolution of a footprint if a more representative footprint is calculated (i.e., not the convex hull).

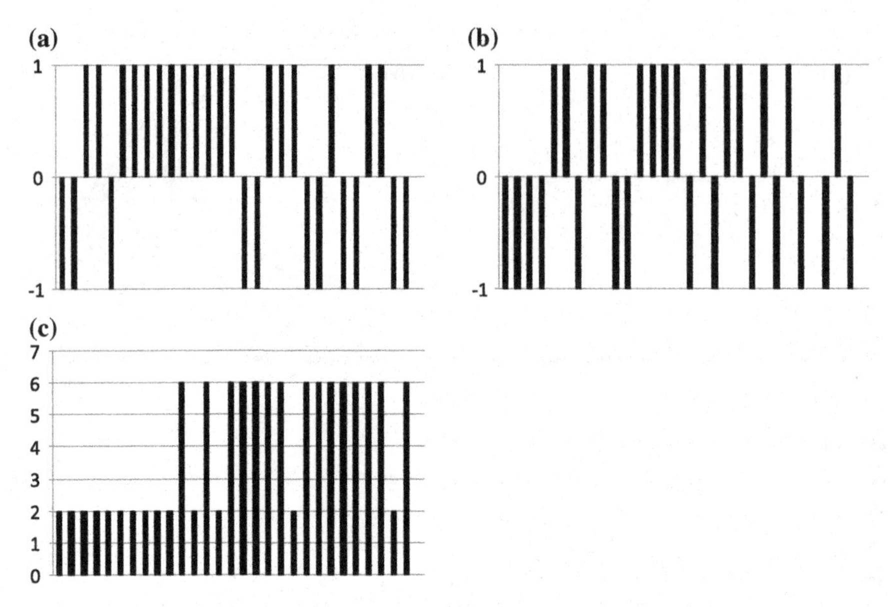

Fig. 3 The segments relating to footprint evolution. **a** Segments relating to change in area of footprint. **b** Segments relating to change in perimeter of footprint. **c** Segments relating to centroid

Existing research has noted the need for the motion of the collective to be examined in relation to the motions of its individuals. The TLA method proposes the use of a string similarity algorithm to capture the level of coordination between the movement of individual members but also between the movement of the collective and that of its members. This method appears novel but the current implementation of the TLA method only includes a simple string similarity algorithm which is computationally expensive. A thorough investigation is currently being undertaken to identify a more suitable algorithm. It is possible that those used in DNA sequence analysis could be more suitable and would also be able to deal with gaps in a dataset (i.e., when the position of an individual is not known). Set-based similarity algorithms can be used to establish the proportion of elements within a set that have a similarity score above a given threshold. Such a method could be used to calculate the proportion of individuals that the collective's motion is similar to; the higher the rating, the more coordinated the motion of the collective to those of its members. Comparing the motion of the individuals relative to each other can be calculated in the same way. Each individual member could be taken as the query string and compared to the remaining individuals resulting in the number of members that are considered to be above the threshold of similarity. This process could be repeated for each member of the collective. The overall rating of similarity between individuals would be the average of the individual similarity scores. Further investigation is needed to see if a set-based algorithm would improve the computational cost. For each similarity algorithm considered for the TLA method, it must be understood how sensitive it would be to noise in the data (e.g., missing segments).

The similarity algorithm used in the current implementation of the TLA method compares the coordination across the entire movement pattern. It would seem more appropriate to identify temporal sequences of common episodes types to establish periods of time when the coordination is high. For example, the Longest Common Substring algorithm could be used which identifies the longest possible substring between a set of strings [2]. This similarity measure could also be used to account for temporal similarities in combination of segments (i.e., if there is any repetition of sequences of segments).

The TLA method produces a consolidated report that provides a summary of the analysis undertaken. Given the assumptions of the proposed method it is unlikely that it would be able to analyse a collective that did not exhibit some form of spatial coherence at some point during its existence. It has been suggested that such a profile could be used to identify the different types of spatial collective. However, this would require a systematic investigation into the different types of spatial collective that exist and the different types of profile that each would generate.

7 Conclusion

The Three-Level Analysis (TLA) method has been proposed to allow collective motion to be analysed in detail taking account of the important features of collective motion. The method suggests the extraction of three aspects of the motion and the combined use of segments and profiles to enable a string similarity algorithm to be used to calculate the level of coordination between the motions of the individuals relevant to each other and the level of coordination between the motion of the collective and that of its members. The consolidated profile that is produced by the proposed method could be used to distinguish the different types of spatial collective. The method has been implemented and applied to a sample spatiotemporal dataset.

Acknowledgments Dr. Antony Galton is acknowledged for his insightful and useful comments.

References

1. W. Ali and B. Moulin. 2D–3D multiagent geosimulation with knowledge-based agents of customers' shopping behaviour in a shopping mall. In David M. Mark and Anthony G. Cohn, editors, *Spatial Information Theory: Proceedings of International Conference COSIT 2005*, Lecture Notes in Computer Science, pages 445–458, Ellicottville, NY, USA, 2005. Springer.
2. I. Alsmadi and M. Nuser. String matching evaluation methods for dna comparison. *Internation Journal of Advanced Science and Technology*, 47:13–32, 2012.
3. M. Andersson, J. Gudmundsson, P. Laube, and T. Wolle. Reporting leaders and followers among trajectories of moving point objects. *Geoinformatica*, 12(4):497–528, 2008.

4. G. Andrienko and N. Andrienko. Extracting patterns of individual movement behaviour from a massive collection of tracked positions. In B. Gottfried, editor, *Workshop on Behaviour Modelling and Interpretation.*, pages 1–16, Germany, 2007. Technologie-Zentrum Informatik.
5. Gennady Andrienko, Natalia Andrienko, Salvatore Rinzivillo, Mirco Nanni, Dino Pedreschi, and Fosca Giannotti. Interactive visual clustering of large collections of trajectories. In *IEEE Visual Analytics Science and Technology (VAST 2009)*, pages 3–10, Atlantic City, New Jersey, USA, October 12–13 2009. IEEE Computer Society Press.
6. S. Dodge, R. Weibel, and E. Forootan. Revealing the physics of movement: Comparing the similarity of movement characteristics of different types of moving objects. *Computers, Environment and Urban Systems*, 33(6):419–434, November 2009.
7. S. Dodge, R. Weibel, and A.K. Lautenschütz. Towards a taxonomy of movement patterns. *Information Visualization*, 7:240–252, 2008.
8. J. A. Dykes and D. M. Mountain. Seeking structure in records of spatiotmeporal behaviour: visualization issues, efforts and application. *Computational Statistics and Data Analysis*, 43 (Data Visualization II Special Edition)(4):581–603, 2003.
9. A. Galton and M. Duckham. What is the region occupied by a set of points? In M. Raubal, H. Miller, A. Frank, and M. Goodchild, editors, *Geographic Information Science: Proceedings of the Fourth International Conferencem GIScience 2006*, pages 81–98, Berlin, 2006. Springer.
10. J. Gudmundson, P. Laube, and T. Wolle. Computational movement analysis. In Wolfgang Kresse (1) W. Kresse and D. Danko, editors, *Springer Handbook of Geographic Information*, chapter C, pages 423–438. Springer, Berlin Heidelberg, 2012.
11. J. Gudmundsson, P. Laube, and T. Wolle. Movement patterns in spatio-temporal data. In *Encyclopedia of GIS*. Springer-verlag, 2007.
12. M. Hadjieleftherioux and D. Srivastava. Weighted set-based string similarity. *IEEE Data Engineering Bulletin*, 33(1):25–36, 2010.
13. D. M. Mountain and J. F. Raper. Modelling human spatio-temporal behaviour: A challenge for location-based services. In *Proceedings of the 6th International Conference on Geocomputation 2001*, Brisbane, Australia, 24–26 September 2001.
14. G. Sinha and D. Mark. Measuring similarity between geospatial lifelines in studies of environmental health. *Journal of Geographical Systems*, 7(1):115–136, 2005.
15. S. Spaccapietra and C. Parent. Adding meaning to your steps. In *Conceptual Modeling ER 2011 Conference, 30th International Conference on Conceptual Modeling*, volume 6998, Brussels, Belgium, October 31 - November 3 2011. Lecture Notes in Computer Science, Springer.
16. M. Thériault, C. Claramunt, and P.Y. Villeneuve. A spatio-temporal taxonomy for the representation of spatial set behaviours. In H.B. Böhlen, C.S. Jensen, and M. Scholl, editors, *Spatio-Temporal Database Management, International Workshop STDBM'99*, pages 1–18. Springer, 1999.
17. Z. Wood. *Detecting and Identifying Collective Phenomena within Movement Data.* PhD thesis, University of Exeter, 2011.
18. Z. Wood and A. Galton. Classifying collective motion. In B. Gottfried and H. Aghajan, editors, *Behaviour Monitoring and Interpretation BMI: Smart Environments*, pages 129–155. IOS Press, Amsterdam, The Netherlands, 2009.
19. Z. Wood and A. Galton. A taxonomy of collective phenomena. *Applied Ontology*, 4(3–4):267–292, 2009.
20. Z. Wood and A. Galton. Zooming in on collective motion. In Mehul Bhatt, Hans Guesgen, and Shyamanta Hazarika, editors, *Proceedings of the International Workshop on Spatio-Temporal Dynamics, co-located with the European Conference on Artificial Intelligence (ECAI-10)*, Lisbon, Prtugal, 2010. ECAI Workshop Proceedings., and SFB/TR 8 Spatial Cognition Report Series.

Sentiment Classification Using Supervised Sub-Spacing

Sadiq Sani, Nirmalie Wiratunga, Stewart Massie and Robert Lothian

Abstract An important application domain for Machine learning is sentiment classification. Here, the traditional approach is to represent documents using a Bag-Of-Words (BOW) model, where individual terms are used as features. However, the BOW model is unable to sufficiently model the variation inherent in natural language text. Term-relatedness metrics are commonly used to overcome this limitation by capturing latent semantic concepts or topics in documents. However, representations produced using standard term relatedness approaches do not take into account class membership of documents. In this work, we present a novel approach called Supervised Sub-Spacing ($S3$) for introducing supervision to term-relatedness extraction. $S3$ works by creating a separate sub-space for each class within which term relations are extracted such that documents belonging to the same class are made more similar to one another. Recent approaches in sentiment classification have proposed combining machine learning with background knowledge from sentiment lexicons for improved performance. Thus, we present a simple, yet effective approach for augmenting $S3$ with background knowledge from SentiWordNet. Evaluation shows $S3$ to significantly out perform the state-of-the-art SVM classifier. Results also show that using background knowledge from SentiWordNet significantly improves the performance of $S3$.

S. Sani (✉) · N. Wiratunga · S. Massie · R. Lothian
Robert Gordon University, Aberdeen, Scotland
e-mail: s.a.sani@rgu.ac.uk

N. Wiratunga
e-mail: n.wiratunga@rgu.ac.uk

S. Massie
e-mail: s.massie@rgu.ac.uk

R. Lothian
e-mail: r.m.lothian@rgu.ac.uk

M. Bramer and M. Petridis (eds.), *Research and Development in Intelligent Systems XXX*, DOI: 10.1007/978-3-319-02621-3_8,

1 Introduction

Sentiment classification has many important applications e.g. market analysis and product recommendation. A common approach is to apply machine learning algorithms in order to classify individual texts (generally called documents) into the appropriate sentiment category (e.g. positive or negative) [15]. Doing this however requires text documents to be represented using a suitable set of features. A popular approach is the Bag-Of-Words (BOW) model where documents are represented using individual terms as features. However, the BOW model is unable to cope with variation in natural language vocabulary (e.g. synonymy and polysemy) which often requires semantic indexing approaches [18].

The general idea of semantic indexing is to discover terms that are semantically related and use this knowledge to identify conceptual similarity even in the presence of vocabulary variation. The result is the generalisation of document representations away from low-level expressions to high-level semantic concepts. Several techniques have been proposed for transforming document representations from the space of individual terms to that of latent semantic concepts. Examples include Latent Semantic Indexing (LSI) which uses singular-value decomposition to exploit co-occurrence patterns of terms and documents in order to create a semantic concept space which reflects the major associative patterns in the corpus [9]. Other more simple approach use statistical measures of term co-occurrence within documents to infer semantic similarity. However, representations produced using these approaches are not optimal for sentiment classification because they do not take into account class membership of documents [5]. We introduce a novel technique called Supervised Sub-Spacing ($S3$) for introducing supervision to term-relatedness extraction. $S3$ works by creating a separate sub-space for each class within which term relations are extracted. The power of $S3$ lies in its ability to modify document representations such that documents that belong to the same class are made more similar to one another while, at the same time, reducing their similarity to documents of other classes. In addition, $S3$ is flexible enough to work with a variety of term-relatedness metrics and yet, powerful enough that it leads to consistent improvements in text classification accuracy when adopted by a local learner such as the nearest neighbour classifier.

Recent works in sentiment classification indicate that machine learning approaches can benefit from using knowledge from sentiment lexicons [8, 13, 14]. Combining the two approaches has a number of benefits. Firstly, it allows machine learning classifiers to utilise general knowledge relevant for sentiment classification, thus, avoiding overfitting the training data. Secondly, supplementing training data with knowledge from sentiment lexicons has the potential to reduce the number of training examples required to build accurate classifiers. However, achieving significant improvements using this combined approach has proved difficult [14]. In this work, we show how background knowledge from a general purpose sentiment lexicon (SentiWordNet [10]) can be used with $S3$. Evaluation shows combining $S3$ with knowledge from SentiWordNet leads to significant improvement in sentiment classification performance.

The rest of this paper is organised as follows, related work is presented in Sect. 2. In Sect. 3, we describe statistical term relatedness extraction and we present a few techniques for extracting term relatedness based on corpus co-occurrence. In Sect. 4, we present our supervised term realtedness extraction approach, $S3$, and describe how background knowledge from SentiWordNet is utilised by $S3$. Section 5 describes the datasets we use for evaluation. In Sect. 6 we present an evaluation of $S3$ on a number of sentiment classification tasks, followed by conclusions in Sect. 7.

2 Related Work

The success of machine learning for sentiment classification largely depends on the performance of the underlying classifier. The state-of-the-art classification algorithm is considered to be SVM, which works by creating higher dimension representation of documents that leads to linear separability of document classes [12]. However, other studies have suggested that instance-based learners such as k Nearest Neighbour (kNN) are equally competitive with SVM performance if a proper data representation is chosen [7]. Several feature transformation approaches have been proposed for generating effective text representations by transforming document representations from the space of individual terms to that of latent concepts or topics that better capture the underlying semantics of the documents. Our Supervised Sub-Spacing ($S3$) algorithm is related to this general class of feature transformation techniques.

A typical example of feature transformation approaches is a generative probabilistic model called latent dirichlet allocation (LDA) where each document in a collection is modelled as an infinite mixture over an underlying set of topics [3]. The set of LDA topics is considered as being representative of the underlying latent semantic structure of the document collection. Another approach is latent semantic indexing (LSI) which uses singular value decomposition to transform an initial term-document matrix into a latent semantic space [9]. A variant of LSI called probabilistic latent semantic analysis (pLSA) uses a latent variable model for general co-occurrence data which associates a unobserved class model variable with each occurrence of a word w in a document [11]. The main limitation of LDA, LSI and pLSA for text classification is that these techniques are agnostic to class knowledge. Thus, the concept/topic spaces extracted by these approaches are not necessarily the best fit for the class distribution of the document collection [1].

Recent approaches have been proposed for introducing supervision into LDA and LSI. For example, a supervised version of LDA called sLDA is presented in [2], where a response variable (class label, real value, cardinal or ordinal integer value) associated with each document is added to the LDA model. Thus the topic model is learnt jointly for the documents and responses such that the resultant topics are good predictors of the response variables. Evaluation of sLDA on regression tasks suggests moderate improvements. An extension of LSI has also been proposed called supervised LSI (SLSI), that iteratively computes SVD on term similarity matrices of separate classes [17]. A separate term-doc matrix is constructed for each class and in each iteration, SVD is performed on each class-specific term-doc matrix. Evaluation results showed SLSI performed better than LSI. However, only marginal gains were achieved with SLSI over BOW. The main limitation of sLDA and SLSI is that they are both tied to specific document transformation techniques i.e. LDA and LSI respectively which limits the portability of these approaches. Our goal with $S3$ is to develop a technique that is flexible enough to be used with a variety of feature term relatedness approaches.

Despite the success of machine learning for sentiment classification, it is understood that sentiment lexicons provide important background knowledge that is useful for sentiment classification [8, 13, 14]. Accordingly, many recent approaches have proposed combining machine learning with background knowledge from sentiment lexicons. An example is a system called *pSenti* which utilises a sentiment lexicon at the feature generation stage to filter out non-sentiment-bearing words such that only words that appear in the sentiment lexicon are used as features [14]. These features are then used to train an SVM classifier. The evaluation compared *pSenti* with a baseline lexicon only approach and also SVM using a standard BOW representation. Results show pSenti to perform better than the lexicon only approach however, not as good as SVM. A second evaluation on a cross domain task where the training and test datasets are obtained from different domains showed *pSenti* to perform better than SVM indicating that *pSenti* avoids overfitting the training data. However, the low performance of *pSenti* compared with SVM makes it non-ideal where high classification accuracy is a priority.

Another approach which builds two generative models: one from a labelled corpus and a second from a sentiment lexicon is presented in [13]. The distributions from the two models are then adaptively pooled to create a composite multinomial Naive Bayes classifier. This approach allows for reducing the amount of labelled training data required and at the same time, produces better quality classifiers than lexicon only models. The pooling approach used employs a linear combination of conditional probabilities from the different generative models. The combined approach was compared to using Naive Bayes classifier built using only the labelled corpus with significant improvement in classification accuracy. Also, evaluation showed the combined approach to reach high levels of accuracy using very little of the training data. However, a weakness of this approach is that it is specific to the Naive Bayes classifier.

An alternative approach which uses a lexical database (SentiWordNet) to extract a set of sentiment features and applies that in combination with content-free and content specific features is presented in [8]. Content-free features include statistical measures of lexical variation, measures of vocabulary richness, statistics of function words, punctuations, part of speech categories as well as the use of greetings, signatures, number of paragraphs and average paragraph length. Content-specific features capture important n-grams where importance is considered as having an occurrence frequency of at least five times in the corpus. Sentiment features are extracted by looking up unigrams in SentiWordNet. Evaluation was conducted on several datasets using SVM with significant improvement. However, the feature engineering effort required in this approach is significantly high compared to our approach. Also, unlike our approach, the use of additional sentiment features leads to a significant increase in the length of feature vectors which in turn increases sparsity in document representations.

3 Term Relatedness Extraction

In this section, we describe how corpus co-occurrence can be used to infer semantic relatedness. These class of approaches are based on the premise that co-occurrence patterns of terms in a corpus are indicative of semantic relatedness such that the more frequently two terms occur together in a specified context, the more semantically related they are [19]. In Sects. 3.1 and 3.2, we present two different approaches for estimating term relatedness from corpus co-occurrence. Later on in this paper we describe how these term-relatedness metrics can be used with $S3$.

3.1 Document Co-Occurrence

Documents are considered similar in the vector space model (VSM) if they contain a similar set of terms. In the same way, terms can also be considered similar if they appear in a similar set of documents. Given a standard term-document matrix D where column vectors represent documents and the row vectors represent terms, the similarity between two terms can be determined by finding the distance between their vector representations. The relatedness between two terms, t_1 and t_2 using the cosine similarity metric is given in Eq. 1.

$$Sim_{DocCooc}(t_1, t_2) = \frac{\sum_{i=0}^{n} t_{1,i} t_{2,i}}{|t_1||t_2|} \quad (1)$$

3.2 Normalised Positive Pointwise Mutual Information (NPMI)

The use of mutual information to model term associations is demonstrated in [6]. Given two terms t_1 and t_2, mutual information compares the probability of observing t_1 and t_2 together with the probability of observing them independently as shown in Eq. 2. Thus, unlike document co-occurrence, PMI is able to disregard co-occurrence that could be attributed to chance.

$$PMI(t_1, t_2) = log_2 \frac{P(t_1, t_2)}{P(t_1)P(t_2)} \quad (2)$$

If a significant association exists between t_1 and t_2, then the joint probability $P(t_1, t_2)$ will be much larger than the independent probabilities $P(t_1)$ and $P(t_2)$ and thus, $PMI(t_1, t_2)$ will be greater than 0. Positive PMI is obtained by setting all negative PMI values to 0. The probability of a term t in any class can be estimated by the frequency of occurrence of t in that class normalised by the frequency of all words in all classes.

$$P(t) = \frac{f(t)}{\sum_{j=1}^{N} \sum_{i=1}^{N} f(t_i, t_j)} \tag{3}$$

PMI values do not lie within the range 0 to 1 as is the case with typical term relatedness metrics [16]. Thus we need to introduce a normalisation operation. We normalise PMI as shown in Eq. 4.

$$Sim_{Npmi}(t_1, t_2) = \frac{PPMI(t_1, t_2)}{-log_2 P(t_1, t_2)} \tag{4}$$

4 Supervised Sub-Spacing

A major limitation of basic approaches to term relatedness extraction on supervised tasks is that term relations are extracted from the entire term-document space instead of being confined to the space of the target class. Consider the example term-document space shown in Fig. 1 where, for the purpose of illustration, terms are shown in the space of *Positive* and *Negative* sentiment classes rather than individual documents. The relation between 'best' and 'fantastic' is likely to be strong because of their proximity within this space. However, terms like 'director' and 'movie'

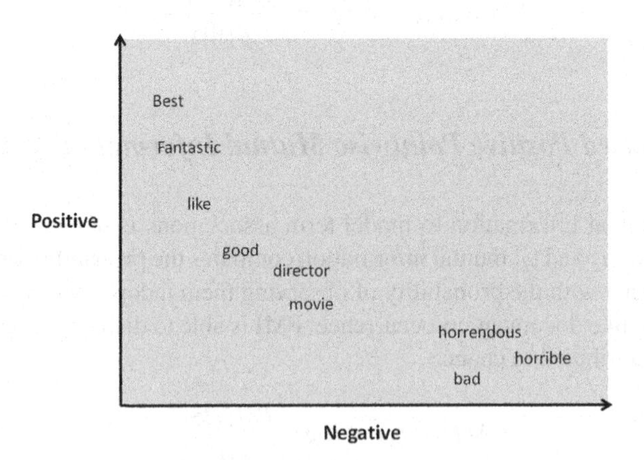

Fig. 1 Two-dimensional visualisation of terms in the space of *positive* and *negative* sentiment classes

are likely to occur in either sentiment class. Thus, establishing a strong association between the terms ‘good’ and ‘director’ (due to their proximity in the entire space) is likely be a major source of noise for documents belonging to the negative sentiment class. However, by extracting term relatedness from separate subspaces, we avoid assigning a strong relation between ‘director’ and ‘good’ in documents belonging to the negative class because of the low occurrence of ‘good’ in the negative class.

*S*3 works by computing term-relatedness in class-partitioned subspaces which has the desired effect of making documents that belong to the same class more similar to one another while making them more dissimilar to documents of other classes. This is achieved using a processing pipeline where the entire term-document space is partitioned into N term-document sub-spaces where N is the number of classes. Hence, documents that belong to the same class can be processed together and separate from documents of other classes. An overview of this approach is shown in Fig. 2.

A standard term-documents matrix D is initially created from the training corpus. The matrix D is then partitioned into N sub-matrices D_1, to D_N for each of N classes in the training corpus such that the sub-matrix D_1 contains only training

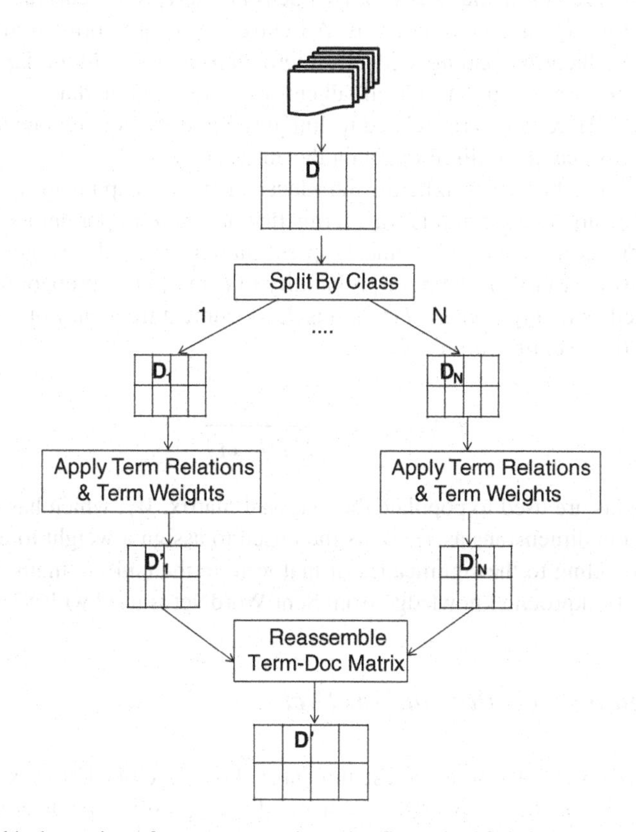

Fig. 2 Graphical overview of text representation using Supervised Sub-Spacing

documents from class 1 and so on. Class-specific term relations and term weights are then applied to each sub-matrix D_i to obtain the transformed matrix D'_i. This transformation process is discussed in detail in Sect. 4.1.

4.1 Document Transformation

Each sub-matrix is transformed by applying term-relations and term-weights to obtain a generalised term-document matrix D' as expressed in Eq. 5 where D_i is the term-document sub-matrix of class i, while T_i and W_i are class-specific term-relation and term-weighting matrices respectively.

$$D'_i = D_i \times T_i \times W_i \tag{5}$$

The first step in this transformation process is to obtain all pairwise term relatedness values to populate a term-term similarity matrix T_i where the row and column dimensions of T_i represent the corpus terms. Each entry in T_i represents the similarity of corresponding row and column terms. All entries in T_i are normalised with the value 1 in any cell corresponding to identical term pairs and 0 to dissimilar. Because any term can be at most similar to itself, all entries on the leading diagonal of T_i are consequently 1. Here, any term related techniques e.g. document co-occurrence or PMI (see Sect. 3), can be utilised to obtain the entries of T_i.

Next, class-specific term weights are introduced in order to capture the importance of terms to the target class. In this way, terms that are more important to the target class have a higher influence in document representations. Here, the weight of a term t_i in class c_j is measured as the prior probability of t_i in c_j. The probability of t_i in c_j is calculated using Eq. 6 where $f(t_i, c_j)$ is the document frequency of t_i in c_j and N is the size of the entire corpus.

$$w(t_i, c_j) = \frac{f(t_i, c_j)}{\sum^N f(t_i, c_k)} \tag{6}$$

Term weights are used to populate the diagonal matrix, W_i, which has the same row and column dimensions as T_i. W_i is then used to assign a weight to each term in D_i corresponding to the significance of that term in the (sub) domain. Next, we describe how background knowledge from SentiWordNet is used with $S3$.

4.2 Combining S3 with SentiWordNet

SentiWordNet is a lexical resource for sentiment analysis developed as an extension to the popular lexical resource, WordNet. SentiWordNet is partitioned into Noun, Verb, Adjective and Adverb dictionaries and entries within each dictionary are grouped into

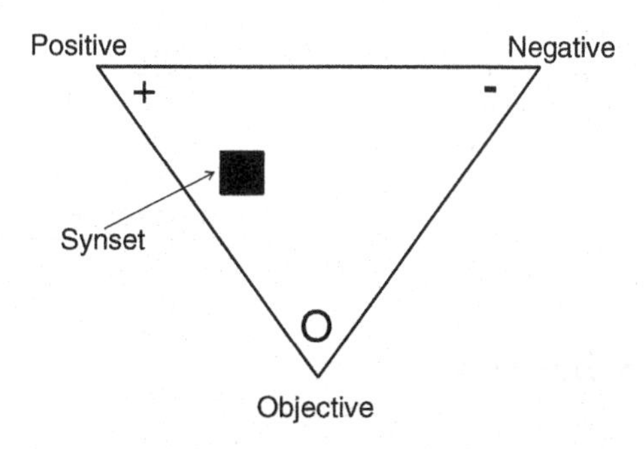

Fig. 3 Representation of the position of a synset in three-dimensional sentiment space as provided by SentiWordNet

Fig. 4 Matching synsets for the term 'fantastic' in SentiWordNet showing both negative and positive sentiment scores for each sense

sense 1:	pos = 0.375	neg = 0.0
sense 2:	pos = 0.75	neg = 0.0
sense 3:	pos = 0.375	neg = 0.375
sense 4:	pos = 0.0	neg = 0.625
sense 5:	pos = 0.375	neg = 0.375

sets of cognitive synonyms called synsets. Each synset in SetiWordNet is associated with scores along three sentiment dimensions, a negative score, a positive score and an objective score, indicating how strongly that entry is associated with the respective sentiment dimension. The positive, negative and objective scores of each entry sum to a total of 1.0. An alternative way of visualising this is in a three dimensional sentiment space where an entry (synset) can be considered as occupying a position in this space as show in Fig. 3.

Our aim here is to obtain a new weight $w(t_i, c_j)'$ for any term t_i by augmenting the $S3$ weight $w(t_i, c_j)$ of t_i with its class-specific sentiment score. Given any lemmatised term t_i, we obtain its sentiment score from SentiWordNet by matching t_i the appropriate synsets in SentiWordNet. Terms are matched to synsets by searching for matching entries in the Noun, Verb, Adverb and Adjective dictionaries in that order. If a matching entry is found in any dictionary then the lookup is abandoned and subsequent dictionaries are not searched. The final sentiment score of term t_i is obtained as the average score of all matching synsets in the target dictionary, along the *positive* and *negative* sentiment dimensions. For example the term 'fantastic' matches 5 synsets in the Noun dictionary as shown in Fig. 4. Thus, the score of fantastic for the Positive class is obtained as the average of the positive scores of all 5 senses. The same approach is used for the Negative class.

Once the class-specific sentiment score $score(t_i, c_j)$ of t_i has been obtained, $w(t_i, c_j)'$ is computed by combining $score(t_i, c_j)$ and $w(t_i, c_j)$ using a linear interpolation approach as shown in Eq. 7.

$$w(t_i, c_j)' = \alpha w(t_i, c_j) + \beta score(t_i, c_j) \tag{7}$$

For the purpose of this work, we use the values $\alpha = \beta = 1.0$.

Table 1 Overview of datasets used for evaluation showing number of documents in each dataset

Dataset	Number of documents
Movie reviews	1,000
Amazon reviews	1,000
Twitter dataset	900
Hotel reviews	1,000

5 Datasets

A summary of the datasets used in our evaluation is provided in Table 1. All datasets contain only the binary sentiment classes *postive* and *negative* with equal distribution of documents between the two classes. We describe these datasets in detail in Sects. 5.1, 5.2, 5.3 and 5.4.

5.1 Movie Reviews

This is a sentiment classification corpus comprising movie reviews from the Internet Movie Database (IMDB) [15]. We used version 1 of this corpus which contains 1,400 reviews, half of which are classified as expressing positive sentiment while the other half is classified as negative. Accordingly, the classification task for this dataset is to determine the sentiment orientation of any given review.

5.2 Amazon Reviews

This is another sentiment classification corpus consisting of customer reviews obtained from the Amazon website. We used version 1 of this dataset which is described in [4]. Four types of products were considered in the dataset: books, DVDs, electronics and kitchen appliances. The original user reviews had a star rating between 1 and 5. We transformed this into binary sentiment classes using the same approach as [4] where reviews with star rating less than 3 are considered negative and those with star rating of 4 and 5 are considered positive.

5.3 Twitter Dataset

This is a collection of 5,513 tweets on four topics: *Apple*, *Google*, *Microsoft* and *Twitter*, available from Sanders Analytics.[1] All tweets have been manually classified into one of three sentiment categories: negative, positive, neutral, including an

[1] http://www.sananalytics.com/lab/twitter-sentiment/

additional uncategorised category for tweets that are not considered to bear any sentiment. We utilise only the positive and negative sentiment classes for our evaluation.

5.4 Hotel Reviews

This is a collection of hotel reviews obtained from the TripAdvisor website as described in [20]. The corpus contains a total of 235,793 reviews, each with a user assigned star rating between 1 and 5. We convert these ratings into binary sentiment classes by tagging reviews with a star rating lower than 3 as negative while reviews with a rating above 3 are tagged as positive. We then randomly select 500 reviews from each of the positive and negative classes to create our evaluation dataset.

6 Evaluation

The aim of our evaluation is two-fold. Firstly, we wish to determine the performance of combining *S*3 with SentiWordNet on sentiment classification tasks. To achieve this we compare sentiment classification performance on document representations obtained using the following strategies.

- BASE: Basic BOW approach without term relatedness
- S3COOC: Supervised term-relatedness extracted using our *S*3 approach with COOC term-relations
- S3NPMI: Supervised term-relatedness extracted using our *S*3 approach with NPMI term-relations
- S3COOCSWN: S3COOC augmented with SWN sentiment scores
- S3NPMISWN: S3NPMI augmented with SWN sentiment scores

We report classification accuracy using a similarity weighted kNN classifier (with k=3) and using the cosine similarity metric to identify the neighbourhood. Our expectation is that in comparison with S3COOC and S3NPMI, S3COOCSWN and S3NPMISWN should lead to better sentiment classification performance. The results for BASE serve as a baseline to measure the improvement achieved using term relatedness.

Secondly, we compare the performance of the two augmented *S*3 representations, S3COOCSWN and S3NPMISWN, to state-of-the-art sentiment classification approach. Thus we include a comparison with the following:

- SVM: Support Vector Machine classifier applied to a standard BOW representation.

For SVM classification, we use the LibSVM package with default parameters. Standard preprocessing operations i.e. lemmatisation and stopwords removal are

Table 2 Results showing classification accuracy in percentage for the different approaches with best results shown in bold

Dataset	BASE	SVM	S3COOC	S3NPMI	S3COOCSWN	S3NPMISWN
Movie reviews	70.7	82.3	83.4	85.0	85.4$^+$	**85.8**$^+$
Amazon reviews	65.9	66.6	78.7	**81.3**	76.8$^-$	81.0
Twitter data	71.6	69.2	82.9	82.7	84.2$^+$	**85.1**$^+$
Hotel reviews	64.5	63.5	68.4	67.3	**70.7**$^+$	68.1$^+$

applied to all datasets. Feature selection is also used to limit our indexing vocabulary to the top 300 most informative terms for all datasets. We report results over 5 runs of 10-fold cross validation. Statistical significance is reported at 95 % using the paired t-test.

6.1 Results

Table 2 presents results of our evaluation. The best results for each dataset are shown in bold. Values in the S3COOCSWN and S3NPMISWN columns with a $^+$ represent significant improvement in classification accuracy compared with their non-lexicon-based counterparts i.e. S3COOC and S3NPMI respectively while values with $^-$ represent a significant depreciation in performance. As expected, the best results, except on the AmazonReviews dataset, are achieved using either S3COOCSWN or S3NPMISWN representations. Also, augmenting $S3$ with SentiWordNet produces significant improvements on all datasets except AmazonReviews where the augmented representation resulted in a decline in classification accuracy. The depreciation in accuracy between S3COOC and S3COOCSWN is statistically significant. This was likely caused by noise from SentiWordNet. Indeed, observation of sentiment scores returned by SentiWordNet reveals some rather unintuitive values. This suggests that further study is perhaps needed in order to determine how to eliminate noisy sentiment scores obtained from SentiWordNet e.g. using a threshold. However, the significant improvements achieved on most datasets indicate that our approach is nonetheless effective for sentiment classification.

Observe that the performance of SVM is quite poor on most datasets with the exception of the MovieReviews dataset. This is perhaps because the decision boundary for these datasets is quite complex and thus a linear separability between the two classes is difficult to achieve. This is also reflected in the poor performance of BASE on the same datasets. Note however that the performance of SVM on the MovieReview dataset is similar to the 82.9 % accuracy reported in [15] which indicates that our SVM classifier delivers standard performance.

7 Conclusion

In this paper, we have presented a novel technique called Supervised Sub-Spacing ($S3$) for introducing supervision into term-relatedness extraction. The effectiveness of $S3$ lies in its ability to transform document representations such that documents that belong to the same class are made more similar to one another while, at the same time, making them more dissimilar to documents of a different class. Also, unlike typical feature transformation approaches e.g. sLDA and SLSI, $S3$ is not tied to any specific model (i.e. LDA and LSI respectively). We demonstrated this by using $S3$ with both COOC and NPMI term relatedness metrics. Evaluation shows our two $S3$-based representations to out perform SVM applied to BOW.

We have also demonstrated how background knowledge from a sentiment lexicon (SentiWordNet) can be utilised with $S3$ to improve sentiment classification performance. Our approach is a simple, yet effective linear interpolation of class-specific term weights derived from corpus statistics and class-specific term scores derived from the lexicon. Evaluation shows combining $S3$ with knowledge from a sentiment lexicon significantly improves sentiment classification accuracy.

Future work will investigate less noisy techniques for obtaining sentiment scores from SentiWordNet e.g using a threshold. Also, we intend to investigate the performance of $S3$ representations used with other types of classifiers e.g SVM.

References

1. Aggarwal, C.C., Zhai, C. (eds.): Mining Text Data. Springer (2012)
2. Blei, D., McAuliffe, J.: Supervised topic models. In: J. Platt, D. Koller, Y. Singer, S. Roweis (eds.) Advances in Neural Information Processing Systems 20, pp. 121–128. MIT Press, Cambridge, MA (2008)
3. Blei, D.M., Ng, A.Y., Jordan, M.I.: Latent dirichlet allocation. J. Mach. Learn. Res. **3**, 993–1022 (2003)
4. Blitzer, J., Dredze, M., Pereira, F.: Biographies, bollywood, boom-boxes and blenders: Domain adaptation for sentiment classification. In: Proceedings of the 45th Annual Meeting of the Association of Computational Linguistics, pp. 440–447. Association for Computational Linguistics, Prague, Czech Republic (2007)
5. Chakraborti, S., Wiratunga, N., Lothian, R., Watt, S.: Acquiring word similarities with higher order association mining. In: Proceedings of ICCBR, pp. 61–76. Springer (2007)
6. Church, K.W., Hanks, P.: Word association norms, mutual information, and lexicography. Computational Linguistics **16**(1), 22–29 (1990)
7. Colas, F., Brazdil, P.: Comparison of SVM and Some Older Classification Algorithms in Text Classification Tasks. pp. 169–178 (2006)
8. Dang, Y., Zhang, Y., Chen, H.: A lexicon-enhanced method for sentiment classification: An experiment on online product reviews. IEEE Intelligent Systems **25**(4), 46–53 (2010)
9. Deerwester, S.C., Dumais, S.T., Landauer, T.K., Furnas, G.W., Harshman, R.A.: Indexing by latent semantic analysis. Journal of the American Society of Information Science **41**(6), 391–407 (1990)
10. Esuli, A., Sebastiani, F.: Sentiwordnet: A publicly available lexical resource for opinion mining. In. In Proceedings of the 5th Conference on, Language Resources and Evaluation (LREC06, pp. 417–422 (2006)

11. Hofmann, T.: Probabilistic latent semantic indexing. In: Proceedings of the 22nd annual international ACM SIGIR conference on Research and development in information retrieval, SIGIR '99, pp. 50–57. ACM, New York, NY, USA (1999)
12. Joachims, T.: Text categorization with support vector machines learning with many relevant features. In: European Conf. Mach. Learning, ECML98, pp. 137–142 (1998)
13. Melville, P., Gryc, W., Lawrence, R.D.: Sentiment analysis of blogs by combining lexical knowledge with text classification. In: Proceedings of the 15th ACM SIGKDD international conference on Knowledge discovery and data mining, KDD'09, pp. 1275–1284. ACM, New York, NY, USA (2009)
14. Mudinas, A., Zhang, D., Levene, M.: Combining lexicon and learning based approaches for concept-level sentiment analysis. In: Proceedings of the First International Workshop on Issues of Sentiment Discovery and Opinion Mining, WISDOM'12, pp. 5:1–5:8. ACM, New York, NY, USA (2012)
15. Pang, B., Lee, L., Vaithyanathan, S.: Thumbs up?: sentiment classification using machine learning techniques. In: Proceedings of the ACL-02 conference on Empirical methods in natural language processing - Volume 10, EMNLP'02, pp. 79–86. Association for Computational Linguistics, Stroudsburg, PA, USA (2002)
16. Rohde, D.L.T., Gonnerman, L.M., Plaut, D.C.: An improved model of semantic similarity based on lexical co-occurence. COMMUNICATIONS OF THE ACM **8**, 627–633 (2006)
17. Sun, J.T., Chen, Z., Zeng, H.J., Lu, Y.C., Shi, C.Y., Ma, W.Y.: Supervised latent semantic indexing for document categorization. Data Mining, IEEE International Conference on 0, 535–538 (2004)
18. Tsatsaronis, G., Panagiotopoulou, V.: A generalized vector space model for text retrieval based on semantic relatedness. In: Proceedings of the Student Research Workshop at EACL 2009, pp. 70–78 (2009)
19. Turney, P.D., Pantel, P.: From frequency to meaning: vector space models of semantics. J. Artif. Int. Res. **37**, 141–188 (2010)
20. Wang, H., Lu, Y., Zhai, C.: Latent aspect rating analysis on review text data: a rating regression approach. In: Proceedings of the 16th ACM SIGKDD international conference on Knowledge discovery and data mining, KDD'10, pp. 783–792. ACM, New York, NY, USA (2010)

Intelligent Agents

On Applying Adaptive Data Structures to Multi-Player Game Playing

Spencer Polk and B. John Oommen

Abstract In the field of game playing, the focus has been on two-player games, such as Chess and Go, rather than on *multi-player* games, with dominant multi-player techniques largely being an extension of two-player techniques to an N-player environment. To address the problem of multiple opponents, we propose the merging of two previously unrelated fields, namely those of multi-player game playing and Adaptive Data Structures (ADS). We present here a novel move-ordering heuristic for a dominant multi-player game playing algorithm, namely the Best-Reply Search (BRS). Our enhancement uses an ADS to *rank* the opponents in terms of their respective threat levels to the player modeled by the AI algorithm. This heuristic, referred to as Threat-ADS, has been rigorously tested, and the results conclusively demonstrate that, while it cannot damage the performance of BRS, it performs better in all cases examined.

1 Introduction

The majority of research in the field of game playing has been focused on two-player games, particularly addressing long-standing traditional games such as Chess [10, 16]. However, multi-player games, such as Chinese Checkers and multi-player forms of Go, have seen comparatively little research. Techniques for multi-player games have been largely limited to extensions of two-player game playing algorithms [7, 11, 15, 16]. Furthermore, the majority of multi-player game playing algorithms

---

S. Polk (✉) · B. J. Oommen
Carleton University, 1125 Colonel By Dr, Ottawa, ON K1S 5B6, Canada
e-mail: andrewpolk@cmail.carleton.ca

B. J. Oommen
University of Agder, Grimstad, Norway
e-mail: oommen@scs.carleton.ca

M. Bramer and M. Petridis (eds.), *Research and Development in Intelligent Systems XXX*, 125
DOI: 10.1007/978-3-319-02621-3_9,

have issues performing on a level comparable with their two-player counterparts, for a variety of reasons [15, 17, 18].

Unrelated to game playing, Adaptive Data Structures (ADS) is a field that deals with reordering data structures dynamically to improve the efficiency of queries [2, 4, 5]. This field has developed a number of techniques to accomplish this task with both a high efficiency and expediency, for a variety of data structures. When we observe that in a multi-player game, we, by definition, encounter multiple opponents, we can deduce that valuable knowledge can *possibly* be gleaned by ranking them according to any number of parameters. Thus, we propose merging the fields of multi-player game playing and ADS to improve existing multi-player game playing strategies, and to possibly develop *entirely* new approaches which are not possible by merely using the art and science of the original domains themselves. In this work, we present the first result of our endeavors to accomplish this: We propose a low-overhead, easy to implement move ordering heuristic for the state-of-the-art multi-player game playing technique called the Best-Reply Search, using adaptive list techniques to rank opponents based on their *relative threat levels*, to achieve improved move ordering. We refer to this new heuristic as "Threat-ADS".

Section 2 *briefly* surveys the background information on two-player and multi-player game playing, as well as ADS, relevant to this work. Section 3 describes the Threat-ADS heuristic and its properties in detail, when applied to the BRS. Section 4 describes the multi-player game models that we have used to test the effectiveness of the Threat-ADS heuristic, as well as the Virus Game, a simple multi-player game we have developed to serve as a highly-configurable testing environment. Sections 5 and 6 describe our experimental design and results, and finally Sects. 7 and 8 contain our analysis and discussion of these results, as well as conclusions and open avenues of further research.

2 Background

The dominant *deterministic* game playing techniques (as opposed to stochastic techniques, such as the UCT algorithm; see [3] for details) for two-player games are founded on the Mini-Max algorithm, based on work by Neumann, and which was first demonstrated for Chess by Shannon [10, 13]. The Mini-Max algorithm provides a natural way to tackle the problem of combinatorial game playing, through the use of game trees. Further, well-known techniques, such as alpha-beta pruning, exist that allows the Mini-Max algorithm to search much farther in the tree than a naive search, without changing its value [6, 16]. The dominant deterministic multi-player game playing algorithms are also extensions of the Mini-Max algorithm to an N-player environment, where $N \geq 2$. Specifically, the two most studied multi-player game playing algorithms are the *Paranoid* and *Max-N* algorithms [11, 15, 16].

In the case of the Paranoid algorithm, all players are considered to be *minimizing* the perspective player, or to have formed a *coalition* against him [15]. Thus, for a three player game, the Paranoid algorithm could be referred to as Max-Min-Min, and for a four player game, a Max-Min-Min-Min scheme. While the Paranoid algorithm is perhaps the most intuitive extension of Mini-Max to multi-player games, and the simplest to implement, the Max-N algorithm is considered the natural extension of Mini-Max to N-person games [7]. The basic philosophy of the Mini-Max algorithm, after all, is not based on minimizing a specific player's score, but instead on maximizing your own score [13]. The Paranoid algorithm's "coalition" approach is therefore unlikely to be an accurate model of reality. In the case of the Max-N algorithm, rather than the heuristic function returning a single value, it instead returns a *tuple* of values of size N, where N is the number of players [7]. At the ith player's turn, he is assumed to choose the move that provides the maximum value in position i in the tuple, and, similar to the Mini-Max or Paranoid algorithms, this value is passed up the tree [7].

The Paranoid and Max-N algorithms, and their variants, have remained the standard for deterministic multi-player game playing for many years. However, very recently, a new Mini-Max style multi-player game playing algorithm has been introduced, which can, in some cases, significantly outperform both the Paranoid and Max-N algorithms, named the Best-Reply Search [11]. The Best-Reply Search is similar to the Paranoid algorithm, in that all opponents only seek to minimize the perspective player. However, instead of allowing *each* opponent to take the most minimizing move, the Best-Reply Search considers only the most damaging move available to *all* opponents, viewed as a group, at each "Min" level of the tree [11]. Thus, the Best-Reply Search functions *identically* to Mini-Max after this grouping is complete. By necessity, the Best-Reply Search has to also consider illegal move states [11]. This is intuitive, as according to our planning, every second action is the perspective player's—something that will clearly not be the case in an actual multi-player game. However, despite this drawback there are significant benefits to using this method, as it can make use of full alpha-beta pruning, unlike Max-N or the Paranoid algorithm, and can achieve farther look-ahead by grouping opponents together into a single level of the game tree. For maximum clarity, Algorithm 1 shows the formal execution of the Best-Reply Search.

Unrelated to game playing, the field of ADS evolved to deal with the well-known problem in data structures that the access frequencies of the elements in the structure are not uniform [2, 4, 5]. As the access probabilities are not known, the data structure must learn them as queries proceed, and *adapt* to this changing information by altering its internal structure to better serve future queries [4]. Individual types of ADS provide *adaptive mechanisms* that lead to this sort of behaviour for the specific data structure. The field of ADS is extensive and covers many different styles of update mechanisms. In this work, however, for a *prima facie* case, we are interested only in one of the oldest and most studied adaptive list mechanisms in the field of ADS, the Move-to-Front rule [1, 2, 4, 9, 14]. As the name suggests, this rule states that when an object is accessed, it is moved to the front of the list. Other traditional adaptive list rules, such as the Transposition rule, Move-Ahead-k, and POS(k) have also been reported in the literature.[1] Some preliminary results involving the latter schemes are

[1] For a complete overview of adaptive list mechanisms, the reader is referred to [1].

Algorithm 1 Best-Reply Search

```
Function Best-Reply Search
Input: Root node n, depth d
Output: Value of n
1: if n is terminal or d ≤ 0 then
2:    return heuristic value of n
3: else
4:    if node is max then
5:       α = −∞
6:       for all child of n do
7:          α = max(α, BRS(child, d − 1))
8:       end for
9:    else
10:      α = ∞
11:      for all opponents in game do
12:         for all child nodes for opponent do
13:            α = min(α, BRS(child, d − 1))
14:         end for
15:      end for
16:   end if
17:   return α
18: end if
End Function Best-Reply Search
```

available, but as we have only studied, in detail, the Move-to-Front rule when applied to Threat-ADS, the results concerning the other ADSs are therefore omitted from this work. More detailed analyses of these are currently being undertaken.

3 Threat-ADS

In this section, we discuss the Threat-ADS heuristic in detail. We begin with a detail of its development and functionality, and then note salient features of the heuristic that assist in its theoretical performance.

3.1 Developing the Threat-ADS Heuristic

It is well known that alpha-beta pruning benefits substantially from efficient *move ordering*, that is, mechanisms that increase the probability that the strongest moves will be investigated first [6]. As the Best-Reply Search operates by grouping all opponents together into a "super-opponent", it is worthwhile to investigate if effective move ordering can be achieved within that specific context. Instead of naively grouping the opponents' moves, we, instead, propose to generate them in order based on the relative threat level each opponent has in relation to the perspective player.

We achieve this through an order derived from an ADS. As we will demonstrate, this heuristic is inexpensive, both in terms of memory and additional processing time. We will refer to this heuristic as "Threat-ADS".

Threat-ADS adds an adaptive list with a size equal to the number of opponents in the game to the memory required for the BRS. The ADS is initialized at startup to contain elements representing each opponent in a random order (we assume that the number of opponents is constant, although this is not a requirement). Thereafter, we invoke this ADS to organize opponent moves when combining them, as in the Best-Reply Search. For example, if in a game, the perspective player, say P1, has three opponents, and our ADS has the order 2, 4, 3, we would arrange the opponents' turns so that all of opponent 2's moves are placed first, then all of opponent 4's moves, and finally all of opponents 3's moves. A single level of a BRS tree using the Threat-ADS heuristic is presented in Fig. 1 for clarity.

The Threat-ADS heuristic must also contain a mechanism by which the ADS can "learn" the relative opponent threat levels as the play proceeds. We model this mechanism as being analogous to the access of an object in a normal ADS. That is, we must "query" a specific opponent from the ADS when we believe that he is in an advantageous position of some sort. Luckily, the Best-Reply Search contains a *natural* mechanism for finding this advantageous position. As a natural part of its execution, the Best-Reply Search finds which opponent has the most threatening move at any given "Min" phase! Thus, we can use this information to query the ADS, and thus update the relative threat levels of the opponents.

3.2 Salient Features of the Threat-ADS Heuristic

As is usual when considering move ordering heuristics [6, 12], Threat-ADS in no way alters the value of the BRS tree, and therefore its application has no impact on the actual decisions made by the BRS algorithm. Furthermore, while traditional move ordering heuristics are applicable to the Best-Reply Search, there is no specified

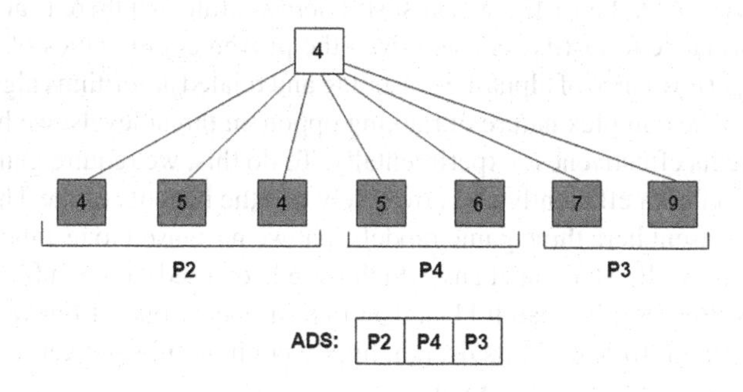

Fig. 1 The operation of a single level of a BRS search tree using the Threat-ADS heuristic, where Player P1 is the perspective player

order in which moves should be collected from each opponent [11]. Therefore, no information is being lost in the application of the Threat-ADS heuristic. There are, of course, more established move ordering heuristics, such as the domain-based heuristics or killer moves – which can also be applied to the Best-Reply Search. While these could have their own specific strengths in their own right, we can demonstrate that the "lightweight" Threat-ADS heuristic is competitive even though its memory and computational footprint is much less.

The Threat-ADS heuristic adds its adaptive list to the memory footprint of the Best-Reply Search, as well as the update and read operations to its execution time. However, the number of players in a game is a constant, or can even be diminuting as in the case of eliminations, and almost always a small number (very rarely above eight). If this number is assumed to be a constant value, then all of Threat-ADS' operations run in constant time, and the list is of constant size. Adaptive list update mechanisms are very efficient, in general, and certainly minimal compared to the vast number of node expansions in a Mini-Max style tree search. We can therefore conclude that the Threat-ADS heuristic only increases the Best-Reply Search's runtime marginally, by a constant factor, and does not influence its asymptotic runtime.

Finally, the Threat-ADS heuristic provides a very useful benefit in that it is able to order moves *without requiring sorting*, which must be contrasted with other move ordering heuristics, such as the History Heuristic [12]. This can be accomplished by generating each opponent's moves during a Min phase of the game tree, in the order of the ADS, and investigating them before continuing. This mechanism is akin to the way the killer moves heuristic operates, i.e., by investigating the killer moves, if they are valid, before even generating the remainder of the moves [12]. This observation, combined with those in the previous paragraph, suggests that one loses essentially nothing by enhancing the BRS as in the Threat-ADS heuristic.

4 Game Models

While we have established that the BRS algorithm loses almost nothing from application of the Threat-ADS heuristic, we must still demonstrate that there is an improvement in performance to justify its use. Given the proven complexities of analyzing average-case performance of alpha-beta pruning and related algorithms algebraically [16], as well as the complex nature of ranking opponent threat levels, we have opted to demonstrate its effectiveness experimentally. To do this, we require game models on which we can both efficiently and effectively test the benefits of the Threat-ADS heuristic. We present here three game models that we have used to test the heuristic, two of which are well-established and which have been used in the publication that introduced the Best-Reply Search [11]. The final model is one of our own design, made to be a straightforward in its design, highly configurable and yet to guarantee an easy-to-implement test environment.

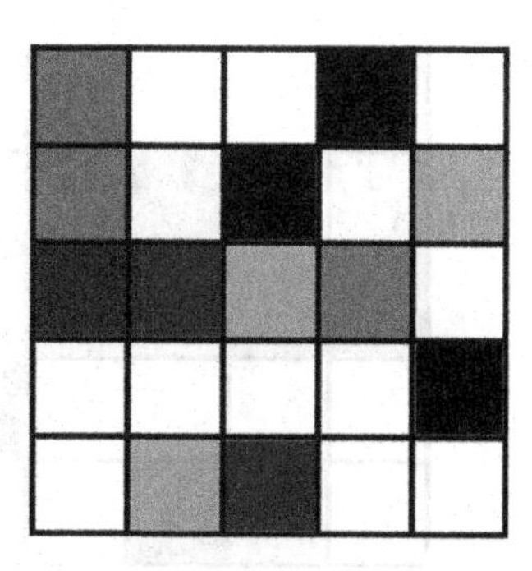

Fig. 2 A possible starting position for the Virus Game

4.1 The Virus Game

To create a simple testbed for the Threat-ADS heuristic, we have, first of all, opted to design our own experimental game. We have named this game the "Virus Game", carrying on the biological analogy found in similar experimental games used in the past [8]. The Virus Game takes place on an N-by-N two-dimensional board, divided into discrete squares, similar to popular combinatorial games like Chess, Checkers, and Go. It can be played by multiple opponents from two to k players, where k can be any whole number. Each of the k players is assigned a colour, and then randomly has a configurable number of pieces placed on the board. To clarify this explanation, a sample starting position for the Virus Game, on a five-by-five board, is shown in Fig. 2. In the context of this work, we used a version on a 5×5 board, with four players, and three starting pieces each, as in Fig. 2.

The play proceeds in a turn-based order around the "table", similar to most multi-player combinatorial games. During a player's turn, he may "infect" any square that he controls, or is *adjacent to* a square he controls, either horizontally or vertically. Only one square may be infected like this, after which the play proceeds to the next player. The player claims the square that he has infected, if he did not already control it. Furthermore, he also claims the squares adjacent to, again horizontally or vertically, *the square that has been infected*. For clarity on how the Virus Game works, a sample first turn of the game shown in Fig. 2 is shown in Fig. 3.

As befits such a simple game, the evaluation function used simply counts the number of squares a player currently controls as the value of that specific game state. As a game, the Virus Game may not appear to be particularly exciting or challenging. Players can easily cancel each others' progress out by simply reclaiming the squares taken by their opponents. In fact, should play come down to two players, optimal play will result in the two of them canceling each others moves perpetually, and the game will never end. However, as in the analyses of such strategies, we are interested in measuring the efficacy of the heuristic, i.e., the *move ordering*, and are not concerned with proceeding till the game terminates. Thus, an easy-to-implement environment like the Virus Game, where players will clearly pose more of a threat to specific opponents than others, is a useful tool in this work.

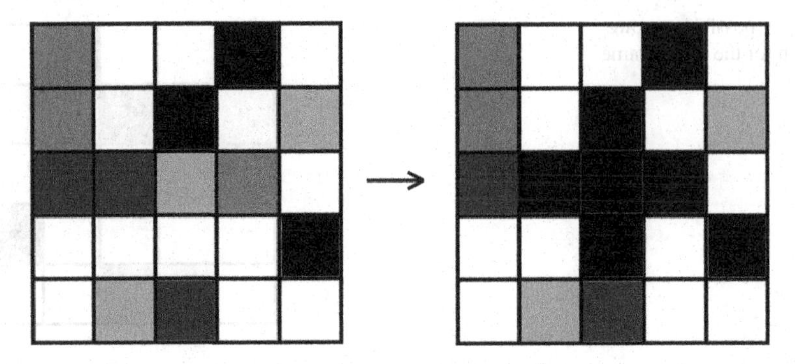

Fig. 3 The figure depicts a possible turn of the Virus Game for the *Black* player. The central square was infected

4.2 Focus

Focus is a commercially-released multi-player abstract strategy game released originally in 1963. It was initially used to test the Best-Reply Search in its original publication [11]. The game takes place on an 8-by-8 board, with the 3 squares in each corner removed. In the four player version of the game, used in this work, each player's pieces are initially arranged according to Fig. 4. Unlike in most games of this type, pieces in Focus may be stacked on top of one another. During a turn, a

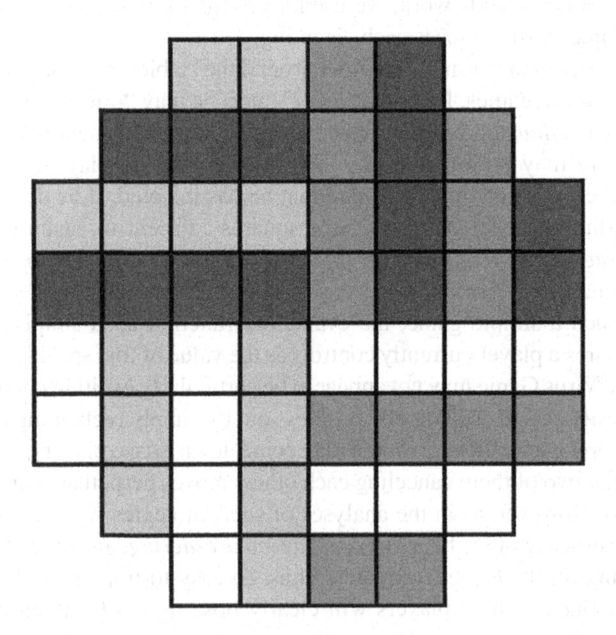

Fig. 4 This figure depicts a starting position of the game Focus

player may move any stack he controls, indicated by the top piece which bears his colour, a number of squares vertically or horizontally equal to the size of the stack. Thus, a stack that is four pieces high may be moved one to four squares left, right, up, or down. Furthermore, players may choose to split a stack they control, moving only the top number of pieces of their choosing. However, they can never move a stack vertically or horizontally more squares than the size of the stack.

When a stack is moved onto another square which already has a stack on it, the stacks are merged, with the moved pieces being placed on top. Stacks have a maximum size of five, and when stacks are merged, the bottom n—5 pieces, where n is the size of the stack, are captured by the player who made the move. A player may, instead of moving a stack, choose to place one of their own pieces that they have captured back onto the board, on any square of their choosing. In the standard version of the game, play ends when all players but one cannot make a legal move, at which point the remaining player is considered to be the winner. However, we make use of the shortened rules introduced in [11], which specifies that in the four-player version of the game, a player has won when he has captured ten opponent pieces, or two pieces from each opponent.

The evaluation function we have invoked is the same as the one in [11], divided into two parts. The first part is the minimum number of pieces the player needs to capture to win the game for either win condition, subtracted from a large number, and multiplied by a thousand. The second is based on the position of the player's pieces in each stack, observing that pieces higher up in stacks are harder to capture. Therefore, for every piece the player has on the board, its height is squared and added to the first part. A small random factor of five points is included to prevent repetition, as in [11].

4.3 Chinese Checkers

Chinese Checkers is a very well-known multi-player abstract strategy game for two, three, four, or six players. In this work, we consider the six player game. Under normal circumstances, each player has ten pieces arranged on a star-shaped board, in each of the "corners". However, we use the smaller version of the board described in [11], where each player has six pieces. This keeps the game's branching factor and processing time to a minimum. The starting position for the six-player version of Chinese Checkers is shown in Fig. 5.

The goal of each player is to move all of his pieces to the opposite "corner" of the board, at which point he is declared the winner. This is accomplished by, on his turn, moving one of his pieces either one space to an adjacent, unoccupied cell, or by "jumping" one of his pieces or an opponents' piece, to an unoccupied square beyond it. Furthermore, jumps may be chained together, one following the other, allowing a single piece to make considerable progress on the board in a single turn. In [11], a table allowing perfect play, in the single player case, was used as the evaluation function for the game. However, as our work is concerned with move ordering, we

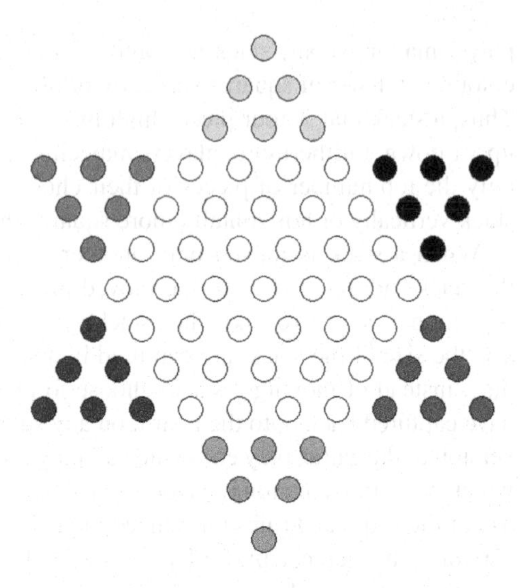

Fig. 5 This figure depicts a starting position of the six-player version of Chinese Checkers

instead opted to use an alternative evaluation function, which takes the total distance the players' pieces are away from the opposite corner of the board, subtracted from a large number, as the value for that player.

5 Experimental Setup

What we are interested in determining is the improvement in tree pruning gained from the use of the Threat-ADS heuristic, which translates to a faster search for a given *ply* depth, potentially allowing the Best-Reply Search to search even deeper in an allotted time frame. Therefore, a logical measure would be the runtime to search to a certain depth, because if it is improved through the use of the Threat-ADS heuristic, there is a net gain. However, runtime is a problematic measure for move ordering heuristics, as it is sensitive to platform and implementation influences [12]. Therefore, we will use the Node Count measure, which counts the total number of nodes within the tree in which the processing occurs, where, for the sake of completeness, we record both the internal and leaf nodes. It has been shown in the past that Node Count is highly correlated to runtime, and that it therefore serves as an effective tool for measuring improvements to alpha-beta search [12]. Given that the Threat-ADS heuristic is able to retain its learning after a turn completes, it is advantageous for us to measure the Node Count over multiple turns, which will enable us to obtain the best possible assessment of its effectiveness.

Thus, as our performance metric, we have measured the total Node Count over the first few turns of each of the games. The number of turns that the games were run for varies between each of the three game models, depending on the execution time and the length of the game. The Virus Game, being relatively simple with a

Table 1 Node count

Game	Threat-ADS	Avg. node count
Virus Game	No	267,256.19
Virus Game	Yes	242,737.96
Focus	No	6,921,645.48
Focus	Yes	6,404,891.12
Chinese Checkers	No	3,486,251.60
Chinese Checkers	Yes	3,284,024.50

low branching factor, was allowed to run for the first ten turns. Chinese Checkers, having a larger branching factor and more complex moves, was run for the first five turns. Finally Focus, with a very large branching factor and relatively short duration using the win conditions from [11] (often under a total of ten turns), was run for three turns. Other than the perspective player, who used the algorithm being tested, all other players made random moves, to save time, as the issue of whether the player won the game or not was rather irrelevant.

Each of these trials was repeated 200 times, for both the Best-Reply Search and the Best-Reply Search using the Threat-ADS heuristic, at a 4-ply search depth. The results are shown in the next section, along with further information showing their statistical properties and the significance of any improvement.

6 Results

Consider Table 1 in which we illustrate the performance of the Threat-ADS heuristic within the context of our three game models. As the reader observes, there is an improvement in Node Count in all cases, when the Threat-ADS heuristic is used alongside the Best-Reply Search. The degree of improvement varies between game models, being around a 6 % reduction in the tree size for Chinese Checkers, and then 8 and 10 % for Focus and the Virus Game, respectively.

To further accentuate the power of the schemes, we have chosen to subject the results to a more formal statistical analysis. To that end, consider Table 2, in which we present the standard deviation of the values from Table 1. We have also included the corresponding certainty values to ensure that the difference before and after the use of the Threat-ADS heuristic did not occur by chance. As is apparent from the table, all cases reject the null hypothesis with at least 95 % certainty.

7 Discussion

As hypothesized, these results demonstrate that a clear improvement in move ordering, within the Best-Reply Search, can be obtained through the use of the Threat-ADS heuristic. This showcases that the ADS is, indeed, capable of identifying the most powerful player, and improving move ordering by placing its moves towards the

Table 2 Statistical details

Game	Threat-ADS	Standard deviation	Certainty (P-value)
Virus Game	No	42, 540.49	$4.61 * 10^{-8}$
Virus Game	Yes	44, 228.66	$4.61 * 10^{-8}$
Focus	No	754, 872.85	$2.53 * 10^{-12}$
Focus	Yes	638, 580.33	$2.53 * 10^{-12}$
Chinese Checkers	No	966, 513.02	0.045
Chinese Checkers	Yes	839, 093.81	0.045

head of the list. All the experiments conducted for all the reported games displayed that an improvement in move ordering was achieved by the Threat-ADS, resulting in a 5–10 % reduction in tree size. One must remember, though, that this advantage is obtained without invoking complex heuristics, such as domain-specific move ordering, and the History Heuristic [12]. Further, it must not be forgotten that this is a reduction of potentially hundreds of thousands of nodes, at only a 4-ply depth, achievable while losing *essentially nothing at all.*

The best results were obtained from the Virus Game, and then from Focus and finally Chinese Checkers, as is apparent from Table 1. It is not surprising that the best results were obtained within the scope of the Virus Game, as it is a territory control game, where players can only target squares adjacent to those they control. Thus, it is natural that players would pose a different level of threat to each other at different points in the game. Though Focus achieved slightly less of an improvement, one must observe that the improvement has a particularly high level of statistical certainty. One further observes that Chinese Checkers did not do as well. Indeed, as we expected, those opponents who start in adjacent "corners" would be most capable of minimizing the strength of the perspective player. Consequently, placing their moves at the head of the list would improve their performance. The reduction of approximately only 6 % is certainly caused by the difficulty of the task of blocking an opponent in Chinese Checkers in *just a few moves*.

We respectfully submit that the results presented here are very promising. They demonstrate the strength of applying an ADS to multi-player game playing, and that this strategy has statistically noticeable improvements. On the other hand, it is very easy to implement and costs the algorithm *essentially nothing*. It can furthermore be easily combined with other move ordering heuristics, as it does not necessitate the use of sorting and may be used to break ties within the framework of the Best-Reply Search. This work, therefore, provides a strong basis for further combinations of list and tree-based ADS and game playing, which we believe to be a novel and pioneering concept.

8 Conclusions and Future Work

In this paper we have submitted a pioneering concept of enhancing multi-player game playing strategies by incorporating into them the science and art of ADS. Indeed, the results presented can be seen to be particularly promising. The fact that the first proposed scheme, the Threat-ADS heuristic, produces statistically significant improvements, even when all the opponents use the same strategy for play, provides strong evidence that it can improve the performance of the Best-Reply Search in the vast majority of cases. Furthermore, as it does not rely on any game-specific knowledge, and it bounds the number of players by a constant, it does not increase the asymptotic execution time of the algorithm. Indeed, it costs essentially nothing for the Best-Reply Search to utilize an ADS-based enhancement.

Perhaps, more importantly than improving on a single Multi-player playing algorithm, the fact that it produces such a consistent improvement over the Best-Reply Search by invoking a relatively simple change, provides a strong justification for further merging the fields of multi-player game playing and ADS.

Currently, we are working on investigating alternate update mechanisms for the ADS. As it stands now, the scheme attempts to converge on the assumption that the threat level is the same at all locations in the tree—which is something that can be certainly improved. Furthermore, we are investigating applying the Threat-ADS heuristic to deeper search trees, as preliminary results suggest that it may provide an even greater reduction in Node Count as the tree grows. We are also considering how we can combine the Threat-ADS heuristic with established move ordering heuristics, such as killer moves and the History Heuristic.

Rather than simply use the ADS paradigm for ordering moves based on opponent threat levels, we suggest that there is also the potential for applying the theory of ADS to many more areas in both multi-player and two-player games. It is our hope that this work will provide a basis for future endeavours to examine other areas in game playing to which ADS may be applied.

References

1. Albers, S., Westbrook, J.: Self-organizing data structures. In: Online Algorithms, pp. 13–51 (1998)
2. Corman, T.H., Leiserson, C.E., Rivest, R.L., Stein, C.: Introduction to Algorithms, 3rd edn., pp. 302–320. MIT Press, Upper Saddle River, NJ, USA (2009)
3. Gelly, S., Wang, Y.: Exploration Exploitation in Go: UCT for Monte-Carlo Go. In: Proceedings of NIPS'06, the 2006 Annual Conference on Neural Information Processing Systems (2006)
4. Gonnet, G.H., Munro, J.I., Suwanda, H.: Towards self-organizing linear search. In: Proceedings of FOCS '79, the 1979 Annual Symposium on Foundations of Computer Science, pp. 169–171 (1979)
5. Hester, J.H., Hirschberg, D.S.: Self-organizing linear search. ACM Computing Surveys **17**, 285–311 (1985)
6. Knuth, D.E., Moore, R.W.: An analysis of alpha-beta pruning. Artificial Intelligence **6**, 293–326 (1975)

7. Luckhardt, C., Irani, K.: An algorithmic solution of n-person games. In: Proceedings of the AAAI'86, pp. 158–162 (1986)
8. Rendell, P.: A universal Turing machine in Conway's Game of Life. In: Proceedings of HPCS'11, the 2011 International Conference on High Performance Computing and Simulation, pp. 764–772 (2011)
9. Rivest, R.L.: On self-organizing sequential search heuristics. In: Proceedings of the 1974 IEEE Symposium on Switching and Automata Theory, pp. 63–67 (1974)
10. Russell, S.J., Norvig, P.: Artificial Intelligence: A Modern Approach, 3rd edn., pp. 161–201. Prentice-Hall, Inc., Upper Saddle River, NJ, USA (2009)
11. Schadd, M.P.D., Winands, M.H.M.: Best Reply Search for multiplayer games. IEEE Transactions on Computational Intelligence and AI in Games **3**, 57–66 (2011)
12. Schaeffer, J.: The history heuristic and alpha-beta search enhancements in practice. IEEE Transactions on Pattern Analysis and Machine Intelligence **11**, 1203–1212 (1989)
13. Shannon, C.E.: Programming a computer for playing Chess. Philosophical Magazine **41**, 256–275 (1950)
14. Sleator, D.D., Tarjan, R.E.: Amortized efficiency of list update and paging rules. Communications of the ACM **28**, 202–208 (1985)
15. Sturtevant, N.: A comparison of algorithms for multi-player games. In: Proceedings of the Third International Conference on Computers and Games, pp. 108–122 (2002)
16. Sturtevant, N.: Multi-player games: Algorithms and approaches. Ph.D. thesis, University of California (2003)
17. Sturtevant, N., Bowling, M.: Robust game play against unknown opponents. In: Proceedings of AAMAS'06, the 2006 International Joint Conference on Autonomous Agents and Multiagent Systems, pp. 713–719 (2006)
18. Sturtevant, N., Zinkevich, M., Bowling, M.: Prob-Maxn: Playing n-player games with opponent models. In: Proceedings of AAAI'06, the 2006 National Conference on Artificial Intelligence, pp. 1057–1063 (2006)

Anytime Contract Search

Sunandita Patra, Satya Gautam Vadlamudi and Partha Pratim Chakrabarti

Abstract Heuristic search is a fundamental problem solving paradigm in artificial intelligence. We address the problem of developing heuristic search algorithms where intermediate results are sought at intervals of time which may or may not be known apriori. In this paper, we propose an efficient anytime algorithm called Anytime Contract Search (based on the contract search framework) which incrementally explores the state-space with the given contracts (intervals of reporting). The algorithm works without restarting and dynamically adapts for the next iteration based on the current contract and the currently explored state-space. The proposed method is complete on bounded graphs. Experimental results with different contract sequences on the Sliding-tile Puzzle Problem and the Travelling Salesperson Problem (TSP) show that Anytime Contract Search outperforms some of the state-of-the art anytime search algorithms that are oblivious to the given contracts. Also, the non-parametric version of the proposed algorithm which is oblivious of the reporting intervals (making it an anytime algorithm) performs well compared to many available schemes.

1 Introduction

Heuristic search has been widely applied over the years in diverse domains involving planning and combinatorial optimization [15]. It is one of the fundamental problem solving techniques of artificial intelligence [14]. A* [6] is the central algorithm around which most other state-of-the-art methods are developed. Owing to the large

S. Patra (✉) · S. G. Vadlamudi · P. P. Chakrabarti
Indian Institute of Technology Kharagpur, Kharagpur, West Bengal 721302, India
e-mail: sunandita.patra@cse.iitkgp.ernet.in

S. G. Vadlamudi
e-mail: satya@cse.iitkgp.ernet.in

P. P. Chakrabarti
e-mail: ppchak@cse.iitkgp.ernet.in

M. Bramer and M. Petridis (eds.), *Research and Development in Intelligent Systems XXX*, 139
DOI: 10.1007/978-3-319-02621-3_10, © Springer International Publishing Switzerland 2013

amount of time required by A* algorithm to produce a solution (which is guaranteed to be optimal) in case of complex problems, several themes are pursued that can report solutions (possibly sub-optimal) quickly. Most prominent amongst them is the class of anytime search algorithms [4]. The objective of anytime search algorithms is to produce a solution quickly and improve upon it as time passes. Several methods are proposed in the literature to address this problem.

In this paper, we explore a new dimension to this problem, where a user applying the anytime algorithms typically checks the progress of the algorithm (for any improved solutions) at certain periods of time rather than continuously monitoring it. Or equivalently, it can be said that he/she expects the anytime algorithm to improve upon its current solution after a certain amount of time. We propose to formulate this behavior formally by taking as input a series of timepoints from the user at which he/she would like to get an update from the anytime algorithm.We call this problem, the *anytime contract search problem*, which takes in a Contract series (a series of timepoints) as input at which improved solutions are sought.Equivalently, one may feed the algorithm during its run by the next contract (at which a better solution is sought), instead of giving the whole series apriori. We solve this problem by intelligently combining the strategies of the anytime algorithms and the contract search algorithms.

The basic contract search problem aims at finding the best possible solution in the given time. In [7], a time constrained search algorithm is proposed for solving this problem based on Weighted A* [10]. Contract search algorithm [2, 3] uses probabilistic models to distribute the contract as node expansions across different levels. Deadline aware search algorithm [5] proceeds in a best-first manner, but ensuring that those nodes are chosen for expansion which can lead to a goal state within the given time.

In this work, we follow the approach of distributing the contract as node expansions at different levels to reach a solution within the given time. We learn dynamically to come up with a chosen distribution of contracts and the algorithm progresses without any restarts ensuring no unnecessary re-expansion of nodes. The key contributions of this work are as follows:

1. The introduction of the anytime contract search problem,
2. An efficient algorithm called Anytime Contract Search (ACTR) to solve this problem, and a modification of the algorithm called Oblivious Anytime Contract Search (OACTR) that makes it usable as a traditional anytime algorithm, and
3. Experimental results comparing the proposed methods with some of the state-of-the-art algorithms such as AWA*, ANA*, etc. on the Sliding-tile Puzzle Problem and the Traveling Salesperson Problem where the proposed algorithms are observed to be outperforming the other methods significantly. Further, the performance of the proposed algorithm is analyzed with respect to different input Contract series distributions.

The rest of the paper is organized as follows: In Sect. 2, we present the proposed methods, namely, anytime contract search algorithm and the oblivious version of it along with some of the important properties satisfied by the algorithms. In Sect. 3, the implementation details and comparison of the proposed methods with some of the state-of-the-art anytime algorithms are presented on the Sliding-tile Puzzle problem and the Traveling Salesperson Problem. Finally, we present a brief discussion pointing to the scope for improvements and future research directions, and conclude, in Sect. 4.

2 Proposed Methods

In this section, we present the Anytime Contract Search algorithm (ACTR) that takes in a series of timepoints at which the user would like to note the progress, and tries to report the best possible solutions at those timepoints.Note that, giving the entire series of timepoints apriori is not mandatory and one may instead dynamically give the contract inputs in a step by step manner during the run of the algorithm after observing the result of the algorithm in the current iteration, as discussed in the Introduction.

The method works by distributing the contract available (till the next report time) into number of nodes to be expanded at each level of the search graph, similar to the idea of Contract Search [2, 3]. However, instead of using probabilistic rank profiles for distributing the contract, our method attempts to learn and adapt the contract distribution scheme based on the feedback from its own previous iterations.Note that, the basic contract search algorithms [2, 3] are not anytime in nature and hence can not be compared with the methods in this work. The previous methods do not have concept of producing multiple solutions but aim for only one best possible solution within the given time.While they use probabilistic rank profiles for their operations, one may either use such approach during the first iteration of the methods proposed in this work, or alternatively, use a simpler beam-search like equal distribution of contracts for different levels.Here, our main focus is not on the initial distribution, rather the later dynamic learning and solving the anytime contract search problem. However, all aspects are open for exploration as discussed in Sect. 4.

Algorithm 1 presents the proposed approach (`ACTR`). It takes as input the search graph, start node, the sequence of timepoints (or just the next timepoint) at which solutions are sought, and also a parameter defining the maximum limit of contract that can be distributed in a given iteration. The last input was induced since we observed that using large contracts directly hampers the learning process of distribution scheme resulting in a poor performance. Separate open lists are maintained for storing the nodes of different levels which are to be expanded, which helps in choosing the most promising node across different levels whose contract limit has not been reached. $ExpCount$ array keeps track of the number of nodes expanded at different levels. $ExpLimit$ array contains the information of maximum number of nodes that can be expanded at different levels as decided by the `DistributeContract` routine. Initially, the $ExpLimit$ value for each level is set to 1 so that the `DistributeContract` routine can allocate equal limits for all levels.

Algorithm 1 Anytime Contract Search (ACTR)

1: **INPUT ::** A search graph, a start node s, maximum contract per iteration, and the Contracts series C of length n.
2: $BestSol \leftarrow infinity$; $g(s) \leftarrow 0$; Calculate $f(s)$; $Level(s) \leftarrow 0$; $OpenList(0) \leftarrow \{s\}$; $ClosedList \leftarrow \phi$; $OpenList(i) \leftarrow \phi$; $\forall i (0 < i < MAX\_DEPTH)$;
3: $ExpCount(i) \leftarrow 0$, $ExpLimit(i) \leftarrow 1$, $\forall i (0 \leq i < MAX\_DEPTH)$;
4: **for** $i \leftarrow 1$ to n and $\exists j\ OpenList(j) \neq \phi$ **do**
5:   **while** $C(i) - time\_elapsed > max\_ctr\_per\_iter$ and $\exists j\ OpenList(j) \neq \phi$ **do**
6:     $ExpLimit \leftarrow$**DistributeContract**$(max\_ctr\_per\_iter, ExpLimit, ExpCount)$;
7:     $BestSol \leftarrow$ **SearchForSolution**$(ExpLimit, BestSol, OpenList, ClosedList, ExpCount)$;
8:   $ExpLimit \leftarrow$**DistributeContract**$(C(i) - time\_elapsed, ExpLimit, ExpCount)$;
9:   $BestSol \leftarrow$ **SearchForSolution**$(ExpLimit, BestSol, OpenList, ClosedList, ExpCount)$;
10: return $BestSol$;

After the initialization (Lines 2 and 3), the method invokes `DistributeContract` routine and `SearchForSolution` routine in tandem according to the given Contracts series and the time-elapsed to come up with the best possible solutions. The *for* loop from Lines 4–9 indicates execution for each of the contracts, as long as there exists at-least one level at which the $OpenList$ is not empty (since all levels being empty would mean that the search is complete and an optimal solution is found, if exists). As mentioned before, for better learning and better performance of the algorithm, when the contract to be distributed is larger than a pre-defined limit per iteration, it is sub-divided into several blocks of the size of the maximum limit (Lines 5–7) followed by the remaining time (Lines 8–9). For example, if the total contract for the current iteration is given to be 10 min and the maximum contract limit per invocation of `SearchForSolution` routine is set to 4 min, then the while loop (Lines 5–7) executes two times with the maximum limit– 4 min, and then Lines 8–9 execute with the remaining time 2 min.

Algorithm 2 DistributeContract

1: **INPUT ::** Contract to be distributed c, $ExpLimit$, $ExpCount$, and a pre-defined tunable parameter $\alpha \in (0, 1)$.
2: **for** $i \leftarrow 0$ to $MAX\_DEPTH - 1$ **do**
3:   $ExpRatio(i) \leftarrow \alpha \times ExpLimit(i) + (1 - \alpha) \times ExpCount(i)$;
4: **for** $i \leftarrow 0$ to $MAX\_DEPTH - 1$ **do**
5:   $ExpRatio(i) \leftarrow ExpRatio(i) / \Sigma_j ExpRatio(j)$;
6:   $ExpLimit(i) \leftarrow ExpLimit(i) + ExpRatio(i) \times c \times node\_expansion\_rate$;
7: return $ExpLimit$;

The `DistributeContract` routine takes as input: the contract to be distributed, the current $ExpLimit$ and $ExpCount$ of various levels, and a pre-defined tunable parameter $\alpha \in (0, 1)$. For deciding the node expansion ratios of different levels, it takes into account the previous node expansion limit assigned and the actual number of node expansions that happened at different levels. While the former represents the promise of nodes at different levels as assessed previously, the latter acts as feedback as to whether the nodes of that level turned out to be globally competitive. Clearly, this is one of the many possible schemes that can be used here, more discussion on which is provided later. Different values of α can be tested with to find the most suited one for the given domain. Lines 2–3 of the `DistributeContract` routine show the decision on the expansion ratios of different levels, and Lines 4–6 show the normalization of the ratios and updating of the node expansion limits based on the given contract and the node expansion rate. Here, note that, $ExpRatio(i)$ is a real number (between 0 and 1), while $ExpLimit(i)$ has to be an integer which is guaranteed by using the *math floor* function while computing the same. Node expansion rate (number of nodes expanded per second (unit of time)) is an user input/parameter which is pre-computed for each domain and a given problem size. One may explore using more complex profiling of the node expansion rates if it varies for nodes belonging to different levels, and adjust the contract distribution accordingly.

Algorithm 3 SearchForSolution

1: **INPUT ::** $ExpLimit$, $BestSol$, $OpenLists$, $ClosedList$, and $ExpCounts$.
2: **while** $\exists i$ such that $OpenList(i) \neq \phi$ and $ExpCount(i) < ExpLimit(i)$ **do**
3: $n \leftarrow$ least f-valued node from all the $OpenLists$ at different levels (i) for which $ExpCount(i) < ExpLimit(i)$;
4: **if** IsGoal(n) **then**
5: **if** $BestSol > f(n)$ **then**
6: $BestSol \leftarrow f(n)$;
7: Move n from its $OpenList$ to $ClosedList$; **continue**;
8: **GenerateChildren**(n); Move n from its $OpenList$ to $ClosedList$;
9: $ExpCount(Level(n)) \leftarrow ExpCount(Level(n)) + 1$;
10: return $BestSol$;

The `SearchForSolution` routine takes the node expansion limits, the current best solution, the lists and the node expansion counts as input and searches for a better solution. It chooses the most promising node from lists of all levels whose node expansion limit has not been reached. The node is checked as to whether it is a goal node, and if it is, the current best solution is updated, otherwise, its children are generated and the corresponding expansion count is updated. The process is continued until either all the $OpenLists$ become empty or the node expansion limits for all levels are reached. Note that when using admissible heuristics, one could terminate this routine when the f-value of the node chosen is greater than or equal to that of the current best solution (for a minimization problem), after pruning such nodes.

Lastly, the `GenerateChildren` routine takes in the node to be expanded and the lists, and generates the children of the node. The children are checked as to whether they are already present in the memory and if so they are updated with the currently known best path from the start node. The lists are updated accordingly as per the new levels of the children.

Algorithm 4 GenerateChildren

1: **INPUT ::** Node n whose children are to be generated, and the lists.
2: **if** $Level(n) = MAX\_DEPTH - 1$ **then**
3: **return**;
4: **for** each successor n' of n **do**
5: **if** n' is not $OpenLists$ and $ClosedLists$ **then**
6: $Level(n') \leftarrow Level(n) + 1$; Insert n' to $OpenList(Level(n'))$;
7: **else if** $g(n') <$ its previous g-value **then**
8: Update $Level(n')$, $g(n')$, $f(n')$; Insert n' to $OpenList(Level(n'))$;

Next, we present a theorem showing the completeness of the proposed method.

Theorem 1. *ACTR is complete and guarantees terminating with an optimal solution, provided MAX_DEPTH is at-least as large as the number of nodes on an optimal solution path and the search is not constrained by the memory or the time available.*

It is easy to observe that the theorem holds true since the algorithm does not discard any promising node within MAX_DEPTH when given enough time.

Note that, here, the algorithm guarantees to find an optimal solution (if exists), even when inadmissible heuristics are used. This is because the algorithm does not prune nodes whose f-values are greater than or equal to that of the best solution, hence, covering the entire search space. However, when using admissible heuristics, the user can prune the nodes with f-values greater than or equal to that of the best solution whenever encountered, and therefore leverage the benefit by reducing the search space to be explored.

Another property satisfied by the algorithm which helps in making it efficient is: ACTR does not re-expand any node unless a better path has been found from the start node to that node.

In the following, we present a version of the proposed algorithm called Oblivious Anytime Contract Search (OACTR) which is oblivious to the input Contract series, and hence acts as a simple anytime algorithm.

Oblivious Anytime Contract Search (OACTR)

Some users may be interested in a simple traditional anytime algorithm which does not ask for Contract series as input, and which is expected to improve upon the solutions as soon as it can.

Algorithm 5 Oblivious Anytime Contract Search (OACTR)

1: **INPUT ::** A search graph, and a start node s.
2: $BestSol \leftarrow infinity$; $g(s) \leftarrow 0$; Calculate $f(s)$; $Level(s) \leftarrow 0$; $OpenList(0) \leftarrow \{s\}$; $ClosedList \leftarrow \phi$; $OpenList(i) \leftarrow \phi$; $\forall i(0 < i < MAX\_DEPTH)$;
3: $ExpCount(i) \leftarrow 0$, $ExpLimit(i) \leftarrow 1$, $\forall i(0 \leq i < MAX\_DEPTH)$;
4: **while** $\exists i$ $OpenList(i) \neq \phi$ **do**
5: $ExpLimit \leftarrow$**DistributeContract**$(MINIMAL\_C, ExpLimit, ExpCount)$;
6: $BestSol \leftarrow$ **SearchForSolution**$(ExpLimit, BestSol, OpenList, ClosedList, ExpCount)$;
7: return $BestSol$;

Algorithm 5 shows a simplified version of Anytime Contract Search which can suit to such requirement. Here, in each iteration, a better solution is sought by using a certain pre-defined minimal contract $MINIMAL\_C$ chosen as per the problem domain. We call this algorithm Oblivious Anytime Contract Search (OACTR) as it just ignores any Contract series inputs. One may also explore other patterns of varying the value of "$MINIMAL\_C$" dynamically for maximizing the performance.

3 Experimental Results

In this section, we present the experimental results comparing the proposed algorithms against several existing anytime algorithms, namely, Beam-Stack search (BS) [17], Anytime Window A* (AWA*) [1], Iterative Widening (IW) [11, 12], Anytime Non-parametric A* (ANA*) [16], and Depth-First Branch and Bound (DFBB) [9]. All the experiments have been performed on a Dell Precision T7500 Tower Workstation with Intel Xeon 5600 Series at 3.47−GHz × 12 and 192-GB RAM.

Anytime performances of different algorithms are usually compared by plotting the average solution costs of all the testcases, obtained at different timepoints. We too use such strategy in our work where we use a metric called $\%Optimal\ Closeness$ which is defined as: $Optimal\ Solution \times 100/Obtained\ Solution$.

It indicates how close the obtained solution is to an optimal solution for a minimization problem. This helps in normalizing the solution costs of different testcases before the average is taken. However, the average value of the quality of the output of an algorithm may become high if it outperforms other algorithms significantly on few corner cases whereas it may be bad in a number of other cases. Such cases can be detected via `Top Count`.

`Top Count` at time t is defined as the number of instances on which a particular algorithm has produced the best solution cost by that time. For example, let two algorithms A_1 and A_2 be compared on 5 testcases at a given time t. Let us assume that A_1 reports better solutions than A_2 in 3 cases, and A_2 finds the better solution in 1 case, and both A_1 and A_2 come up with the same solution on the remaining testcase. Then, the Top Counts for the given algorithms A_1 and A_2 become 4 and 2 respectively. This indicates that a given algorithm is dominating in so many number of cases at that particular time. Note that, on any instance, multiple algorithms may produce the best solution in a given time, and so, the sum of the top counts of various algorithms at a given time can be greater than the total number of testcases, as shown in the above example. This measure gives a complementary picture to the traditional comparison of the average anytime performances.

In each problem domain, we first compare the proposed algorithm with the existing anytime algorithms at different timepoints of the Contract series given by user using Top Count measure. Note that, while the proposed algorithm is run afresh in each case as per the Contract series, other algorithms are run only once as they are oblivious to the given contract series. Next, we show the comparison of Oblivious Anytime Contract search algorithm with the existing ones. Next, results showing the average anytime performances of all the algorithms using the *% Optimal Closeness* measure are presented. Finally, we show a comparative analysis of the proposed algorithms run with different Contract series inputs amongst themselves to test their adaptiveness towards the fed inputs.

We used a constant value of $\alpha = 0.5$ (the learning parameter in contract distributions) in all our experiments since it was observed to be the most suitable with the considered setting. And, the maximum contract per iteration is set to 8 min in all the experiments.

3.1 Sliding-Tile Puzzle Problem

We have chosen the 50 24-puzzle instances from [8, Table II] for our experiments. Manhattan distance heuristic is used as the heuristic estimation function (which underestimates the actual distance to goal). All algorithms explore up-to a maximum depth of 1000 levels.

Firstly, we run ACTR with the Contract series: $\{4, 12, 20, 28, 36, 44, 52, 60\}$ which is an arithmetic progression (AP) having uniformly distributed intervals. We compare the obtained results with that of the existing algorithms such as AWA*, Beam-Stack search (BS), Iterative Beam search (IW), ANA*, and DFBB which are oblivious to the given Contract series. Table 1 shows the results comparing the top counts at the timepoints of the AP Contract series. BS_{500} indicates the results of Beam-stack search algorithm when run with beam-width = 500. The beam-width value is chosen after experimenting with several values such that neither the initial solution of the algorithm is delayed (due to large beam-width) nor the anytime performance is hampered (due to small beam-width). It can be observed that the proposed algorithm dominates the other algorithms at the given timepoints. Note that, here the Top Count value corresponding to ACTR decreases for the last two time contracts. This is because one of the other algorithms must have come up with a better solution than ACTR on one of the testcases by that time. Equivalently, ACTR might have come up with a better solution on one of the testcases and the other algorithm(s) might have bettered it on two other testcases, etc. The point to note is that, Top Count value for an algorithm need not be monotonically increasing, and is dependent on the relative performance of the algorithms at the time point under consideration.

Next, we run ACTR with other Contract series distributions such as a Geometric Progression (GP) and a Randomly generated series (RD). Tables 2 and 3 show the comparison of the obtained results in terms of top count with respect to that of the existing algorithms. Note that, while ACTR is run afresh with the given Contract series, the other algorithms being oblivious to the given Contract series are not effected by it. Only the analysis of Top Counts is carried out with respect to a different set of timepoints in each case. It can be seen that the proposed algorithm outperforms the others in these cases as well. Note that, it is expected that the first columns (corresponding to 4 min.) of Tables 1 and 3 should match, however, the minor discrepancy may be due to ACTR producing a better solution than others

Table 1 Comparison of ACTR with the existing algorithms on the 50 (sliding-tile) 24-puzzle benchmarks when the contract series input is an AP series

	Top count versus **time** (min)							
Algorithm	**4**	**12**	**20**	**28**	**36**	**44**	**52**	**60**
ACTR	**48**	**48**	**49**	**49**	**49**	**49**	**48**	**47**
AWA*	3	4	4	4	4	3	3	4
BS_{500}	1	0	0	0	0	0	0	0
IW	0	0	0	0	0	1	1	1
ANA*	0	0	0	0	0	0	0	0
DFBB	0	0	0	0	0	0	0	0

Table 2 Comparison of ACTR with the existing algorithms on the 50 (sliding-tile) 24-puzzle benchmarks when the contract series input is a GP series

	Top count versus **time** (min)						
Algorithm	**1**	**2**	**4**	**8**	**16**	**32**	**64**
ACTR	**47**	**48**	**47**	**48**	**48**	**46**	**45**
AWA*	6	3	5	3	4	7	7
BS_{500}	2	0	0	1	1	1	1
IW	1	0	0	1	1	1	1
ANA*	0	0	0	0	0	0	0
DFBB	0	0	0	0	0	0	0

Table 3 Comparison of ACTR with the existing algorithms on the 50 (sliding-tile) 24-puzzle benchmarks when the contract series input is generated randomly

	Top count versus **time** (min)							
Algorithm	**4**	**10**	**23**	**26**	**31**	**49**	**54**	**60**
ACTR	**49**	**49**	**48**	**48**	**48**	**48**	**47**	**46**
AWA*	3	3	4	4	3	5	5	6
BS_{500}	1	0	0	0	0	1	0	0
IW	0	0	0	0	0	1	1	1
ANA*	0	0	0	0	0	0	0	0
DFBB	0	0	0	0	0	0	0	0

slightly before the 4 min time limit during the Random-series input execution, and producing a better solution than others slightly after the 4 min time limit during the AP-series input execution. They are expected to perfectly match if the measure is in terms of node expansions, instead of time (which may get effected slightly during different runs of the same algorithm).

Clearly, in the previous cases, the proposed algorithm is expected to gain advantage over the other algorithms as it mends itself as per the given Contract series. Now, we present the comparison of the Oblivious ACTR (OACTR) which is similar to traditional anytime algorithms that do not take any Contract series as input.

Here, OACTR takes as input a minimal contract to be used in each iteration which is set to 1000 node expansions. Table 4 shows the comparative results in terms of top count at uniformly distributed time intervals. The outstanding performance of OACTR suggests that the basic underlying anytime framework proposed is of significance.

Table 5 shows the comparison of average of the anytime performances of the algorithms in terms of $\%\ Optimal\ Closeness$. ACTR_{AP} and ACTR_{GP} denote the runs of ACTR algorithm with the input Contract series being the AP and the GP series shown before. One may note that the ACTR versions outperform the other algorithms and amongst them the performance of the version fed with the AP series input stands out better. Figure 1 shows the pictorial view of the same comparison of anytime performances across all the timepoints in the given time window.

Till now, we have shown the comparison of ACTR with the existing algorithms where it is clearly outperforming the others. Now, we compare the different runs of ACTR algorithm fed with different Contract series inputs amongst themselves to analyze their behavior. Table 6 shows the comparison of different runs of ACTR fed with different inputs, the AP series, the GP series, the Random series, and the default series (OACTR), when sampled at the intervals of the AP series. As one

Table 4 Comparison of OACTR with the existing algorithms on the 50 (sliding-tile) 24-puzzle benchmarks at uniformly distributed time intervals

	Top count versus **time** (min)							
Algorithm	**4**	**12**	**20**	**28**	**36**	**44**	**52**	**60**
OACTR	**46**	**47**	**46**	**44**	**43**	**44**	**43**	**44**
AWA*	6	5	7	9	8	8	7	6
BS_{500}	2	1	1	1	1	1	1	1
IW	1	2	1	1	0	1	1	1
ANA*	0	0	1	1	1	1	1	1
DFBB	0	0	0	0	0	0	0	0

Table 5 Comparison of the average anytime performances on the 50 puzzle benchmarks at uniformly distributed time intervals

	$\%OptimalCloseness$ versus **time** (min)							
Algorithm	**4**	**12**	**20**	**28**	**36**	**44**	**52**	**60**
ACTR_{AP}	87.5	89.7	**90.7**	**91.4**	**91.6**	**91.7**	**91.8**	**91.8**
ACTR_{GP}	**88.0**	**90.1**	**90.7**	91.1	91.2	91.4	91.5	**91.8**
OACTR	85.5	88.3	89.3	90.0	90.3	90.5	90.6	90.8
AWA*	77.7	80.2	81.6	82.4	82.6	82.8	83.0	83.3
BS_{500}	72.6	74.2	75.6	76.1	76.6	76.9	77.2	77.5
IW	73.4	77.0	77.6	78.0	78.2	79.3	79.5	79.9
ANA*	62.4	67.2	68.6	71.0	71.8	72.1	72.4	72.7
DFBB	0.0	0.0	0.0	0.0	0.0	0.0	0.0	0.0

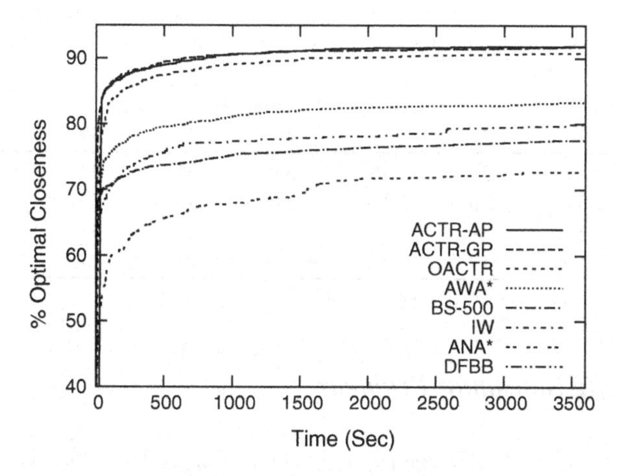

Fig. 1 Comparison of the average anytime performances on the 50 puzzle benchmarks

Table 6 Comparison of different runs of ACTR on the 50 (sliding-tile) 24-puzzle benchmarks at contract series of the AP series

	Top count versus **time** (min)							
Algorithm	**4**	**12**	**20**	**28**	**36**	**44**	**52**	**60**
$ACTR_{AP}$	26	33	**34**	**36**	**37**	**37**	**35**	32
$ACTR_{GP}$	**36**	**34**	32	33	34	34	34	**35**
$ACTR_{RD}$	26	27	25	27	31	32	31	32
OACTR	13	15	19	23	26	26	26	26

Table 7 Comparison of different runs of ACTR on the 50 (sliding-tile) 24-puzzle benchmarks at contract series of the GP series

	Top count versus **time** (min)						
Algorithm	**1**	**2**	**4**	**8**	**16**	**32**	**64**
$ACTR_{AP}$	31	27	26	31	**38**	**37**	33
$ACTR_{GP}$	**34**	**34**	**36**	**35**	35	33	**35**
$ACTR_{RD}$	32	24	26	34	25	28	33
OACTR	12	14	13	16	20	25	26

would expect/want, the ACTR run fed with the AP series input does well in most cases. Similarly, in Table 7, we observe that the ACTR run with the GP series input does well compared to others as the output sampling is measured against the GP series timepoints.

However, in Table 8, we observe that the ACTR runs with the AP series and the GP series inputs continue to dominate while we expect the run with the random series input to do well as the output sampling is measured against the random series timepoints. More viewpoints on such behavior are presented in Sect. 4.

Table 8 Comparison of different runs of ACTR on the 50 (sliding-tile) 24-puzzle benchmarks at contract series of the random series

	Top count versus **time** (min)							
Algorithm	**4**	**10**	**23**	**26**	**31**	**49**	**54**	**60**
$ACTR_{AP}$	26	31	**36**	**35**	**37**	**36**	**34**	32
$ACTR_{GP}$	**36**	**36**	32	33	33	33	**34**	**35**
$ACTR_{RD}$	26	31	26	25	27	32	32	32
OACTR	13	17	21	24	24	27	27	26

3.2 Travelling Salesman Problem

The first 50 symmetric TSPs (when sorted in increasing order of their sizes) from the traveling salesman problem library (TSPLIB) [13] are chosen for our experiments. These range from burma14 to gr202 where the numerical postfixes denote the size of the TSPs. Minimum spanning tree (MST) heuristic is used as the heuristic estimation function (which is an under-estimating heuristic).

In the initial state, some city c is chosen as the starting point and in each successive state the next city n to be visited is chosen (which is not already visited) till all cities are visited. TSP was often looked as a tree search problem, however, careful examination suggests that if two states denote paths from c to n through a same set of cities S, then only the best of the two need to be pursued further and the other one can be admissibly pruned.This modification makes the search space of TSP a graph with the state being represented by $\{c, n, S\}$ rather than the traditional way of using the path as the signature of a state (which can just be maintained as an attribute in the current state space).

Note that, a duplicate node can only exist in the same level as that of the new node under consideration (as the number of cities covered must be equal). Also, all the goal nodes are at a fixed depth, which helps reducing the number of goal node checks. Goal nodes at a known maximum depth m also means that when a expansion limit is reached at a particular level $l(< m)$ as per ACTR, none of the nodes belonging to the levels $< l$ need to be expanded in that iteration.

We repeat the exercise done in the case of Sliding-tile Puzzle Problem experiments in here as well. Firstly, we compare the performance of the proposed algorithm with that of the other algorithms when fed with different distributions of the timepoints across the time-window. Table 9 shows the comparison when ACTR is fed with the AP series input. BS_{100} indicates the results of Beam-stack search algorithm when run with beam-width = 100. The beam-width value is chosen after experimenting with several values such that neither the initial solution of the algorithm is delayed (due to large beam-width) nor the anytime performance is hampered (due to small beam-width). It can be noted that the proposed algorithm once again outperforms the other algorithms.

Table 9 Comparison of ACTR with the existing algorithms on the 50 TSP benchmarks when the contract series input is an AP series

	Top count versus **time** (min)							
Algorithm	**4**	**12**	**20**	**28**	**36**	**44**	**52**	**60**
ACTR	**36**	**40**	**43**	**41**	**42**	**41**	**41**	**41**
AWA*	28	31	27	27	27	28	28	28
BS_{100}	23	24	24	24	25	25	25	25
IW	21	18	19	19	19	20	19	19
ANA*	14	15	15	17	17	17	17	17
DFBB	10	10	11	11	11	11	11	11

Table 10 Comparison of ACTR with the existing algorithms on the 50 TSP benchmarks when the contract series input sequence is a GP series

	Top count versus **time** (min)						
Algorithm	**1**	**2**	**4**	**8**	**16**	**32**	**64**
ACTR	**40**	**40**	**44**	**44**	**43**	**43**	**43**
AWA*	25	27	28	27	25	26	26
BS_{100}	23	24	24	25	24	25	25
IW	19	19	18	17	19	18	18
ANA*	13	13	14	15	17	17	17
DFBB	10	10	10	11	11	11	11

Table 11 Comparison of ACTR with the existing algorithms on the 50 TSP benchmarks when the contract series input is generated randomly

	Top count versus **time** (min)							
Algorithm	**4**	**10**	**23**	**26**	**31**	**49**	**54**	**60**
ACTR	**37**	**41**	**43**	**41**	**42**	**41**	**41**	**41**
AWA*	28	31	27	27	27	27	28	28
BS_{100}	24	24	24	24	25	25	25	25
IW	20	17	19	19	19	19	19	19
ANA*	14	15	15	17	17	17	17	17
DFBB	10	10	11	11	11	11	11	11

Similarly, Tables 10 and 11 also indicate that the proposed algorithm comes on top when fed with the GP series input and the Random series input respectively.

Table 12 shows the comparison of the Oblivious Anytime Contract Search algorithm (OACTR) when sampled at uniformly distributed timepoints. The algorithm is run with a minimal contract of n in each iteration where n is the number of cities of the TSP instance. This strategy also results in best performance amongst the competing algorithms.

Now, we present the comparison of average anytime performances of the algorithms in terms of $\%\ Optimal\ Closeness$. Table 13 shows the comparison at uniformly distributed time intervals. $ACTR_{AP}$ and $ACTR_{GP}$ denote the runs of ACTR

Table 12 Comparison of OACTR with the existing algorithms on the 50 TSP benchmarks at the uniformly distributed time intervals

	Top count versus **time** (min)							
Algorithm	**4**	**12**	**20**	**28**	**36**	**44**	**52**	**60**
OACTR	**34**	**40**	**43**	**43**	**43**	**42**	**42**	**42**
AWA*	27	30	26	25	26	27	27	27
BS_{100}	25	24	25	24	25	25	25	25
IW	19	19	20	19	19	20	19	19
ANA*	14	15	15	17	17	17	17	17
DFBB	11	10	11	11	11	11	11	11

Table 13 Comparison of the average anytime performances on the 50 TSP benchmarks at uniformly distributed time intervals

	% Optimal Closeness versus **time** (min)							
Algorithm	**4**	**12**	**20**	**28**	**36**	**44**	**52**	**60**
$ACTR_{AP}$	97.8	98.2	98.4	98.4	98.4	98.4	98.5	98.5
$ACTR_{GP}$	**98.2**	**98.4**	**98.5**	**98.5**	**98.5**	**98.6**	**98.6**	**98.6**
OACTR	97.9	98.2	98.4	**98.5**	**98.5**	98.5	98.5	**98.6**
AWA*	97.5	97.8	97.8	97.8	97.8	98.0	98.0	98.0
BS_{100}	86.1	97.6	97.7	97.7	97.7	97.7	97.7	97.7
IW	97.2	97.8	97.8	98.0	98.0	98.0	98.1	98.1
ANA*	91.5	92.1	92.7	92.8	92.8	92.8	92.8	92.8
DFBB	90.0	90.3	90.5	90.6	90.7	90.7	90.7	90.8

algorithm with the input Contract series being the AP and the GP series shown before. One may note that the ACTR versions outperform the other algorithms and amongst them the performance of the version fed with the GP series input stands out better. Figure 2 shows the pictorial view of the same comparison of anytime performances

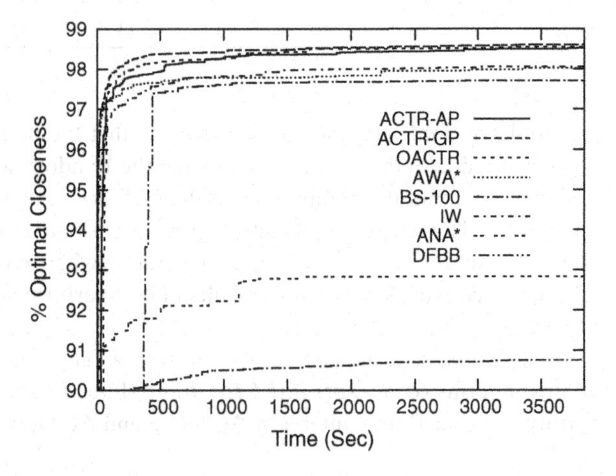

Fig. 2 Comparison of the average anytime performances on the 50 TSP benchmarks

across all the timepoints in the given time window. One may also note that the quality of the solutions reported is as high as 98.5 % on an average.

Finally, we present the comparison of different runs of ACTR when fed with different Contract series inputs. Table 14 shows the comparison when the outputs are sampled at the time intervals that match the given AP series input. We observe that all the runs are very competitive in this domain without a clear winner. An interesting thing to note here is that the oblivious version (OACTR) performs quite well in this domain.

Tables 15 and 16 show the comparison when the outputs of the algorithms are sampled at the intervals of the GP series and the random series, respectively. Once again, we note that there is no clear winner and the run with the GP series input seems to hold a bit of an advantage at some timepoints. This brings us to the discussion on the future scope for studying this problem and improving the proposed algorithms, which we discuss ahead.

4 Discussion and Conclusion

We build our discussion on the experimental observations of Tables 6, 7, and 8 and Tables 14, 15, and 16 where we observed that for some timepoints, the algorithm with the corresponding input series is not the best performer, which is against the expectations. There are several factors which contribute to such deviations which need to be studied and tuned in future. Broadly, the two major factors which impact

Table 14 Comparison of different runs of ACTR on the 50 TSP benchmarks at contract series of the GP series

	Top count versus **time** (min)							
Algorithm	**4**	**12**	**20**	**28**	**36**	**44**	**52**	**60**
$ACTR_{AP}$	35	39	**42**	39	42	42	**43**	40
$ACTR_{GP}$	38	**41**	41	**45**	**45**	42	42	43
$ACTR_{RD}$	**39**	38	**42**	41	42	**43**	42	41
OACTR	35	38	37	40	41	41	**43**	**44**

Table 15 Comparison of different runs of ACTR on the 50 TSP benchmarks at contract series of the GP series

	Top count versus **time** (min)						
Algorithm	**1**	**2**	**4**	**8**	**16**	**32**	**64**
$ACTR_{AP}$	32	32	36	38	38	38	35
$ACTR_{GP}$	**38**	33	**41**	**43**	40	**40**	38
$ACTR_{RD}$	32	**35**	36	41	40	39	36
OACTR	31	30	35	39	**41**	39	**40**

Table 16 Comparison of different runs of ACTR on the 50 TSP benchmarks at contract series of the random series

	Top count versus **time** (min)							
Algorithm	**4**	**10**	**23**	**26**	**31**	**49**	**54**	**60**
$ACTR_{AP}$	**45**	38	39	41	41	40	42	42
$ACTR_{GP}$	37	**40**	**45**	**46**	**46**	41	42	42
$ACTR_{RD}$	38	39	41	42	42	**42**	41	41
OACTR	35	**40**	39	39	41	**42**	**43**	**43**

the performance of the algorithm in this case are: (1) the maximum contract per iteration, and (2) the contract distribution scheme.

Regarding the maximum contract per iteration which is kept as a constant in our experiments (= 8 min), the same can be either increased gradually, in an AP series or a GP series, or can be learned dynamically as the algorithm progresses. This aspect needs to be studied further in future and how the same can be adapted to different domains.

The contract distribution scheme is a major aspect of the algorithm. In this paper, our scheme builds on the previous distributions and the number of nodes expanded so far. However, there is ample scope here to study entirely different and novel schemes and choose the best one. We have studied some other basic schemes such as, one which always distributes the contract equally amongst all levels, which did not perform well compared to the ones presented in the paper. The given scheme uses a pre-defined constant value for α which may also be learned/tuned during the run.

Coming up with a good dynamically learning framework for the contract distribution is an interesting problem. Also, one may explore other options inspired by the existing contract search algorithms that are parameter-free.

In conclusion, we proposed the problem of optimizing anytime performance of an algorithm according to a given input Contract series. An efficient algorithm is presented which distributes the contracts at various stages across different levels in terms of node expansion limits. Experimental results indicate that the proposed algorithm and its variations outperform some of the existing anytime algorithms consistently on the Sliding-tile Puzzle Problem and Traveling Salesperson Problem domains. This also highlights the strength underlying the proposed framework in terms of traditional anytime performance. Several interesting future research directions are also identified.

References

1. Aine, S., Chakrabarti, P.P., Kumar, R.: AWA* - A window constrained anytime heuristic search algorithm. In: M.M. Veloso (ed.) IJCAI, pp. 2250–2255 (2007).
2. Aine, S., Chakrabarti, P.P., Kumar, R.: Contract search: Heuristic search under node expansion constraints. In: ECAI, pp. 733–738 (2010).

3. Aine, S., Chakrabarti, P.P., Kumar, R.: Heuristic search under contract. Computational Intelligence **26**(4), 386–419 (2010).
4. Dean, T., Boddy, M.: An analysis of time-dependent planning. In: Proceedings of 6th National Conference on Artificial Intelligence (AAAI 88), pp. 49–54. AAAI Press, St. Paul, MN (1988).
5. Dionne, A.J., Thayer, J.T., Ruml, W.: Deadline-aware search using on-line measures of behavior. In: SOCS (2011).
6. Hart, P.E., Nilsson, N.J., Raphael, B.: A formal basis for the heuristic determination of minimum cost paths. IEEE Transactions on Systems Science and Cybernetics **4**(2), 100–107 (1968).
7. Hiraishi, H., Ohwada, H., Mizoguchi, F.: Time-constrained heuristic search for practical route finding. In: H.Y. Lee, H. Motoda (eds.) PRICAI98: Topics in Artificial Intelligence, *Lecture Notes in Computer Science*, vol. 1531, pp. 389–398. Springer, Berlin Heidelberg (1998).
8. Korf, R.E., Felner, A.: Disjoint pattern database heuristics. Artif. Intell. **134**(1–2), 9–22 (2002).
9. Lawler, E.L., Wood, D.E.: Branch-and-bound methods: A survey. Operational Research **14**(4), 699–719 (1966).
10. Likhachev, M., Gordon, G.J., Thrun, S.: ARA*: Anytime A* with provable bounds on suboptimality. In: Advances in Neural Information Processing Systems 16. MIT Press, Cambridge, MA (2004).
11. Lowerre, B.: The Harpy Speech Recognition System. PhD thesis, Carnegie Mellon University (1976).
12. Norvig, P.: Paradigms of Artificial Intelligence Programming: Case Studies in Common Lisp, 1st edn. Morgan Kaufmann Publishers Inc., San Francisco, CA, USA (1992).
13. Reinelt, G.: TSPLIB - A traveling salesman problem library. ORSA Journal on Computing **3**, 376–384 (1991).
14. Russell, S.J., Norvig, P.: Artificial Intelligence: A Modern Approach, 2 edn. Pearson, Education (2003).
15. Sturtevant, N.R., Felner, A., Likhachev, M., Ruml, W.: Heuristic search comes of age. In: AAAI (2012).
16. van den Berg, J., Shah, R., Huang, A., Goldberg, K.Y.: Anytime nonparametric A*. In: AAAI (2011).
17. Zhou, R., Hansen, E.A.: Beam-stack search: Integrating backtracking with beam search. In: Proceedings of the 15th International Conference on Automated Planning and Scheduling (ICAPS-05), pp. 90–98. Monterey, CA (2005).

Diagnosing Dependent Action Delays in Temporal Multiagent Plans

Roberto Micalizio and Gianluca Torta

Abstract Diagnosis of Temporal Multiagent Plans (TMAPs) aims at identifying the causes of delays in achieving the plan goals. So far, approaches to TMAP diagnosis have relied on an assumption that might not hold in many practical domains: action delays are independent of one another. In this paper we relax this assumption by allowing (indirect) dependencies among action delays. The diagnosis of a given TMAP is inferred by exploiting a qualitative Bayesian Network (BN), through which dependencies among actions delays, even performed by different agents, are captured. The BN, used to compute the heuristic function, drives a standard A* search, which finds all the most plausible explanations. Results of a preliminary experimental analysis show that the proposed Bayesian-based heuristic function is feasible.

1 Introduction

The execution of Multiagent Plans (MAPs)—plans carried on by a team of cooperating agents—has to consider that plan actions might deviate, for a number of reasons, from their expected, nominal behavior. Detecting and isolating these discrepancies (i.e., failures) is essential to properly react to them. In other words, to attempt any plan repair procedure, one should first infer a *plan diagnosis* that explains why action failures have occurred.

Plan diagnosis has been addressed by several works (see e.g., [5, 6]) that consider action failures due to unexpected events such as faults in the functionalities of the

---

R. Micalizio (✉) · G. Torta
Università di Torino, Turin, Italy
e-mail: micalizio@di.unito.it

G. Torta
e-mail: torta@di.unito.it

M. Bramer and M. Petridis (eds.), *Research and Development in Intelligent Systems XXX*, DOI: 10.1007/978-3-319-02621-3_11,

agents, or unpredictable changes in the environment. These works, however, do not consider temporal failures; for instance, actions that take longer than expected to be completed.

The temporal dimension has been taken into account in MAP diagnosis only recently in [7, 9]. These approaches, however, rely on the assumption that action delays are independent of one another. Since this assumption can be too strong in many practical situations, in this paper we consider the problem of diagnosing action delays that, at least in some cases, can be dependent of one another. To capture dependencies among action delays, we extend the Temporal Multiagent Plan (TMAP) concept we have already introduced in [7]. In particular, within a TMAP we model not only the actions to be performed by agents, but also the agents themselves and the shared objects they use. As in other approaches to diagnosis of MAPs (see e.g., [6]), agents are modeled as a set of *devices*; each device has its own (health) state—known as *behavioral mode* in the Model-Based Diagnosis (MBD) literature—that can change unpredictably over time.

The basic idea is that the state of a device can affect the duration of all the actions using that device. For instance, when the mobility of an agent is not properly working, all the movement actions of that agent will take longer than expected. Similarly, also the states of the shared objects can affect the duration of the actions; e.g., moving a heavy pack may require more time than moving a light pack.

Indirect dependencies between action delays are therefore captured by means of agents' devices and shared objects, that nondeterministically affect the duration of the actions using them. To deal with such nondeterministic relations, we propose the adoption of Bayesian Networks (BNs). A peculiarity of our BNs is that they use qualitative probabilities (i.e., *ranks*). This idea is inspired by the work by Goldszmidt and Pearl [3, 4], where they introduce the notion of ranks as a means to represent, in a qualitative way, the expectation that a given event is likely to happen or not.

The paper is organized as follows: Sect. 2 introduces some background notions about BNs; Sect. 3 extends the notion of TMAP previously introduced in [7] with a BN that captures the device-to-action and object-to-action dependencies; Sects. 4 and 5 provide further details on the BN discussing, respectively, its structure and the qualitative, conditional probabilities associated with its nodes. In Sects. 6 and 7 we define the Delayed Action eXecution (DAX) diagnostic problem and discuss how it can be solved by using the BN as the heuristic function of an A* search. Finally, in Sect. 8 we present a preliminary experimental evaluation of the approach, and then conclude.

2 Background

In this section we briefly introduce some basic notions on *Bayesian Networks* (BNs).

Definition 1 (Causal Network) A causal network $\mathcal{N}$ is a tuple $\langle \mathcal{V}, \mathcal{E} \rangle$ such that:

- $\mathscr{V}$ is a set of variables over which the network is defined; each variable $v \in \mathscr{V}$ has a finite domain $dom(v)$, and represents a network node.
- $\mathscr{E}$ is a set of directed edges of the form $\langle v, v' \rangle$, where $v, v' \in \mathscr{V}$, meaning that the value of variable v can influence the value of variable v'. We say that v is a *parent* of v', and that v' is a *child* of v. Edges in $\mathscr{E}$ cannot form loops, so the resulting network must be a Directed Acyclic Graph (DAG).

In the rest of the paper we will denote as $pa(v)$ the subset of variables in $\mathscr{V}$ that are parents of variable v in the network $\mathscr{N}$. When, for a given node v, the set $pa(v)$ is $\emptyset$, then v is a *root node* of the network. A network can have more than one root (i.e., a network is not necessarily a tree).

Definition 2 (Bayesian Network). A Bayesian Network BN is a pair $\langle \mathscr{N}, \Theta \rangle$ such that:

- $\mathscr{N}=\langle \mathscr{V}, \mathscr{E} \rangle$ is a causal network with nodes $\mathscr{V}$ and edges $\mathscr{E}$,
- Θ is a mapping function that associates each variable $v \in \mathscr{V}$ with its conditional probability table (CPT) $P(v|pa(v))$. Of course, when v is a root node ($pa(v) = \emptyset$), the function Θ associates v with its unconditional—*a priori*—probability $P(v)$.

3 Temporal MAPs: Basic Concepts

In [7] we have proposed a framework for modeling Temporal MAPs that supports the diagnosis of anomalous action durations. In that work, however, we assume that action durations are independent of one another. Since in this paper we want to relax such an assumption, we extend the previous notion of TMAPs as follows.

Definition 3 (Temporal MAP). A TMAP P is a tuple $\langle T, O, M, A, R, BN \rangle$ where:

- T is the team of agents involved in P. For each agent $i \in T$, we denote as $DEV^i : \{dev_1^i, \ldots, dev_n^i\}$ the set of its internal devices.
- $O : \{obj_1, \ldots, obj_m\}$ is a the set of objects available in the environment.
- M is a set of execution modes to be associated with actions (see below). Each mode $m \in M$ is a pair $\langle label, range \rangle$, where $label$ is the mode's name, and $range$ is a time interval $[l, u]$ corresponding to the possible durations of an action when it is executed in that specific mode.
- A is the set of action instances, each action $a \in A$ is assigned to a specific agent in T. The notation a_k^i will refer to the k-th action assigned to agent i. Each action a_k^i is characterized by the following relations:
  - $devices(a_k^i) \subseteq DEV^i$ is the subset of devices used by agent i to perform a_k^i,
  - $objects(a_k^i) \subseteq O$ is the subset of objects agent i requires to complete a_k^i.
  - $modes(a_k^i) \subseteq M$ is a relation that associates action a_k^i in A with a set of modes in M. We assume that each a_k^i has always an expected, nominal mode, named nor, in the set $modes(a_k^i)$.

- R is a partial order relation over A: $e \in R$ is a tuple $\langle a, a' \rangle$ (where $a, a' \in A$), meaning that the execution of a must precede the execution of a'; we say that a is a *predecessor* of a', and that a' is a *successor* of a.
- $BN : \langle \mathcal{N}, \theta \rangle$ is a Bayesian Network encoding the causal dependencies binding actions in P to shared objects and to agents' devices, and associating each of these dependencies with a plausibility measure (see below).

This definition extends the notion of TMAP given in [7] in two ways. First, it specifies what devices characterize each agent, and what shared objects are present in the environment. Second, it makes explicit the causal dependencies binding actions to devices and to shared objects by means of the Bayesian network BN.

Example 1. Let us consider a simple scenario in which two agents, Z and W, cooperate for moving a block $BLK1$. The two agents are assigned with three actions each, denoted as $Z1$, $Z2$, and $Z3$ for agent Z, and similarly $W1$, $W2$, and $W3$ for agent W.

Figure 1 shows a possible TMAP for our scenario: nodes represent action instances, and edges between nodes represent precedence constraints; for instance, $Z2$ can start only when $Z1$ has already terminated; $W2$, instead, can only start when both $W1$ and $Z1$ have been completed. Besides the precedence relations specified in Fig. 1, we also need to specify the modes associated with the actions. For simplicity we suppose that each action has three modes: normal $\langle nor, [2, 3] \rangle$, moderate delay $\langle md, (3, 4] \rangle$, and heavy delay $\langle hd, (4, 6] \rangle$. Moreover, we assume that each agent is characterized by only one device: the *mobility*. The health state of *mobility* can be: *ok* (the device is working properly), slow sw, or very slow vsw. The last two states represent situations in which the mobility is malfunctioning, and hence it is likely to cause some amount of delay in the actions that require the mobility device. Also block $BLK1$ can cause delays; in fact, $BLK1$ can either be *light* (l) or *heavy* (h), and an action moving a heavy block may take longer than expected. Note that, even though the state of the block does not change over time, we assume that it is not known during plan execution. All these relations are specified by the last important item we have to provide for modeling the TMAP, namely the Bayesian network that will be the topic of the next two sections.

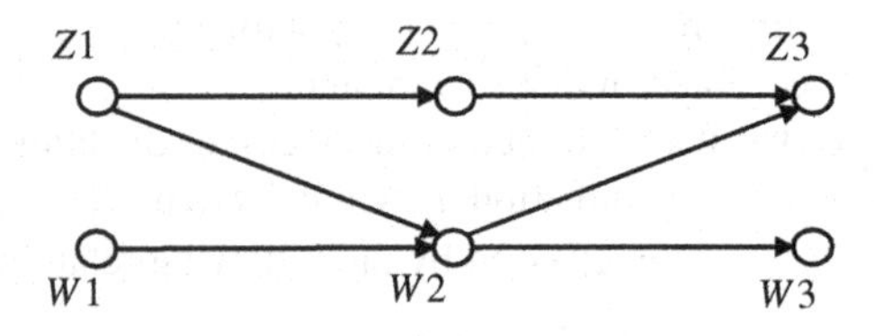

Fig. 1 The TMAP to be performed by agents A and B

4 Building the Bayesian Network: Structure

Given a TMAP $P = \langle T, O, M, A, R, BN\rangle$, the causal dependencies device-to-action and object-to-action are captured by the Bayesian network $BN = \langle \mathscr{N}, \theta\rangle$. The structure of the causal network $\mathscr{N}$ deserves particular attention. The network must in fact ensure that all the possible elements conditioning an action are taken into account without introducing loops. We guarantee this condition by constructing the network $\mathscr{N} = \langle \mathscr{V}, \mathscr{E}\rangle$ as follows:

- $\mathscr{V}$ is the set of variables (i.e., network nodes), and it is partitioned into three disjoined subsets: $\mathscr{A}$, $\mathscr{O}$ and $\mathscr{D}$.
  - $\mathscr{A}$ is the set of action-variables; that is, for each action $a_k^i \in A$ a corresponding variable α_k^i is added in $\mathscr{A}$. The domain of α_k^i is the set of mode labels associated with a_k^i; i.e., $dom(\alpha_k^i) = \{label | \langle label, range\rangle \in modes(a_k^i)\}$.
  - $\mathscr{O}$ is the set of object-variables: for each $obj_j \in O$ a corresponding variable ω_j is defined in $\mathscr{O}$; $dom(\omega_j)$ is set of possible states of object obj_j.
  - $\mathscr{D}$ is the set of device-variables, through these variables we represent the status of the devices of all the agents. However, in the case of devices, the status can evolve over time. Thus, two (subsequent) actions using the same device could be affected in different ways as they use the device at different times. To capture the dynamics of a device, we create a device-variable for each pair (action, used-device). More formally, for each action $a_k^i \in A$ and for each device dev_l^i such that $dev_l^i \in devices(a_k^i)$, we add in $\mathscr{D}$ a device-variable $\delta_{l,k}^i$ representing the status of device dev_l^i when used to perform action a_k^i. The domain of each device-variable $\delta_{l,k}^i$ corresponds to the set of operating modes of device dev_l^i; namely, $dom(\delta_{l,k}^i) = \{ok, dgr_1, \ldots dgr_n\}$, where *ok* represents dev_l^i nominal behavior, whereas the other values represent its degraded operating modes.
- $\mathscr{E}$ is, as usual, the set of edges between variables in $\mathscr{V}$. To guarantee that no loop will be formed, we impose that only three types of edges can be defined:
  - $DA = \{\langle \delta_{l,k}^i, \alpha_k^i\rangle\}$ is the set of *device-to-action* edges. These edges are used to model how the state of a device can influence the duration of an action using that device;
  - $OA = \{\langle \omega_j, \alpha_k^i\rangle\}$ is the set of *object-to-action* edges modeling how the state of an object influences the duration of an action;
  - $DD = \{\langle \delta_{l,j}^i, \delta_{l,k}^i\rangle\}$ is the set of *device-to-device* edges. This type of edges models how the status of a device can evolve over time. In particular, we impose that $\langle \delta_{l,j}^i, \delta_{l,k}^i\rangle$ can be added in DD iff these three conditions are satisfied:
    1. a_j^i precedes a_k^i in the plan,
    2. no other action a_h^i placed between a_j and a_k in the plan uses device dev_l^i.

Since there is a one-to-one mapping between actions in A and variables in $\mathscr{A}$, and between objects in O and variables in $\mathscr{O}$, we will henceforth use roman letters

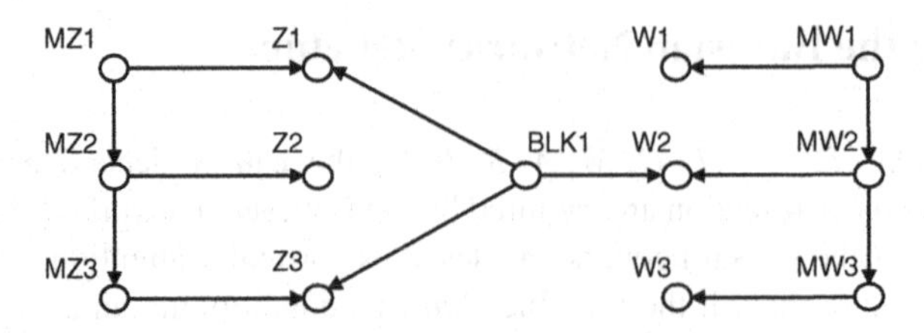

Fig. 2 The Bayesian network associated with the example TMAP P

also for the variables within $\mathscr{A}$ and $\mathscr{O}$ to simplify the notation. While we keep δ to denote variables in $\mathscr{D}$ which are not direct maps to the variables in DEV.

Example 2. Let us now build the Bayesian network BN for the TMAP P in Example 1. In building BN, we assume that:

1. Actions $Z1$, $Z2$, and $Z3$ require agent Z's *mobility*;
2. Actions $W1$, $W2$, and $W3$ require agent W's *mobility*;
3. Block $BLK1$ is used, in the order, by actions $Z1$, $W2$, and $Z3$.

The resulting Bayesian network is given in Fig. 2. Nodes represent variables mentioned within $\mathscr{A}$, $\mathscr{O}$, and $\mathscr{D}$. In particular, nodes in $\mathscr{D}$ are not directly mapped one-to-one to agents' devices as the state of a device can change over time. For instance, agent Z uses the mobility device during its three actions. Such a device, however, could affect the three actions in different ways; for instance, the mobility could be *ok* during $Z1$, then it might evolve into sw during $Z2$, and into vsw during $Z3$. To capture the dynamic behavior of the mobility device, we create a variable $MZ1$ for the mobility of agent Z during action $Z1$, a variable $MZ2$ for its mobility during $Z2$ and a variable $MZ3$ for its mobility during $Z3$. We link these three new variables with dependency edges in such a way that the value of $MZ2$ depends on the value of $MZ1$, and that the value of $MZ3$ depends on the value of $MZ2$. Similarly, variables $MW1$, $MW2$, and $MW3$ are created to capture the possible evolution of agent W's mobility during the execution of its actions.

Of course, each action depends on the devices and objects it uses. For instance, action $Z1$ uses the block, so an edge from $BLK1$ to $Z1$ is added to the BN. At the same time, $Z1$ needs also the mobility, and hence an edge from $MZ1$ to $Z1$ is included.

5 Building the Bayesian Network: Qualitative CPTs

As stated in the background section, given a Bayesian network $BN=\langle \mathscr{N}, \Theta \rangle$, component Θ is a function associating each node v in the network with its conditional probability table (CPT); where $CPT(v)$ encodes the conditional probability $P(v|pa(v))$.

In some situations, however, the full power of continuous probabilities might be unnecessary and somewhat awkward to deal with. In fact, although probabilities could be learnt, the learning process could require a number of—possibly time-consuming and expensive—training sessions. A possible alternative to continuous probabilities is the adoption of *qualitative probabilities*, or *ranks*. Ranks have been introduced in [3, 4] as a means for supporting belief revision in the context of non-monotonic reasoning. They are interpreted as order-of-magnitude approximations of probabilities, and can be used to model in a qualitative way the degree of "surprise" of observing a given behavior or state in a system.

In [4] a precise mapping between probabilities and ranks is discussed. For the purpose of this work, it is sufficient to say that:

1. A ranking function κ associates an event e with a nonnegative integer; the lower the rank $\kappa(e)$, the more plausible, or "normal", the event e is; when $\kappa(e) = 0$, the event e is considered normal (i.e., highly probable).
2. Given two events e_1 and e_2, it holds $\kappa(e_1|e_2) = \kappa(e_1 \wedge e_2) - \kappa(e_2)$.

Note that the latter formula corresponds to Bayes rule in classic probability, where $p(e_1|e_2)$ is given by $p(e_1 \wedge e_2)/p(e_2)$.

From our point of view, an event e is an assignment of values to a subset of variables in the BN; namely, $e{:}\langle v_1 = val_1, \ldots, v_k = val_k\rangle$ (where: $i : 1, \ldots, k$, $v_i \in \mathcal{V}$, and $val_i \in dom(v_i)$). Given a variable $v \in \mathcal{V}$, with parents $pa(v) = \{v_1, \ldots, v_k\}$, the table $CPT(v)$ associated with v is obtained by specifying $\kappa(v = val|v_1 = val_1, \ldots, v_k = val_k)$ for each value $val \in dom(v)$, and for every possible tuple in $dom(v_1) \times \cdots \times dom(v_k)$.

Example 3. To complete our TMAP model, we just need to provide the CPTs for each node mentioned in the Bayesian network. First of all, we can identify the roots of the network, $MZ1$, $BLK1$, and $MW1$; for these nodes we have just to provide *a priori* likelihoods as they are not affected by any other node in the network.

For instance, Table 1 shows the *a priori* likelihoods for node $MZ1$; it states that the mobility of agent Z during its first action $Z1$ is likely to be *ok*, whereas modes sw and vsw are, progressively, more unlikely. The *a priori* likelihood associated with node MW1 is similar. Table 2 shows another example of *a priori* likelihoods for the weight of block $BLK1$: normally it is expected that $BLK1$ is light—in fact value l has rank 0.

Table 3 reports how variable $MZ2$ causally depends on $MZ1$. For instance, when $MZ1$ is *ok* it is highly probable that $MZ2$ is *ok* too ($MZ2 = ok$ has rank 0), but $MZ2$ can still evolve into sw (rank 1) or vsw (rank 2). On the other hand, in this example we assume that faults are not intermittent; that is, an impaired device cannot

Table 1 Node $MZ1$'s CPT

$MZ1$	*ok*	*sw*	*vsw*
	0	1	2

Table 2 Node $BLK1$'s CPT

$BLK1$	*l*	*h*
	0	1

evolve back to a normal mode. For this reason we say that when $MZ1$ is sw, then $MZ2 = ok$ has rank ∞. We treat the "very slow" case in a similar way: once the mobility is in the *very slow* state (vsw), it cannot evolve back to the *slow* state (sw); thus, when $MZ1 = vsw$ the only possible value for $MZ2$ is vsw.

Finally, in Table 4 we show the CPT associated with action node $Z3$. From the Bayesian network in Fig. 2 it is apparent that node $Z3$ causally depends on nodes $MZ3$ and $BLK1$; namely, the delay possibly accumulated during action $Z3$ may depend on the state of the mobility device (node $MZ3$), and on the weight of block $BLK1$. Thus, when the mobility device is *ok* and the block is light, $\langle MZ3 = ok, BLK1 = l\rangle$, it is highly probable that $Z3$ is *nor*, meaning that action $Z3$ will be completed on-time. On the other hand, when the mobility is not *ok* or when the block is not *light*, action $Z3$ may gather some delay. Due to space reasons we leave out the remaining CPTs; these, however, are just variations of the given examples.

6 Delayed Action Execution Diagnostic Problem (DAX)

We can now describe how the Bayesian network is used in the solution of a Delayed Action eXecution Diagnostic problem, or DAX for short. Intuitively, given a TMAP P, a DAX problem arises during the execution of P when we observe that at least one action has been completed with some delay. To formalize this intuitive notion, we need a current diagnostic hypothesis H (e.g., all actions are initially assumed normal), and a set of observations Obs. Each observation $obs \in Obs$ is a pair $\langle oe, t\rangle$, where oe is the observed event (i.e., either the start or the end of an action in A); whereas t is the time at which oe occurs. More formally:

Table 3 Node $MZ2$'s CPT

		$MZ1$		
		ok	sw	vsw
$MZ2$	ok	0	∞	∞
	sw	1	0	∞
	vsw	2	1	0

Table 4 Node Z3's CPT

		$\langle MZ3, BLK1\rangle$					
		$\langle ok, l\rangle$	$\langle sw, l\rangle$	$\langle vsw, l\rangle$	$\langle ok, h\rangle$	$\langle sw, h\rangle$	$\langle vsw, h\rangle$
Z3	*nor*	0	1	2	0	2	∞
	md	1	0	0	1	1	2
	hd	2	1	0	1	0	0

Definition 4 (DAX Problem). A tuple $\langle P, H, Obs \rangle$ is a DAX problem instance iff:

$$P \cup H \cup Obs \vdash \bot \tag{1}$$

Namely, hypothesis H is inconsistent with Obs given that TMAP P is being executed.

To make the definition of a DAX problem more precise, we have to clarify two points: (1) what a diagnostic hypothesis looks like, and (2) what we mean when we require that the diagnostic hypothesis is consistent with the observations, and how we can assess this property. We address the first point in the following of this section by defining a solution to a DAX problem. The second point, instead, will just be sketched in the next section, since it has been deeply addressed in [7] with a solution based on Simple Temporal Networks (STNs) [2].

Definition 5 (Diagnostic Hypothesis). A diagnostic hypothesis H is a mapping that associates each action variable $a_k^i \in A$ with a mode in $dom(a_k^i)$.

Note that, since A and $\mathscr{A}$ coincide, a diagnostic hypothesis is also an assignment to some of the variables within the Bayesian network BN associated with the TMAP P. This is important because in solving a DAX problem we will use a (possibly partial) diagnostic hypothesis as an evidence within BN (see next section).

Definition 6 (DAX Solution). Let $DP = \langle P, H, Obs \rangle$ be a DAX problem, a solution to DP is a new diagnostic hypothesis H' such that $P \cup H' \cup Obs \nvdash \bot$.

In other words, the new hypothesis H' is consistent with the observations received so far. This means that H' is a *consistency-based* explanation [8] for the observations about the agents' behaviors.

In general, given a DAX problem, there exist several alternative hypotheses H' that satisfy Definition 6; i.e., many explanations, or diagnoses, are possible. Not all the diagnoses, however, have the same relevance, and preference criteria are often used to select a subset of preferred diagnoses. In MBD it is often assumed that faults are independent of one another. Relying on this assumption, one can conclude that a diagnosis assuming the least number of faults is more preferable than a diagnosis that assumes more faults. These diagnoses are referred to as *minimum cardinality diagnoses* within the MBD literature. In our scenario, however, faults (i.e., action anomalous durations) are *not* independent of one another. It follows that minimum cardinality diagnoses are not meaningful in our setting.

In order to take into account the dependencies among action durations, we need to consider a different preference criterion. A natural choice is to prefer diagnoses that are more plausible than others, taking into account the dependencies captured by the Bayesian network BN associated with the TMAP at hand.

Definition 7 (Preferred DAX Solution). Given a DAX problem DP=$\langle P, H, Obs \rangle$, a Preferred DAX Solution is a new diagnostic hypothesis H^{Pref} such that:

- $P \cup H^{Pref} \cup Obs \nvdash \bot$ (i.e., H^{Pref} satisfies Definition 6), and
- $\kappa(H^{Pref})$ is minimal.

The first condition states that the new hypothesis H^{Pref} is consistent with the observations received so far. The second condition states that H^{Pref} is one of the most plausible explanations as its rank is minimal. In other words, for any other diagnostic hypothesis H' that satisfies Definition 6, $\kappa(H^{Pref}) \leq \kappa(H')$. Note that to compute the rank of a diagnostic hypothesis we can easily adapt the standard algorithms for inferring most probable explanations (MPE) in Bayesian networks [1] to our rank-based qualitative BNs.

In the next section we describe how the BN network can be used to guide an A*-search process leading to the most plausible (i.e., minimal ranked) preferred solutions H^{Pref}.

7 Detecting and Solving a DAX Problem

Detecting a DAX Problem. As pointed out above, a DAX problem arises whenever the current diagnostic hypothesis H is no longer adequate for explaining the received observations. A first issue to face is that observations involve a metric time (e.g., $obs_{Z1} = \langle start(Z1), 3\rangle$ meaning that action $Z1$ was observed to start at time 3), but hypothesis H just mentions qualitative behavioral modes associated with actions (e.g., $nor(Z1)$ action $Z1$ is assumed to be normal). To verify whether observation obs_{Z1} is consistent with assumption $nor(Z1)$, in [7] we have proposed a solution based on Simple Temporal Networks (STNs). More precisely, given a TMAP P and the current diagnostic hypothesis H, we build a corresponding STN $S_{\langle P,H\rangle}$. Then, to verify whether a set Obs of observed events are consistent with the current hypothesis, it is sufficient to add them to $S_{\langle P,H\rangle}$, yielding a new STN $S_{\langle P,H,Obs\rangle}$, and then running any *all-pairs shortest paths* algorithm, such as Floyd-Warshall or Johnson. In fact, it has been shown in [2] that if the shortest path between any two nodes in the STN results to be a negative cycle, then the set of constraints encoded by the STN is inconsistent. In our case, a negative cycle in $S_{\langle P,H,Obs\rangle}$ would mean that the current hypothesis H is not consistent with the observations Obs; so it is sufficient to verify this condition to detect a DAX problem.

Solving a DAX problem. To solve a DAX problem we adopt an A* search strategy, similar to the one we have proposed in [7]. The main difference is in the heuristic function we use to drive the search. To explain this function it is sufficient to focus on the information maintained within each node of the A* search. A node *nd* on the fringe of the A* search represents, in general, an intermediate step towards a solution; that is, it encodes a partial solution. In our case, this partial solution is encoded via two main pieces of information:

- H is a (partial) assignment of modes to a (subset) of actions in A (i.e., it is a partial solution);
- $\mathcal{D}oms = \{\langle a, dom(a)\rangle | a \notin H\}$ is the set of actions whose mode has not been assigned yet (they are not mentioned within H). Note that for each action $a \in \mathcal{D}oms$, we also keep the domain $dom(a)$ of this action; that is, the set of modes

still available for a. In fact, as the search process goes on, it is possible that the domains of these variables are pruned by the observations and by the assignments made so far.

It must be noted that during the search process, it is necessary to maintain the consistency of the partial solution H with the received observations Obs. To this end, the A* algorithm also maintains, for each node nd, an STN $S_{\langle P,H,Obs \rangle}$ encoding the TMAP P, the partial hypothesis H, and the observations Obs. This STN ensures that only the action modes consistent with the observations Obs and the assignments in H are actually kept within $\mathcal{D}oms$. Details about this point are discussed in [7].

As usual in any A* search, each node nd is assessed by a heuristic function $f(nd) = g(nd) + h(nd)$: $g(nd)$ estimates the cost of the current partial hypothesis H; whereas $h(nd)$ estimates the cost of completing H by assigning modes to the actions mentioned in $\mathcal{D}oms$. In [7] we assume that faults (i.e., action delays) are independent of one another. Relying on this assumption, the two components $g(nd)$ and $h(nd)$ can be computed as summations on the ranks associated with action modes. More precisely, $g(nd)$ is the summation of the ranks of the modes assigned to the actions in the partial solution H. Whereas $h(nd)$ is the summation of the *minimal* ranks of modes that can be assigned to actions in $\mathcal{D}oms$. The minimality of such ranks guarantees that $f(nd)$ is admissible.

In this paper, however, we have relaxed the assumption of independence of faults. Indeed, to take into account dependencies among action delays, we need a heuristic function that, by exploiting the Bayesian network BN, estimates the most plausible way for completing a partial solution H. To this end, we note that, given an evidence e (i.e., an assignment of values to a subset of variables in a Bayesian network), the Most Probable Explanation (MPE) of such an event is defined as

$$MPE(e) = \arg\max_{x} P(x, e).$$

In other words, $MPE(e)$ is the most probable assignment of values to all the variables in the Bayesian network not mentioned in e. Let $\hat{x}$ be $MPE(e)$, by definition of conditional probability we have that

$$P(\hat{x}, e) = P(e) \cdot P(\hat{x}|e). \tag{2}$$

In our rank-based framework, Eq. 2 can be rewritten as

$$\kappa(e, \hat{x}) = \kappa(e) + \kappa(\hat{x}|e). \tag{3}$$

Now it is easy to see that function $\kappa(e, \hat{x})$ in Eq. 3 is exactly the heuristic valuation function $f(nd)$ we are looking for. In fact, $\kappa(e)$ corresponds to $g(nd)$; namely, the current hypothesis H within node nd is seen as the evidence e. On the other hand, $\hat{x}$ is the best way to complete the hypothesis H, and hence $\kappa(\hat{x}|e)$ corresponds to $h(nd)$.

Property 1 Let nd be a node generated during the A* search, and let H be the partial solution encoded in nd, then the heuristic function

$$f(nd) = \kappa(H) + \kappa(\hat{x}|H),$$

(where $\hat{x} = MPE(H)$), is *admissible*.

Intuitively, this property follows from the observation that $MPE(H)$ is the most probable explanation for H. In our rank-based setting, this means that $MPE(H)$ is the explanation with the least possible rank; namely, $MPE(H)$ underestimates the actual rank of a solution including the assignments already made in H, and hence $f(nd)$ is admissible.

Property 1 assures that the A* search will eventually find an optimal solution, consisting of a complete assignment of modes to actions in A (i.e., $\mathscr{A}$), whose rank is minimal, and hence that is (one of) the most plausible explanation(s).

Example 4. Let us now illustrate how an A* search can solve a DAX problem. Let us assume that agents Z and W are executing the TMAP P modeled in the previous examples, and that after a while we receive the observation $obs = \langle start(Z3), 8\rangle$. Observation obs is inconsistent with the hypothesis that all the actions performed so far are normal. In fact, when all the actions preceding $Z3$ are normal, the earliest and latest start times for action $Z3$ are, respectively, 4 and 6. This means that to explain obs, we must assume at least one anomalous action preceding $Z3$.

To solve the detected DAX problem we start A* with the special node $\langle \emptyset, \mathscr{D}oms\rangle$: the current partial solution H is empty, and all the action-variables are in $\mathscr{D}oms$ (the domain of each of these variables is complete). The search proceeds by selecting a variable, say $Z1$, and generates as many children nodes as there are modes in $dom(Z1)$. Thus, three new nodes are created: $n1 : \langle\{Z1 = nor\}, \mathscr{D}oms \setminus \{Z1\}\rangle$, $n2 : \langle\{Z1 = md\}, \mathscr{D}oms \setminus \{Z1\}\rangle$, $n3 : \langle\{Z1 = hd\}, \mathscr{D}oms \setminus \{Z1\}\rangle$. As mentioned above, our A* algorithm keeps, for each node, a corresponding STN $S_{\langle P,H,Obs\rangle}$ encoding the plan P, the partial solution H, and the received observations (see [7]). Such an STN is used to verify whether the partial solution H is consistent with Obs. In addition, tightening the STN constraints may have the beneficial effect of pruning the domains of some variables in $\mathscr{D}oms$. In the most fortunate case, the domain of a variable could be reduced to a single value. In our simple example, all the three partial solutions are consistent with obs.

Thus, the algorithm proceeds by evaluating these three nodes with the Bayesian-based heuristic function. In our simple example, we have that $f(n1) = 0$, $f(n2) = 1$, and $f(n3) = 2$. For instance, $f(n1)$ is zero because the rank of the partial hypothesis $Z1 = nor$ is zero, this is summed to the rank of the most plausible assignments that can be given to the variables in $\mathscr{D}oms \setminus \{Z1\}$, which is still zero. The other two nodes, $n2$ and $n3$, are evaluated in a similar way.

Since none of the three nodes contains a complete solution, the search algorithm selects the new top node $n1$, and proceeds as before by selecting a variable and generating children nodes. From node $n1$, the algorithm finds the optimal solution

Table 5 Preliminary results

	#sols	#nodes	Time all sols (s)	Time/sol (s)
2Ags	3.1	38.9	32.6	10.6
3Ags	2.8	44.5	41.1	18.0
4Ags	5.5	72.1	86.3	21.3

Columns show avg # of solutions, avg # of visited nodes, avg time to find all solutions (sec), avg time per solution

$sol1 : \{Z1 = nor, Z2 = nor, Z3 = nor, W1 = md, W2 = md, W3 = md\}$. Note that the solution is a complete assignment of modes to all the actions; in particular, the modes of actions $Z3$ and $W3$ are the most plausible modes predicted by the BN given the only available observation. The rank of solution $sol1$ is 1, and it explains the delayed start of $Z3$ with a *moderate delay* affecting the actions $W1$ and $W2$. A second optimal solution $sol2 : \{Z1 = md, Z2 = md, Z3 = md, W1 = nor, W2 = nor, W3 = nor\}$ is derived form node $n2$ in subsequent iterations of the algorithm. These are the only two possible optimal solutions for this simple example as any other explanation has rank greater than 1.

8 Implementation and Experimental Results

We developed the Bayesian-based heuristic function as an extension of the Perl program implementing the A* search discussed in [7]. In particular, Bayesian inferences are performed by transforming the Bayesian network in a join-tree and then applying rank-based versions of standard algorithms for propagating the evidence and computing MAPs (see [1]).

We conducted a first set of tests to check the feasibility of the approach; tests have run on a Intel i7 M640 processor at 2.80 GHz with 8 GB of RAM. In the same blocks world domain of our example, we defined three test sets, *2Ags*, *3Ags* and *4Ags*, each one containing 20 DAX problems with 2, 3 and 4 agents, respectively. Each DAX problem consists of a TMAP and a set of timed observations. Each TMAP is randomly generated and contains 10 actions per agent; in addition, up to 4 precedence links are randomly inserted between actions of different agents in order to make the plan structure more complex. The blocks the agents have to handle are half light and half heavy.

In the tests, we simulated an anomalous execution of each TMAP by randomly injecting faults in the agents' devices, which in turns induced action delays. On average, 6 actions per DAX were affected by the fault. We assessed the approach by using a very scarce observability rate as only up to three timed observations per DAX problem were provided to the system in order to infer *all* the minimal diagnoses.

Table 5 shows the preliminary results of our experiments. These results show that the search process is properly driven by the heuristic function. The low number of explanations, in fact, demonstrates that the search process does take into account the dependencies among action delays. Otherwise, the number of possible solutions for explaining the failure of the actions affected by the injected fault (six on average), as independent delays would necessarily be greater. These data show also that the computation of the heuristic function is still relatively costly (about 1 s is spent for processing one node). We believe that such a computational cost can be reduced by optimizing our implementation in future developments.

9 Conclusion

The paper has addressed the problem of diagnosing action delays in the execution of Temporal Multiagent Plans (TMAPs). The main contribution of the work is the ability of dealing with dependent action delays. In particular, the paper has extended the notion of TMAP introduced in [7] with a Bayesian network, which models indirect, nondeterministic dependencies among actions, even assigned to different agents.

The paper has focused on how the Bayesian network can be built relying on the precedence constraints defined between actions, and on the models of objects and agents. Indirect dependencies among actions are in fact defined in terms of the *devices* the agents are equipped with, and of the *shared objects* the agents may use in performing their activities.

An important characteristic of the proposed Bayesian network is that it uses *ranks* (i.e., qualitative probabilities) in the CPTs associated with the network's nodes. The adoption of ranks eases the problem of supplying the CPTs since we do not need to run a huge number of simulations to learn the real probability values. Ranks are in fact a qualitative measure of the degree of "surprise" that we can associate with events. Defining ranks is therefore intuitive when the normal behavior of a device (or action) is more plausible than the abnormal ones.

The paper has pointed out that the Bayesian network can be exploited in the heuristic function of an A* search algorithm. In particular, any node of the search tree is evaluated by computing the MPE of the partial hypothesis encoded by the node itself. This novel heuristic function allows us to drive the search taking into account the possible dependencies existing among action delays. Thus the method extends previous approaches (see e.g., [7, 9]) which rely on the assumption of independent action delays.

Besides improving the computation of the Bayesian-based heuristic function, in future works we intend to use the Bayesian network more actively. In fact, the network could suggest to the search process not only the most promising node to consider next, but also what variable should be picked up when children nodes have to be created in the node expansion.

Another interesting future development is the identification of *primary causes*. The current solution, in fact, just explains the received observations by tagging each performed action with a mode. The Bayesian network, however, represents a richer model binding action modes to objects and devices, and these can be considered as the primary causes of the action delays. Primary causes may provide a human user with more understandable explanations as they make explicit that action delays are actually related with one another. For instance, a single fault in an agent device could be used to explain the delay of a number of actions assigned to that agent.

References

1. Darwiche, A.: Modeling and Reasoning with Bayesian Networks. Cambridge University Press (2009)
2. Dechter, R., Meiri, I., Pearl, J.: Temporal constraint networks. Artificial Intelligence **49**, 61–95 (1991)
3. Goldszmidt, M., Pearl, J.: Rank-based systems: a simple approach to belief revision, belief update, and reasoning about evidence and actions. In: Proc. KR92, pp. 661–672 (1992)
4. Goldszmidt, M., Pearl, J.: Qualitative probabilities for default reasoning, belief revision, and causal modeling. Artificial Intelligence **84**, 57–112 (1996)
5. de Jonge, F., Roos, N., Witteveen, C.: Primary and secondary diagnosis of multi-agent plan execution. Journal of Autonomous Agent and MAS **18**, 267–294 (2009)
6. Micalizio, R., Torasso, P.: Monitoring the execution of a multi-agent plan:dealing with partial observability. In: Proc. ECAI08, pp. 408–412 (2008)
7. Micalizio, R., Torta, G.: Diagnosing delays in multi-agent plans execution. In: Proc. ECAI12, pp. 594–599 (2012)
8. Reiter, R.: A theory of diagnosis from first principles. Artificial Intelligence **32 (1)**, 57–96 (1987)
9. Roos, N., Witteveen, C.: Diagnosis of simple temporal networks. In: Proc. ECAI08, pp. 593–597 (2008)

Representation and Reasoning

Conditional Preference-Nets, Possibilistic Logic, and the Transitivity of Priorities

D. Dubois, H. Prade and F. Touazi

Abstract Conditional Preference-nets (CP-nets for short) and possibilistic logic with symbolic weights are two different ways of expressing preferences, which leave room for incomparability in the underlying ordering between the different choices. Relations can be expressed between the two settings. A CP-net can be mapped to a weighted set of formulas, one per node of the CP-net, and appropriate constraints between symbolic weights of formula are defined according to the observed priority of father nodes over children nodes in the CP-net. Thus, each potential choice can be associated with a vector of symbolic weights which acknowledges the satisfaction, or not, of each node formula. However, this local priority between father and children nodes in the CP-net does not seem to be transitive. It may happen that the same pair of vectors in the possibilistic representations of two CP-nets correspond to decisions that are comparable in one CP-net structure and incomparable in the other. This troublesome situation points out the discrepancies between the two representation settings, the difficulties of an exact translation of one representation into the other, and questions the faithfulness of the preference representation in CP-nets from a user's point of view. This note provides a preliminary discussion of these issues, using examples

D. Dubois · H. Prade · F. Touazi (✉)
IRIT, University of Toulouse, 118 rte de Narbonne, Toulouse, France
e-mail: dubois@irit.fr

H. Prade
e-mail: prade@irit.fr

F. Touazi
e-mail: faycal.touazi@irit.fr

M. Bramer and M. Petridis (eds.), *Research and Development in Intelligent Systems XXX*, 175
DOI: 10.1007/978-3-319-02621-3_12, © Springer International Publishing Switzerland 2013

1 Introduction

"CP-nets" [3] are a popular framework for expressing conditional preferences, based on a graphical representation. They are based on the idea that users' preferences generally express that, in a given context, a partially described situation is strictly preferred to another antagonistic partially described situation, in a ceteris paribus way. It has been observed that preferences associated with a father node in a CP-net are then more important than the preferences associated with children nodes. This has motivated the idea of approximating CP-nets by means of a possibilistic logic representation of preferences. More precisely, each node of a CP-net is associated with preferences represented by one possibilistic logic formula, to which is attached a symbolic weight, the priority in favour of father node preferences being echoed by constraints between such symbolic weights [5, 9]. Thus, each interpretation is associated with a vector of symbolic weights which acknowledges the satisfaction, or not, of each node formula. This use of symbolic weights leaves room for a possible incomparability of interpretations, as in CP-nets.

Recently, it has been suggested that depending on the order defined between the vectors of symbolic weights, one may approximate any CP-net from below and from above in the framework provided by the symbolic possibilistic logic representation [8]; see Sect. 3. This state of facts prompts us to compare the CP-net and the symbolic weighted logic settings. They are based on two different principles, from the point of view of the representation of preferences. On the one hand, the ceteris paribus principle in CP-nets enables a conditional expression of preferences, usually in the form of a directed acyclic graph where nodes represent variables. On the other hand, in the weighted logic representation, the preferences are expressed as goals having priority weights between which inequality constraints may exist. This induces a partial order between symbolic priority weights, i.e. a transitivity property applies, while in CP-nets, the priority of a father node with respect to its children nodes does not seem to simply extend to the grand children nodes. However, it is worth emphasizing that the symbolic possibilistic logic setting leaves total freedom to the user to indicate the relative priorities between goals, which is not always possible in CP-nets [5, 9]. The relative priorities considered here are only dictated by the will to remain close to CP-nets. This note lays bare new discrepancies between the two representation settings beyond those already pointed out in [8].

The paper is organized as follows. First, short backgrounds on possibilistic logic, and then on CP-nets are respectively provided in Sects. 2 and 3 where the possibilistic logic approximations of a CP-net from below and from above are also recalled. Section 2 starts the comparative discussion by exhibiting two examples of CP-nets that order two interpretations differently, while they have the same encoding as a pair of possibilistic logic vectors of symbolic weights. This raises the questions of the behavior and of the faithfulness of the two representations.

2 Possibilistic Logic

We consider a propositional language where formulas are denoted by $p_1, ..., p_n$, and Ω is its set of interpretations. Let $B^N = \{(p_j, \alpha_j) \mid j = 1, \ldots, m\}$ be a possibilistic logic base where p_j is a propositional logic formula and $\alpha_j \in L \subseteq [0, 1]$ is a priority level [6]. The logical conjunctions and disjunctions are denoted by $\wedge$ and $\vee$. Each formula (p_j, α_j) means that $N(p_j) \geq \alpha_j$, where N is a necessity measure, i.e., a set function satisfying the property $N(p_1 \wedge p_2) = \min(N(p_1), N(p_2))$. A necessity measure is associated with a possibility distribution π as follows: $N(p_j) = \min_{\omega \notin M(p_j)}(1-\pi(\omega)) = 1-\Pi(\neg p_j)$, where Π is the possibility measure associated with N and $M(p)$ is the set of models induced by the underlying propositional language for which p is true.

The baseB^N is associated with the following possibility distribution:

$$\pi_B^N(\omega) = \min_{j=1,m} \pi_{(p_j,\alpha_j)}(\omega)$$

on the set of interpretations, where $\pi_{(p_j,\alpha_j)}(\omega) = 1$ if $\omega \in M(p_j)$, and $\pi_{(p_j,\alpha_j)}(\omega) = 1-\alpha_j$ if $\omega \notin M(p_j)$. This means that an interpretation ω is all the more possible as it does not violate any formula p_j having a higher priority level α_j. So, if $\omega \notin M(p_j)$, $\pi_B^N(\omega) \leq 1 - \alpha_j$, and if $\omega \in \bigcap_{j \in J} M(\neg p_j)$, $\pi_B^N(\omega) \leq \min_{j \in J}(1 - \alpha_j)$. It is a description "from above" of π_B^N, which is the least specific possibility distribution in agreement with the knowledge base B^N. Thanks to the min-decomposability of necessity measures with respect to conjunction, a possibilistic base B^N can be transformed into a base where the formulas p_i are clauses (without altering the distribution π_B^N). We can still see B^N as a conjunction of weighted clauses, i.e., as an extension of the conjunctive normal form.

3 CP-Nets and Their Approximation in Possibilistic Logic

A CP-net [3] is graphical in nature, and exploits conditional preferential independence in structuring the preferences provided by a user. It is reminiscent of a Bayes net; however, the nature of the relation between nodes within a network is generally quite weak, compared with the probabilistic relations in Bayes nets. The graph structure captures statements of qualitative conditional preferential independence.

Definition 1 *A CP-net $\mathcal{N}$ over a set of Boolean variables $V = \{X_1, \cdots, X_n\}$ is a directed graph over the nodes $X_1, \cdots, X_n$, and there is a directed edge from X_i to X_j if the preference over the value X_j is conditioned on the value of X_i. Each node $X_i \in V$ is associated with a conditional preference table $CPT(X_i)$ that associates a strict preference ($x_i > \neg x_i$ or $\neg x_i > x_i$) with each possible instantiation u_i of the parents of X_i (if any).*

A complete preference ordering satisfies a CP-net $\mathcal{N}$ iff it satisfies each conditional preference expressed in $\mathcal{N}$. In this case, the preference ordering is said to be *consistent* with $\mathcal{N}$. A CP-net induces a partial order $\succ_{\mathcal{N}}$ between interpretations that can be linked by a chain of worsening flips [3]. We denote by $Pa(X)$ the set of direct parent variables of X, and by $Ch(X)$ the set of direct successors (children) of X. The set of interpretations of a group of variables $S \subseteq V$ is denoted by $Ast(S)$, with $\Omega = Ast(V)$. Given a CP-net $\mathcal{N}$, for each node $X_i, i = 1, \ldots, n$, each entry in a conditional preference table CPT_i is of the form $\phi = u_i : \star x_i > \star\neg x_i$, where $u_i \in Ast(Pa(X_i))$, $\star$ is blank if the preference is $x_i > \neg x_i$ and is $\neg$ otherwise.

Example 1 *We use an example inspired from [4].* Figure 1a *illustrates a CP-net about preferences over cars: one prefers minivans* (m) *to sedans* (s) *(s_1). For minivans, one prefers Chrysler* (c) *to Ford* (f)*(s_2), and the converse for sedans (s_3). Finally, for Ford, one prefers black* (b) *cars to white* (w) *ones (s_4), while for Chrysler cars it is the converse (s_5). The CP-net Ordering is given in* Fig. 1b, *and* Table 1 *gives a schema instance of $Car(category, make, color)$.*

In possibilistic logic, these CP-net preference statements are encoded by constraints of the form $N(\neg u_i \vee \star x_i) \geq \alpha > 0$, where N is a necessity measure [5]. It is equivalent here to a constraint on a conditional necessity measure $N(\star x_i|u_i) \geq \alpha$, hence to $\Pi(\neg\star x_i|u_i) \leq 1-\alpha < 1$, where $\Pi(p) = 1-N(\neg p)$ is the dual possibility measure associated with N. It expresses that having $\neg\star x$ in context u_i is somewhat not satisfactory, as the possibility of $\neg\star x$ is upper bounded by $1-\alpha$, i.e., satisfying $\neg\star x$ is all the more impossible as α is large.

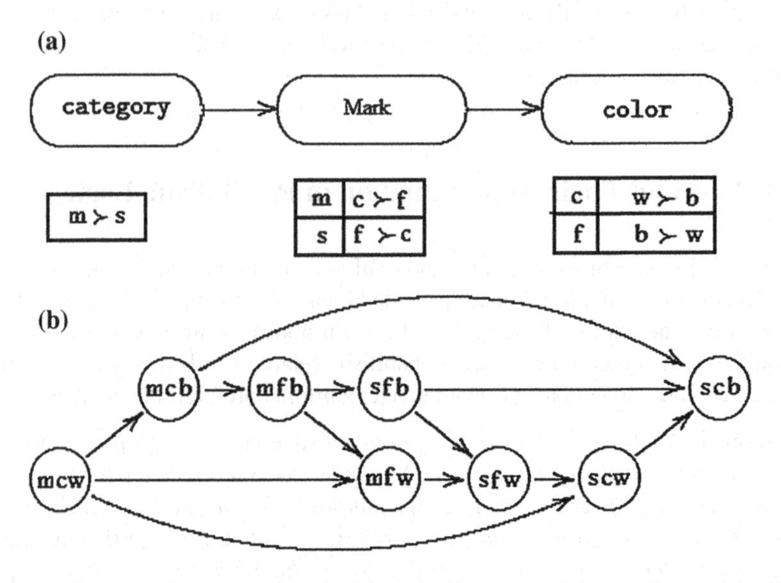

Fig. 1 The CP-net and its preference tables for the car Example, and the resulting partial order

Table 1 The different car instances and the user's preferences

Interpretations	Category	Make	Color		
mcw	Minivan	Chrystler	White	s_1	The user prefers minivan (m) cars to sedan (s) cars
mcb	Minivan	Chrystler	Black	s_2	For minivans, he prefers Chrystler (c) to Ford (f)
mfw	Minivan	Ford	White	s_3	For sedans, he prefers Ford to Chrystler
mfb	Minivan	Ford	Black	s_4	Among Ford cars, he prefers the black (b) ones to the white (w) ones
scw	Sedan	Chrystler	White	s_5	Among Chrystler cars, he prefers the white ones to the black ones
scb	Sedan	Chrystler	Black		
sfw	Sedan	Ford	White		
sfb	Sedan	Ford	Black		

A partially ordered possibilistic base $(\Sigma, \succeq_\Sigma)$ is built from a CP-net in the following way:

- Each entry of the form $u_i : \star x_i > \star\neg x_i$ in the table CPT_i of each node $X_i, i = 1, \ldots, n$ is encoded by the possibilistic logic clause $(\neg u_i \vee \star x_i, \alpha_i)$, where $\alpha_i > 0$ is a symbolic weight (whose value is unspecified). This formula is put in Σ.
- Since the same weight is attached to each clause built from CPT_i, the set of weighted clauses induced from CPT_i is thus equivalent to the weighted conjunction $\phi_i = (\bigwedge_{u_i \in Ast(Pa(X_i))}(\neg u_i \vee \star x_i), \alpha_i)$, one per variable (and thus one per node in the CP-net).
- Additional constraints over symbolic weights are added. Each weight α_i attached to a node X_i, is supposed to be strictly smaller than the weight of each of its parents. This contributes to define a partial order between symbolic weights, which is denoted $\succeq_\Sigma$.

For each interpretation ω, we associate a vector $\boldsymbol{\omega}(\Sigma)$ is obtained as follows. For each weighted formula ϕ_i in the possibilistic base Σ satisfied by ω, we put 1 in the ith component of the vector, and $1 - \alpha_i$ otherwise, in agreement with possibilistic logic semantics [6]. Thus, by construction, we use the set of levels $L = \{1, 1 - \alpha_i | i = 1, \ldots, n\}$, with $1 \succeq_\Sigma 1 - \alpha_i, \forall i$. The vector $\boldsymbol{\omega}(\Sigma)$, associated with interpretation ω, has a specific format. Namely its ith component (one per CP-net node) lies in $\{1, 1 - \alpha_i\}$ for $i = 1, \ldots, n$.

It has been shown that CP-nets can be approximated from above in the possibilistic logic setting by using the *leximin* order between the vectors associated with the interpretations. In the case of a totally ordered scale, the leximin order is defined by first reordering the vectors in an increasing way and then applying the lexicographic order to the reordered vectors. Since we deal with a partial order, the reordering of vectors is no longer unique, and the definition is generalized by:

Definition 2 **(leximin)** *Let* $\mathbf{v}$ *and* $\mathbf{v}'$ *be two vectors having the same number of components. First, delete all pairs* (v_i, v'_j) *such that* $v_i = v'_j$ *in* $\mathbf{v}$ *and* $\mathbf{v}'$ *(each deleted component can be used only one time in the deletion process). Thus, we get two non overlapping sets* $r(\mathbf{v})$ *and* $r(\mathbf{v}')$ *of remaining components, namely* $r(\mathbf{v}) \cap r(\mathbf{v}') = \emptyset$. *Then,* $\mathbf{v} \succ_{leximin} \mathbf{v}'$ *iff* $\min(r(\mathbf{v}) \cup r(\mathbf{v}')) \subseteq r(\mathbf{v}')$ *(where* min *here returns the set of minimal elements of the partial order between the priority weights).*

Then we have the following result (see [8] for the proof):

Proposition 1 *Let* $\mathcal{N}$ *be an acyclic CP-net. Let* $(\Sigma, \succeq_\Sigma)$ *be its associated partially ordered base. Then:* $\forall \omega, \omega' \in \Omega, \omega \succ_{\mathcal{N}} \omega' \Rightarrow \boldsymbol{\omega}(\Sigma) \succ_{leximin} \boldsymbol{\omega}'(\Sigma)$.

Besides, it has been also conjectured [8, 9] that CP-nets can be approximated from below in the same possibilistic logic setting by using the *symmetric Pareto* order, denoted by $\succ_{SP}$, between the vectors. Namely we have

$$\forall \omega, \omega' \in \Omega, \omega \succ_{SP} \omega' \Rightarrow \omega \succ_{\mathcal{N}} \omega'$$

where $\mathcal{N}$ is an acyclic CP-net and $\succ_{SP}$ the partial order defined as follows:

Definition 3 **(symmetric Pareto)** *Let* $\mathbf{v}$ *and* $\mathbf{v}'$ *be two vectors having the same number of components,* $\mathbf{v} \succ_{SP} \mathbf{v}'$ *if and only if there exists a permutation* σ *the components of* $\mathbf{v}'$*, yielding vector* $\mathbf{v}'^{\sigma}$*, such that* $\mathbf{v} \succ_{Pareto} \mathbf{v}'^{\sigma}$ *(where as usual,* $\mathbf{v} \succ_{Pareto} \mathbf{v}'$ *if and only if* $\forall i, v_i \geq v'_i$ *and* $\exists j, v_j > v'_j$*).*

We have also shown that for a particular class of CP-nets where each node has at most one child node, the possibilistic logic approach using the *symmetric Pareto* order exactly captures the CP-net ordering (see [8] for the proof):

Proposition 2 *Let* $\mathcal{N}$ *be an acyclic CP-net where every node has at most one child node. Let* $(\Sigma, \succeq_\Sigma)$ *be its associated partially ordered base. Let* $\succ_{SP}$ *be the symmetric Pareto partial order associated to* $(\Sigma, \succeq_\Sigma)$*. Then,*

$$\forall \omega, \omega' \in \Omega, \omega(\Sigma) \succ_{SP} \omega'(\Sigma) \text{ iff } \omega \succ_{\mathcal{N}} \omega'.$$

In the general case, there are arguments to conjecture that

$$\omega(\Sigma) \succ_{SP} \omega'(\Sigma) \text{ implies } \omega \succ_{\mathcal{N}} \omega'.$$

See [8] for a partial proof.

4 Comparative Discussion About Preference Representation

CP-nets and the possibilistic logic approach handle preferences in their own ways. Even if the above results only provide upper and lower approximations of a CP-net, the possibilistic logic encoding of each preference node in the CP-net offers a basis for discussing the orderings obtained in both settings. If we consider Example 1, we notice that mfw and sfb are incomparable for the CP-net, but if we look at the preference constraints violated by each one, we notice that:

- mfw violates the preference of a child node and grandchild node.
- sfb violates the preference of a father node.

We know that in CP-nets, when comparing two choices, where one violates the preference of a father node and the other violates the preference of children nodes (irrespectively of the number of the violated children nodes), the second one is preferred to the first one (where the father node preference is violated). Moreover, we also know that the priority of a child node is more important than the one of the grandchild node, so it seems obvious that mfw should be preferred to sfb. Indeed, violating the preferences of one child node and one grandchild node should be less important than violating the preferences of two children nodes, which, in turn, is less important than violating the preference associated to a father node. But the CP-net does not acknowledge this matter of fact.

Based on this observation, we further investigate the differences between the CP-net and the possibilistic logic approaches, using the following new example:

Example 2 *Let us consider the two CP-nets $\mathcal{N}_a$ and $\mathcal{N}_b$* (see Fig. 2) over the four variables $\{X, Y, S, T\}$. Their possibilistic logic encoding are:

- $\Sigma_a = \{\phi_1, \phi_2, \phi_3, \phi_4\}$
  where $\phi_1 = (x, \alpha_1)$, $\phi_2 = ((\neg x \vee y) \wedge (x \vee \neg y), \alpha_2$
  $\phi_3 = ((\neg x \vee s) \wedge (x \vee \neg s), \alpha_3), \phi_4 = ((\neg x \vee t) \wedge (x \vee \neg t), \alpha_4)$;
- $\Sigma_b = \{\phi_1, \phi_2, \phi'_3, \phi'_4\}$
  with $\phi'_3 = ((\neg y \vee s) \wedge (y \vee \neg s), \alpha_3), \phi'_4 = ((\neg y \vee t) \wedge (y \vee \neg t), \alpha_4)$.

Let us now consider the three interpretations $\omega = x\bar{y}\bar{s}t$, $\omega' = x\bar{y}s\bar{t}$ and $\omega'' = \bar{x}\bar{y}\bar{s}\bar{t}$.

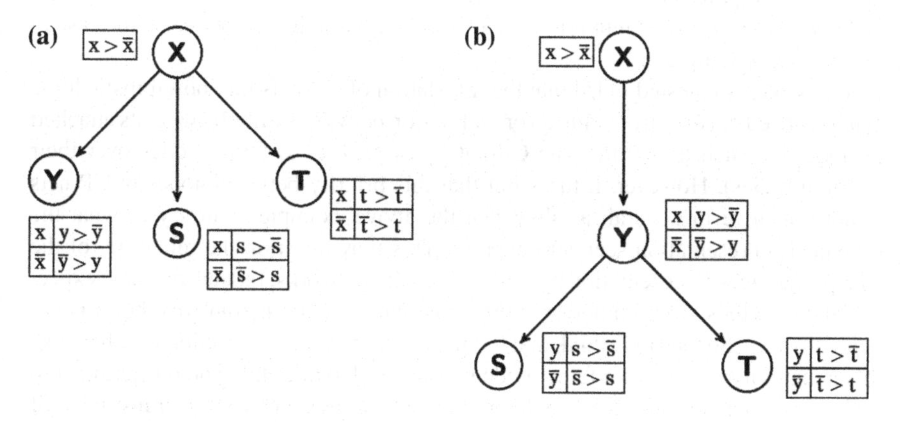

Fig. 2 The two CP-nets $\mathcal{N}_a$ and $\mathcal{N}_b$ of Example 2

Note that their associated vectors $\omega(\Sigma_a) = \omega'(\Sigma_b) = (1, 1-\alpha_2, 1-\alpha_3, 1)$, $\omega'(\Sigma_a) = \omega(\Sigma_b) = (1, 1-\alpha_2, 1, 1-\alpha_4)$ and $\omega''(\Sigma_a) = \omega''(\Sigma_b) = (1-\alpha_1, 1, 1, 1)$ are the same for both bases Σ_a and Σ_b (interchanging ω and ω').

These three interpretations not only have the same weight vectors in both cases, but also the same constraints on the symbolic weights are used when comparing the vectors, namely $\alpha_1 > \alpha_2$, $\alpha_1 > \alpha_3$ and $\alpha_1 > \alpha_4$. However, both CP-nets do not classify these interpretations in the same way:

- In CP-net $\mathcal{N}_a$: ω, ω' are preferred to ω'', but are incomparable.
- In CP-net $\mathcal{N}_b$: all three interpretations are incomparable.

But, in the possibilistic logic setting, since the three interpretations are associated with the same triples of vectors in both cases, we obtain the same partial order over vectors: The total incomparability using the *symmetric Pareto* order, which is in agreement with the CP-net $\mathcal{N}_b$; and if the *leximin* order is used, an identical comparability to the one of CP-net $\mathcal{N}_a$ is obtained, but not to that of the CP-net $\mathcal{N}_b$.

This situation is troublesome for CP-nets, since interpretations with the same vectors (the same patterns of preference violations), lead to different conclusions.

So, when comparing interpretations, CP-nets take into account some further information not directly expressed by a graphical structure (priority of father nodes over children nodes) nor preference tables alone, but which comes from their interaction via the ceteris paribus principle. Is it legitimate with respect to preference representation? What is this extra information?

In the CP-net $\mathcal{N}_b$, let us now consider the four interpretations $\omega_1 = xy\bar{s}\bar{t}$, $\omega_2 = x\bar{y}\bar{s}\bar{t}$, $\omega_3 = \bar{x}\bar{y}\bar{s}\bar{t}$ and $\omega_4 = x\bar{y}s\bar{t}$. The CP-net yields $\omega_1 \succ_{\mathcal{N}_b} \omega_2 \succ_{\mathcal{N}_b} \{\omega_3, \omega_4\}$ but leaves ω_3 and ω_4 incomparable. Thus, in the CP-net, it is true that violating preferences for two grandchildren nodes (in ω_1), is less important than violating those for one child node (in ω_2), which is less important than violating a father node (ω_3). Based on a transitivity argument, it is natural to conclude that violating preferences of one child and one grandchild node (ω_4), should be less than violating a father node preference (ω_3). However, the CP-net does not conclude in this case and leaves the two interpretations ω_3 and ω_4 incomparable, as it was also the case in our original Example 1.

It has been suggested in [5] that the translation of CP-nets into possibilistic logic can be carried out by introducing an order relation between symbolic weights attached to nodes (in such a way that the CP-net gives priority to father nodes over their children nodes). However, it turns out that this priority between nodes in CP-nets is not transitive in general, as shown in the above example, contrary to what the graphical representation seems to express. This may question the reliability of the user preferences representation by directed acyclic networks. Indeed, one may expect that the priorities between nodes in the graphical structure is transitive but it is not generally true. So, one may think that CP-nets contain extra information which blocks this transitivity in the case a father node is compared to one child and one grandchild node. However, the possibilistic logic framework presupposes the transitivity of priorities. Indeed, let us consider ϕ_1, ϕ_2 and ϕ_3' three possibilistic logic formulas encoding preferences in Σ_b. Let us denote "more important" by $\mapsto$; since we have

Table 2 Differences between key features of both approaches

	Possibilistic logic	CP-nets
Representation	A partially ordered set of logical formulas	A directed acyclic graph of variables
Principle	Possibilistic penalty for violations	Ceteris paribus
Dominance test	Vector comparison	Worsening flips
Expressive power	Can handle partial description of preferences	Do not allow a partial description of preferences
Transitivity of priorities	Yes	No

$\phi_1 \mapsto \phi_2$ and $\phi_2 \mapsto \phi'_3$, $\phi_1 \mapsto \phi'_3$ should be deduced, which expresses naturally how the user preferences are. So, if we consider the previous example, it is clear that in the possibilistic logic setting, it is better to violate the preferences of one child node and one grandchild node (ω_4) than to violate a preference of one parent node (ω_3), i.e., we have $\omega_1(\Sigma_b) \succ_{SP} \omega_2(\Sigma_b) \succ_{SP} \omega_4(\Sigma_b) \succ_{SP} \omega_3(\Sigma_b)$.

Recently, some papers have focused on the reconciliation of these two approaches [9], but there are some difficulties for an exact translation of one approach into the other, as shown here (see also [8]). Several questions remain, especially the one of determining the precise differences between these two settings, whether or not all information implicitly encoded in a CP-net can be expressed in a weighted propositional logic format, and which setting is best in agreement with user preferences.

The following table summarizes the comparison between the two approaches Table 2.

5 Conclusion

The possibilistic logic framework is able to deal with preference representation in a modular way, leaving the the user free to state relative priorities between goals. This representation takes advantage of the logical nature of the framework and constitutes an alternative to the introduction of a preference relation inside the representation language, as in, e.g., [2]. Moreover, we have shown that the weighted logic representation does not suffer from problems of lack of transitivity between priorities as it is the case in CP-nets, thanks to the use of an order relation between the symbolic weights. Still much remains to be done. First, the question of an exact representation of any CP-net remains open, even if the discrepancies between the two representation settings look more important than expected. Moreover, an attempt has been made recently [7] for representing more general CP-theories [10] in the possibilistic logic approach (by introducing further inequalities between symbolic weights in order to take into account the CP-theory idea that some preferences hold irrespective of the values of some variables), where the leximin order seems to provide an upper approximation. This remains to be confirmed and developed further. Comparing CP-nets with the possibilistic counterpart of Bayesian nets, namely possibilistic nets [1] would be also of interest, since these possibilistic nets translate into possibilistic logic bases (however in the case of completely ordered weights).

References

1. Benferhat, S., Dubois, D., Garcia, L., Prade, H.: On the transformation between possibilistic logic bases and possibilistic causal networks. Int. J. Approx. Reasoning pp. 135–173 (2002).
2. Bienvenu, M., Lang, J., Wilson, N.: From preference logics to preference languages, and back. In: F. Lin, U. Sattler, M. Truszczynski (eds.) Proc. 12th Inter. Conf. on Principles of Knowledge Represent. and Reasoning (KR'10), Toronto, May 9–13, pp. 414–424 (2010).
3. Boutilier, C., Brafman, R.I., Domshlak, C., Hoos, H.H., Poole, D.: CP-nets: A tool for representing and reasoning with conditional ceteris paribus preference statements. J. Artif. Intell. Res. (JAIR) pp. 135–191 (2004).
4. Brafman, R.I., Domshlak, C.: Database preference queries revisited. In: Technical Report TR2004-1934, Cornell University, Computing and Information Science (2004).
5. Dubois, D., Kaci, S., Prade, H.: Approximation of conditional preferences networks "CP-nets" in Possibilistic Logic. In: Proc. 15th IEEE Inter. Conf. on Fuzzy Systems (FUZZ-IEEE), Vancouver, July 16–21, p. (electronic medium). IEEE (2006).
6. Dubois, D., Prade, H.: Possibilistic logic: a retrospective and prospective view. Fuzzy Sets and Systems pp. 3–23 (2004).
7. Dubois, D., Prade, H., Touazi, F.: Handling partially ordered preferences in possibilistic logic - A survey discussion. Proc. ECAI'12 Workshop on Weighted Logics for AI pp. 91–98 (2012).
8. Dubois, D., Prade, H., Touazi, F.: Conditional preference nets and possibilistic logic. In: L.C. van der Gaag (ed.) Proc. 12th Europ. Conf. Symbolic and Quantitative Approaches to Reasoning with Uncertainty (ECSQARU'13), Utrecht, July 8–10, LNCS, vol. 7958, pp. 181–193. Springer (2013).
9. Kaci, S., Prade, H.: Relaxing ceteris paribus preferences with partially ordered priorities. In: Proc. 9th Eur Conf on Symbolic and Quantitative Approaches to Reasoning with Uncertainty (ECSQARU'07) Hammamet, Oct. 30-Nov. 2, pp. 660–671 (2007).
10. Wilson, N.: Computational techniques for a simple theory of conditional preferences. Artificial Intelligence pp. 1053–1091 (2011).

Using Structural Similarity for Effective Retrieval of Knowledge from Class Diagrams

Markus Wolf, Miltos Petridis and Jixin Ma

Abstract Due to the proliferation of object-oriented software development, UML software designs are ubiquitous. The creation of software designs already enjoys wide software support through CASE (Computer-Aided Software Engineering) tools. However, there has been limited application of computer reasoning to software designs in other areas. Yet there is expert knowledge embedded in software design artefacts which could be useful if it were successfully retrieved. While the semantic tags are an important aspect of a class diagram, the approach formulated here uses only structural information. It is shown that by applying case-based reasoning and graph matching to measure similarity between class diagrams it is possible to identify properties of an implementation not encoded within the actual diagram, such as the domain, programming language, quality and implementation cost. The practical applicability of this research is demonstrated in the area of cost estimation.

1 Introduction

There is no "exact science" in dictating how functional and non-functional requirements of a software application are translated into a software design. Software design is an experience-driven process and the structure of a software artefact stems from a developer's or systems architect's experience, personal preference, understanding of the business requirements, target technologies or may even be reused in part from

M. Wolf (✉) · J. Ma
University of Greenwich, London, UK
e-mail: m.a.wolf@greenwich.ac.uk

J. Ma
e-mail: j.ma@greenwich.ac.uk

M. Petridis
University of Brighton, Brighton, UK
e-mail: m.petridis@brighton.ac.uk

M. Bramer and M. Petridis (eds.), *Research and Development in Intelligent Systems XXX*, DOI: 10.1007/978-3-319-02621-3_13,

existing designs. Giving the same non-trivial requirements to a number of developers or architects, would result, without doubt, in largely different software designs.

For a person comparing software designs depicting applications with similar functionality, a prime indicator for identifying similarity would be the semantic relationship between the different elements. The problem with automating this comparison is that the names chosen for elements composing a software design could be in different languages, abbreviations, words joined together or be meaningful only to the designer. This research therefore ignores the semantic aspect of software artefacts and focuses entirely on their structure.

It is not possible to determine the functionality of an application or system strictly based on its design structure. If each composing element is stripped of its semantic description, it can represent anything and is therefore stripped of much of its meaning. However, there is meaning in how composing elements are related to one another and how they are structured internally. Comparing software designs merely from a structural perspective, within a context, can provide valuable information. This work presents an approach to extract meaningful knowledge from class diagrams in the absence of semantic information using an automated retrieval process.

There are contexts within which the structural similarity of software designs is more relevant than the information provided by semantic comparison. One such example is cost estimation of software development. Cost estimation of software is considered more difficult than in other industries, as it generally involves creating new products [3]. Different approaches exist to software cost estimation; some are based on formal models, such as parametric or size-based models, while others rely on expert estimation. The approach presented here uses case-based reasoning by providing a collection of software designs and corresponding cost it took to implement each design, and then predict the effort necessary to implement a new target design based on the most similar designs in the case base. A software design is defined by its composing elements, which in the case of a UML class diagram are its classes and interfaces. A class/interface is further defined by its internal members, which could be operations, constructors and/or attributes. Beyond the composing elements, an essential aspect of defining a software design is the relationships between the elements. By reducing structurally complex data, as present in a class diagram, to a graph-based notation, it is possible to compare two class diagrams taking into account not just the sum of its composing elements, but also their relationships by applying a graph matching algorithm.

In this research, case-based reasoning and graph matching are combined in order to measure similarity between software designs so that this information can be applied to automated cost estimation and identification of application properties, such as implementation quality, programming language and functionality.

Section 2 of this paper presents the methodologies and techniques that were applied, as well as providing justification for their application. Section 3 outlines the experiments performed and an evaluation of the results obtained. Section 4 of this paper presents the background to this work and identifies related research. Finally, Sect. 5 summarises the key contributions and discusses how this work could evolve.

2 CBR Approach for Measuring Similarity of UML Class Diagrams

The approach taken in this research achieves retrieval of expert knowledge based on the contextualisation of software designs. It is by relating designs to each other that one can extract knowledge. Viewing a design in the context of others makes it necessary to compare them.

Case-Based Reasoning was chosen as the paradigm for extracting knowledge from the class diagrams as it is a method which works well with complex domains and when no algorithmic method or model is available for evaluation.

2.1 Complex Structural Similarity

In principle, when cases share the same structure (contain the same features), the process of measuring the similarity between two cases is straightforward. (1) Each feature in the source case is matched to the corresponding feature in the target case; (2) the degree of match is computed and (3) multiplied by a coefficient which represents the importance of the feature. (4) The results are added to obtain the overall match score. This process is expressed in the following formula, which shows the similarity between a target (C_t) case and a source (C_s):

$$\sigma\left(C_t, C_s\right) = \frac{\sum_{i=1}^{n} \omega_i * \sigma\left(f_i^t, f_i^s\right)}{\sum_{i=1}^{n} \omega_i} \quad (1)$$

where f_i^t is value for the feature and ω_i is the weight (importance) attributed to the feature. In this formula, the result is normalised.

Due to the fact that the cases are represented as a complex hierarchical structure, measuring the similarity between two cases becomes much more complicated.

Class diagrams are used in object-oriented software design to depict the classes and interfaces of a software design and the way these relate to one another. Apart from the classes and their relationships, class diagrams often show additional information, such as certain class-specific properties, attributes and operations, which in turn have associated modifiers. Every feature of a class diagram can be seen as being fully contained in another. For instance, a class diagram contains a class, the class contains an operation, the operation contains a parameter and the parameter contains a data type. This concept of containment makes it possible to regard a class diagram as a hierarchical tree-like structure, as can be seen in Fig. 1.

As a UML class diagram consists of classes which are connected using relationships, they can easily be represented as graphs of nodes (classes) and arcs (relationships). To measure similarity between class diagrams, represented as graphs, requires graph matching. A full search graph matching algorithm has been adapted to be applied to UML class diagrams. Given the graph representations of two UML

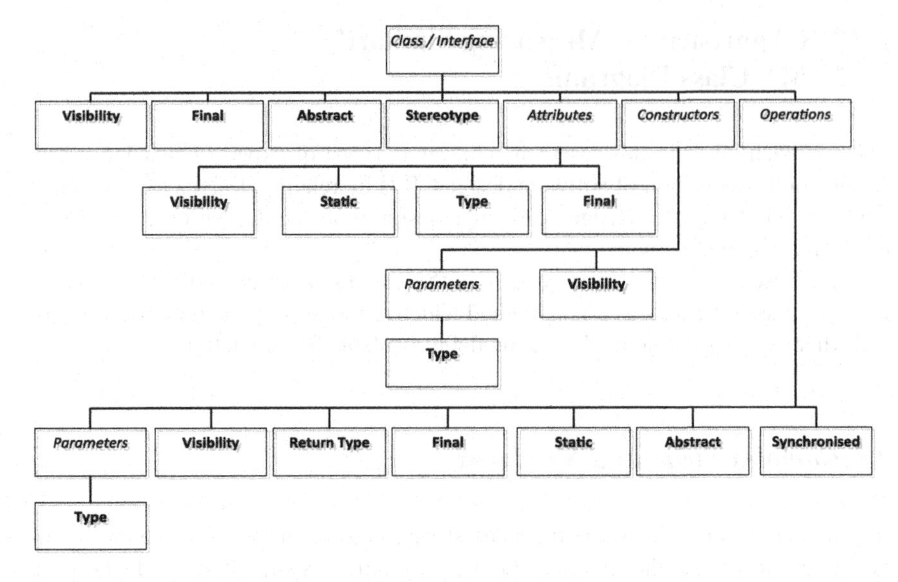

Fig. 1 Hierarchical structure of composing elements of a class

class diagrams, CD_t and CD_s, the algorithm returns the Maximum Common Subgraph ($MCS(CD)_t, CD_s$) present in both graphs.

The graph matching algorithm maps the maximum common connected subgraph based on matches between individual elements of each graph, where a minimum threshold similarity value is satisfied.

The overall similarity between the two case graphs (G_t, G_s) is then defined as:

$$\sigma(G_t, G_s) = \frac{\left(\sum_{\substack{matches\\ C_t, C_s\\ in\\ MCS}} \sigma(C_t, C_s)\right)^2}{count(G_t).count(G_s)} \tag{2}$$

where count (G_t) represents the number of nodes (classes) in graph G_t.

The algorithm used in this research is based on the full recursive search of all elements in the graph representations where both target and source diagrams are being compared. It finds a bijective match to maximize similarity between class diagrams with different number of classes, methods, arguments, etc.

2.2 *Weight Optimisation and Genetic Algorithms*

Many features are compared in order to establish the overall similarity between two classes. It is quite possible that different features contribute to the overall similarity in uneven shares. To account for uneven contribution, a weighting scheme is used to assign weights proportionally. This way, the weights can be adapted individually to reflect the importance of the feature. The features of a class are arranged in a hierarchical structure, which is reflected by the weights, creating different levels of weights.

An important question is to what extent the optimisation of weights can be generalised. Can a weight setting obtained from a particular set of designs be applied to other cases?

According to [11] it is not possible to use a weight optimisation method that would obtain an optimal weight setting which could be used for all tasks, since each task requires a different bias for optimal performance. A possible solution to this would be the creation of several different profiles according to the specific characteristics of different designs.

A key issue in calculating the class similarity effectively is to identify what the weight setting should be in order to successfully match up the correct classes. There is no established norm or convention for measuring similarity between classes. This is an abstract activity and even stating whether a match is good or bad is not always straightforward. A human expert could identify similarities between given designs using a heuristic approach, thus the key lies in classifying the characteristics that would make a human expert identify the similarity and adjust the weights for those features accordingly. However, this could vary from person to person or even diagram to diagram. It is also possible that a human expert would measure some features intuitively without being able to express them in rules.

In order to solve this problem this research employs weight optimisation. This process can be automated by training a system to automatically identify optimal weight settings, given the desired matching results. An expert assigns values from a predefined scale to class matches (similar to a grading scheme).

Given a set of desired/undesired matches, the system can apply different weight settings, run comparisons between all the different classes and keep a score of the points obtained from matching the classes against one another. An analysis of the scores from these comparisons makes it possible to adopt the best weight settings. This weight optimisation algorithm is outlined in Fig. 2.

Rather than randomly changing weights and hoping to chance upon a good setting, it would be more efficient to try to improve a setting by incrementally manipulating the values, verifying the results and making adjustments accordingly. This was achieved in this approach by means of a genetic algorithm.

Running the genetic algorithms to obtain the highest score did not identify a weight setting which would create higher scores than those obtained when using the default weights. While the different weight settings would result in different class similarity values, the granularity of the differences was not enough to modify the maximum common subgraph and thereby alter the score.

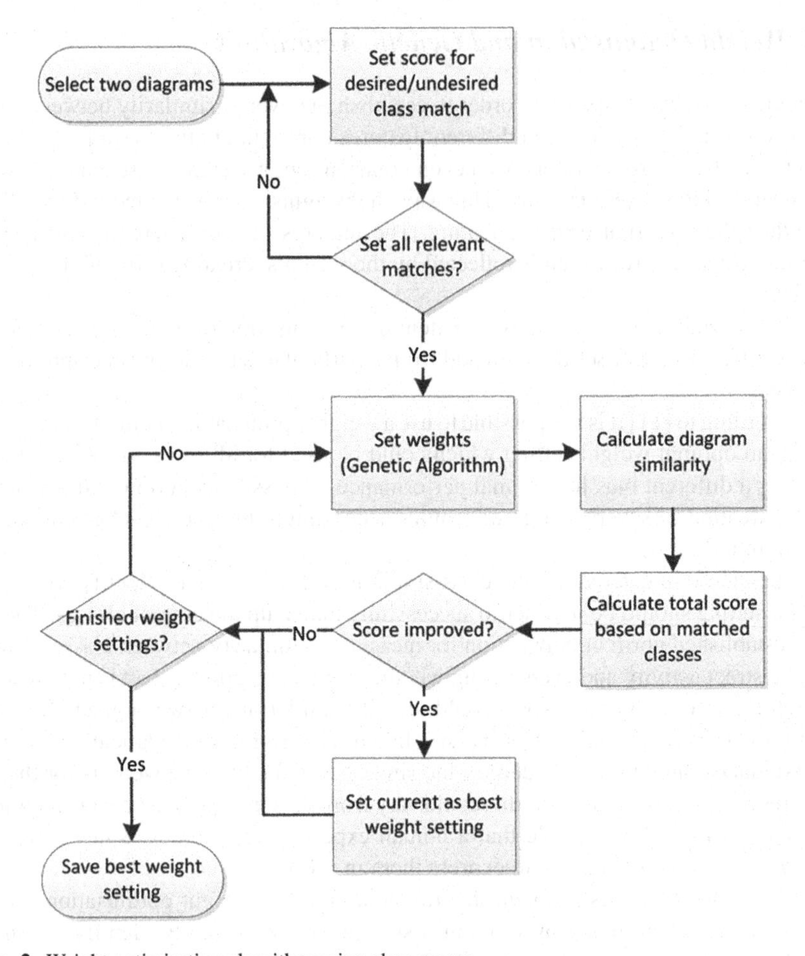

Fig. 2 Weight optimisation algorithm using class scores

A slightly different approach to this weight optimisation problem was adopted which asks the expert to identify the most similar diagrams from a set (see Fig. 3). The system applies different weight settings and adopts the setting which yielded the highest result.

Having generated five different profiles of weights, each one was applied in turn to measure similarity between all class diagrams in the case-base. In the first set of experiments, each weight setting was used on the entire training set of cases. So while a weight profile was optimised for one pair of class diagrams, it would be applied across the board. The reasoning behind this experiment was to determine whether weight profiles could be generalised, i.e. whether an optimised weight profile obtained from a particular experiment would yield good results from across the case-base.

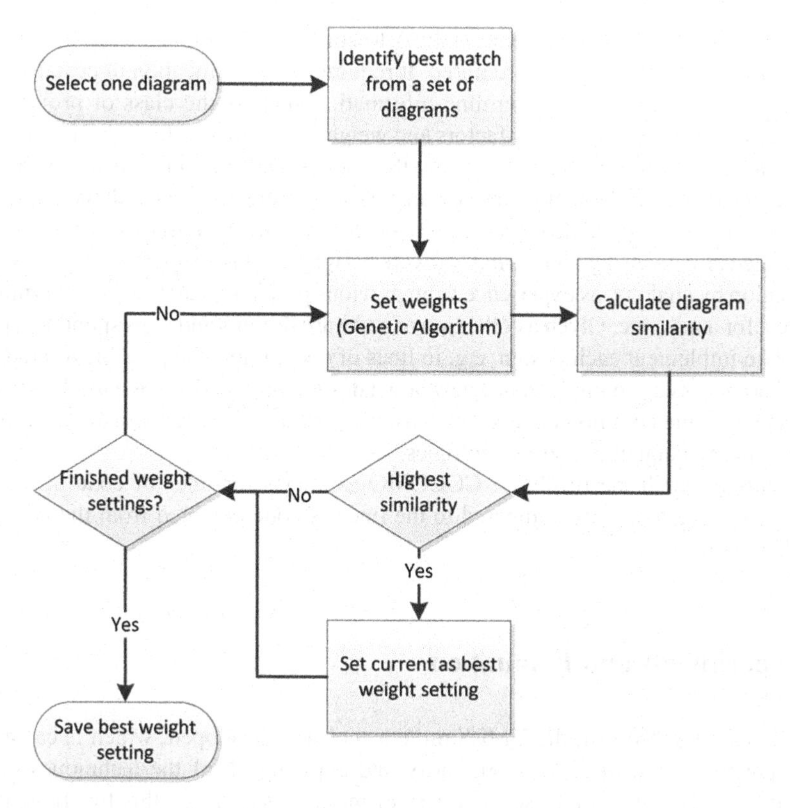

Fig. 3 Weight optimisation algorithm using diagram matching

2.3 Cost Estimation

Software development projects have a notorious reputation for being late and over budget, yet much of the cost estimation for projects is done based on experience of project managers and estimates done on the "back of an envelope". The estimation process also needs to account for numerous factors, such as system size, complexity, experience of the development team, etc. There is also a tendency to create very optimistic estimates, as this makes it more probable for a project to be approved or attract funding.

As with any type of estimation, cost estimation works most efficiently when backed by historical and statistical information. Thus, a variety of different approaches exist to software cost estimation, which take into account historical and statistical information, or which use models which were created using historical data. Some are based on formal models, such as parametric or size-based models, while others rely on expert estimation. The most common approaches are formal estimation models, such as COCOMO (COnstructive COst MOdel), established by Boehm [2], which apply formulas derived from historical data.

COCOMO uses function point analysis to compute program size, development time and the number of people required. It requires the classification of cost drivers using a scale of ratings to determine information such as the class of project. It also requires setting complexity factors and weight assignment, all of which must be performed by a human expert. This means that the estimation is very much dependent on the judgement of the expert and remains prone to human errors and biases [15].

A major motivation of this research is to determine whether cost estimation can be automated effectively by comparing software designs. This is an approach which uses estimation by analogy, as experience from previous projects is analysed to determine the cost for a new one. Given a collection of software designs and corresponding cost it took to implement each design, e.g. in lines of code, it may be possible to predict the effort necessary to implement a new target design, provided that the implementation shares some common context, such as, complexity of the functionality, similar programming languages and technologies.

In this research the results of COCOMO estimates for lines of code of a target software system were compared to the lines of code obtained from the nearest neighbours retrieved.

3 Experiments and Evaluation

To evaluate the research, the UMLSimilator tool was developed, which is capable of importing cases, measuring similarity and applying all of the techniques presented here. The tool comprises a variety of modules to support this functionality. An overview of the architecture of the tool and its main modules has been outlined in Fig. 4.

As the purpose of the application is to retrieve knowledge from structural information of software design artefacts, the content of the case-base consists of representations of UML class diagrams. To ensure that there is enough data and variety within the case-base, just over one hundred cases were obtained and stored. Diversity was introduced by using results from a number of different teaching assignments, thus assuring a good assortment of cases. Another positive consequence of using a set of teaching assignments is the fact that it provides diverse implementations of identical requirements, making it possible to validate results and evaluate provenance. Finally, this approach provided a measure of quality in form of the grade achieved.

The case-base used for the experiments consists of 101 class diagrams obtained from five different coursework assignments. All of the assignments were implemented using an object-oriented language (.NET or Java). Using different assignments means that the cases come from five different domains—Software Cataloguing, Project Management, Car Repair Shop, Stock Management and Project Bidding. The largest class diagram contains 40 classes, the smallest 3 classes and the average number of classes per class diagram across the entire case-base is 10.75.

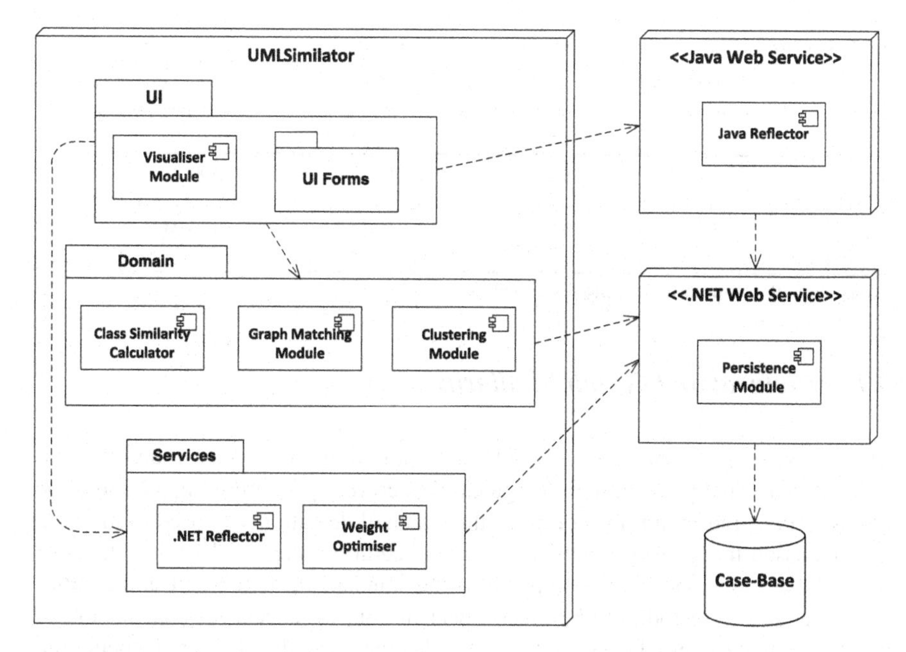

Fig. 4 UMLSimilator architecture

Given that comparing class diagrams is an abstract process, this poses the question of how the competence of the results could be measured. As additional information was known about the class diagrams, this data could be used to measure whether the system identifies information such as the domain from which a case was sourced, the programming language it was built in, the grade it was awarded or the cost of the implementation.

The majority of class diagram comparisons executed very fast, with the average being 1.54 s. However, more complex diagrams could take a very long time to compute. Performing a run of all cases using a particular weight setting took several hours. The entire set of 70 runs of experiments required over 184 h of processing time to compute, the majority of which was taken up by the graph similarity calculations.

Extensive tests were carried out which used the structural information encoded in class diagrams, but ignored the semantic details. A suite of experiments was devised using classes only (without graphs), graphs by calculating the maximum common subgraph and with optimised weights.

Leave-one-out cross-validation was used to validate the results and given that the case-base is not very large, it was decided to repeat this process for a considerable amount of cases. Thus, 50 % of the cases were randomly selected from the case-base and used as the training data, one at a time.

Table 1 Comparison of structural and graph similarity results

	% Matching domain	% Matching programming language	Standard deviation of grade	Standard deviation of lines of code
Structural similarity	74.16	85.40	11.57	509.87
Graph similarity[a]	85.51	93.16	10.84	470.94

[a] Minimum class similarity threshold set to 60 %

3.1 Structural and Graph Similarity

The first sets of experiments used only structural similarity, which found that the target domain and programming language was correctly identified in 74 and 85 % of cases. For grades and lines of code, the standard deviation was used to measure how focussed the nearest neighbours were. In addition, the standard deviations of the nearest neighbours were compared to the standard deviation across the entire case-base and were found to be much lower in both instances. An analysis of the number of nearest neighbours used revealed that the lower the number, the better the results.[1]

The impact of choosing cases from the same domain as the target case was verified (provenance). It was found that overall, this improved results, but only very marginally.

Graph similarity was measured using minimum thresholds of 20, 40, 60 and 80 %, with the best overall results achieved using 60 %. The results improved in every category over structural similarity (see Table 1).

Scoring of class matches did not work, so an alternative approach was taken which required an expert to identify the closest match for a given diagram from the same domain. The genetic algorithm would then run experiments with different weights and select the one giving the highest results. The weight profiles obtained using this technique, were first applied across the entire case-base, where they yielded poorer results than the default weights. However, when each weight profile was applied only to cases from its domain, the results improved in over 63 % of cases.

3.2 Cost Estimation

Five class diagrams were randomly selected (one from each domain). The estimated lines of code for implementing each diagram were calculated using COCOMO. The COCOMO estimates were compared against the actual lines of code it took to

[1] Tests were carried out with one, three and five nearest neighbours.

implement each of the class diagrams and to the lines of code obtained using the nearest neighbour retrieval applying only structural similarity (Fig. 5).

The structural similarity estimates performed better than COCOMO for two of the diagrams, but COCOMO was more accurate for the remaining three.

The next set of experiments compared COCOMO estimates to those achieved using graph similarity. The graph similarity improved the results achieved using only class structure similarity. The estimates were more accurate than COCOMO for three diagrams (Fig. 6).

The final set of experiments applied provenance, by choosing only the nearest neighbours from the same case domain (Fig. 7).

The results were similar to those achieved only with graph similarity, but using provenance improved the results further and COCOMO performed better in only one instance. The total difference of estimated lines of code and the actual lines of code was also reduced to 3766 which is much lower than the 9238 lines of code difference with COCOMO.

In an industrial setting, the provenance approach could be used to discriminate between development teams, companies, technologies used in a software development project or even classification by types of applications.

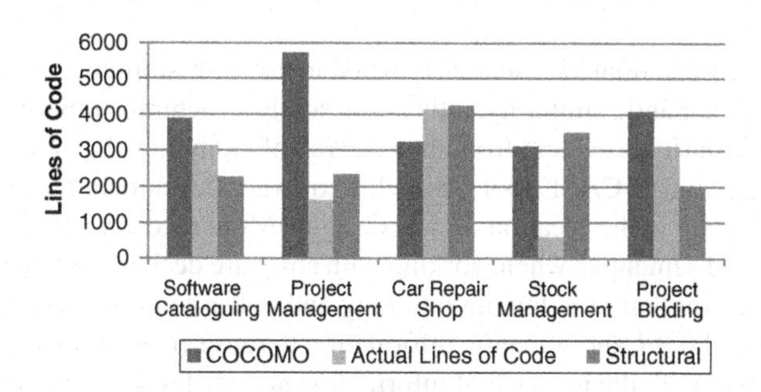

Fig. 5 Comparison of actual lines of code, COCOMO and structural similarity estimates

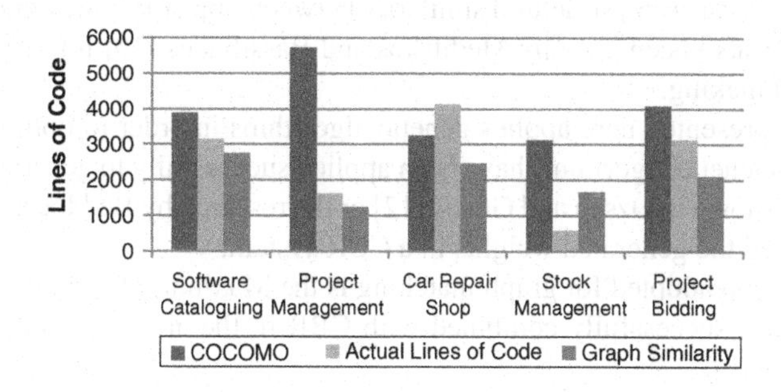

Fig. 6 Comparison of actual lines of code, COCOMO and graph similarity estimates

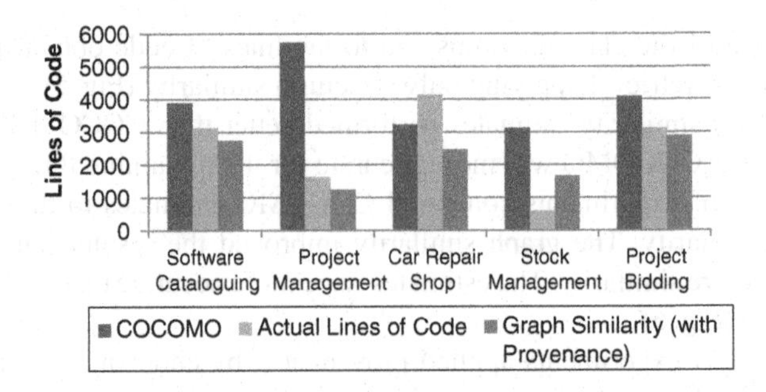

Fig. 7 Comparison of actual lines of code, COCOMO and graph similarity estimates using provenance

4 Related Work

The work presented in this paper continues the research outlined in a previous publication [16]. A number of areas are related to this research, namely, case-based reasoning, structural and graph similarity, retrieval of software knowledge and cost estimation.

The most-closely related research is work measuring similarity between class diagrams using semantic similarity, rather than solely structural similarity. The merit of semantic similarity is evident from the research of Gomes et al. [6], who achieved good results creating a CASE tool that helps software developers reuse previously developed software in the creation of new designs. More recently, Robles et al. [14] have also applied semantics when working with software designs, but their approach is based on the creation of a domain ontology and doesn't apply CBR. Grabert and Bridge combined CBR and semantic information to reuse code snippets (examplets) [7], which works with the retrieval of information at code level as opposed to design level, as is applied in the research presented here.

The idea of measuring structural similarity between object-oriented models using data types has also been used by Meditskos and Bassiliades [10] for semantic web service matchmaking.

The work presented here applies genetic algorithms in order to optimise coefficient values. Genetic algorithms have been applied successfully to generate attribute weights as is shown in Özşen and Güneş [12] or the research by Beddoe and Petrovic [1], which used the generated weights in a CBR system.

The algorithm adopted for graph matching is the Maximum Common Subgraph, which has been successfully combined with CBR in the area of metal casting by Petridis et al. [13].

Much of the research for software cost estimation using case-based reasoning is based on the Desharnais data set [4] with cases containing, among other, data such as project details, project length, team experience, programming language and entities [8, 9]. None of that research, however, takes into consideration the software design artefacts. Given that a software design artefact depicts the intended structure of the software product, it can contain information useful to the estimation of its development.

5 Conclusions and Future Work

The research experiments have provided results which demonstrate that it is possible to retrieve useful knowledge from class diagrams based only on their structural information. Some of the techniques and approaches employed have been more successful than others, but overall the results achieved have been positive, especially in the area of software cost estimation. A software application was developed to automate this process by implementing the proposed algorithms and procedures.

There a number of ways in which this research could evolve. The problem of finding the maximum common induced subgraph of two graphs is known to be NP-hard [5]. The implication of this is that the algorithm doesn't scale to graphs having large numbers of nodes. Experiments with this algorithm found that cases with low complexity (small number of classes and relationships) computed quite fast, but with an increase of the complexity of the graphs, the time required to calculate the maximum common subgraph would increase drastically. This means that the approach came close to reaching the limit of where it could be practicably applied and to increase the case-base or include more complex cases, the current algorithms would need to be optimized to increase performance.

It would also be valuable to field test the method for cost estimation presented in this work in a commercial setting and with a larger number of test cases.

Another context within which the structural similarity of software designs is more relevant than the information provided by semantic comparison is plagiarism detection. Within software development, a person committing plagiarism may intentionally change the semantic values of elements in order to avoid detection. By measuring similarity between software designs purely on a structural level, copying someone's code and changing the names or order of classes, operations and attributes, would still be detected. A known case of plagiarism within the case-base made it possible to do a preliminary evaluation of the suitability of this approach to plagiarism detection. The results were very good, with the plagiarised cases being paired as the highest matches in 90 % of all 70 sets of experiments. Further experiments should be carried out with a larger number of cases.

References

1. Beddoe, G. Petrovic, S. (2006) Determining feature weights using a genetic algorithm in a case-based reasoning approach to personnel rostering. European Journal of Operational Research, Vol. 175, Issue 2, pp. 649–671.
2. Boehm, B. Abts, C. Brown, A. W. Chulani, S. Clark, B. K. Horowitz, E. Madachy, R. Reifer, D. J. Steece, B. (2000) Software Cost Estimation with COCOMO II, Englewood Cliffs, NJ:Prentice-Hall.
3. Briand, L. Wieczorek, I (2002) Resource Estimation in Software Engineering, Encyclopedia of Software Engineering, J. J. Marcinak. New York, John Wiley & Sons: 1160–1196.
4. Desharnais, J. M. (1989) Analyse statistique de la productivitie des projets informatique a partie de la technique des point des fonction. University of Montreal.
5. Garey, M. R. Johnson, D. S. (1987) Computers and Intractability: A Guide to the Theory of NP-Completeness, Freeman.
6. Gomes, P. Gandola, P. Cordeiro, J. (2007) Helping Software Engineers Reusing UML Class Diagrams, in Proceedings of the 7th International Conference on Base-Based Reasoning (ICCBR'07) pp. 449–462, Springer, 2007.
7. Grabert, M. Bridge, D.G. (2003) Case-Based Reuse of Software Examplets, Journal of Universal Computer Science, Vol. 9, No. 7, pp. 627–641.
8. Huang, Z. (2009) Cost Estimation of Software Project Development by Using Case-Based Reasoning Technology with Clustering Index Mechanism. In Proceedings of the 2009 Fourth international Conference on innovative Computing, information and Control, ICICIC. IEEE Computer Society, pp. 1049–1052, Washington, DC.
9. Li, Y. F. Xie, M. Goh, T. N. (2009) A study of mutual information based feature selection for case based reasoning in software cost estimation. Expert Systems with Applications: An International Journal, Volume 36, Issue 3, pp. 5921–5931, Pergamon Press, Tarrytown, NY.
10. Meditskos, G. Bassiliades, N. (2007) Object-Oriented Similarity Measures for Semantic Web Service Matchmaking, in Proceedings 5th IEEE European Conference on Web Services.
11. Mitchell, T. M. (1990) The need for biases in learning generalizations, In Readings in machine learning, San Mateo, CA, Morgan Kaufmann.
12. Özşen, S. Güneş, S. (2009) Attribute weighting via genetic algorithms for attribute weighted artificial immune system (AWAIS) and its application to heart disease and liver disorders problems, Expert Systems with Applications, Vol. 36, Issue 1, pp. 386–392.
13. Petridis, M. Saeed, S. Knight, B. (2007) A Generalised Approach for Similarity Metrics Between 3D Shapes to Assist the Design of Metal Castings using an Automated Case Based Reasoning System, in Proceedings of the 12th UK CBR workshop, Peterhouse, December 2007, CMS press, pp. 19–29, UK.
14. Robles, K. Fraga, A. Morato, J. Llorens, J. (2012) Towards an ontology-based retrieval of UML Class Diagrams, Information and Software Technology, Vol. 54, Issue 1, January 2012, pp. 72–86, Elsevier.
15. Valerdi, R. (2007) Cognitive Limits of Software Cost Estimation. In Proceedings of the First international Symposium on Empirical Software Engineering and Measurement, Empirical Software Engineering and Measurement. IEEE Computer Society, pp. 117–125, Washington, DC.
16. Wolf, M. Petridis, M. (2008) Measuring Similarity of Software Designs using Graph Matching for CBR, in Proceedings of the Artificial Intelligence Techniques in Software Engineering Workshop, 18th European Conference on Artificial Intelligence.

Formulating the Temporal Causal Relationships Between Events and Their Results

J. Ma, M. Petridis and B. Knight

Abstract We introduce in this paper a formalism for representing flexible temporal causal relationships between events and their effects. A formal characterization of the so-called (most) General Temporal Constraint (GTC) is formulated, which guarantees the common-sense assertion that "the beginning of the effect cannot precede the beginning of its causal event". It is shown that there are actually in total 8 possible temporal causal relationships which satisfy the GTC. These include cases where, (1) the effect becomes true immediately after the end of the event and remains true for some time after the event; (2) the effect holds only over the same time over which the event is in progress; (3) the beginning of the effect coincides with the beginning of the event, and the effect ends before the event completes; (4) the beginning of the effect coincides with the beginning of the event, and the effect remains true for some time after the event; (5) the effect only holds over some time during the progress of the event; (6) the effect becomes true during the progress of the event and remains true until the event completes; (7) the effect becomes true during the progress of the event and remains true for some time after the event; and (8) where there is a time delay between the event and its effect. We shall demonstrate that the introduced formulation is versatile enough to subsume those existing representative formalisms in the literature.

J. Ma (✉) · B.Knight
University of Greenwich,London SE10 9LS, Uk
e-mail: j.ma@greenwich.ac.uk
e-mail: b.knight@greenwich.ac.uk

M. Petridis
University of Brighton,Brighton BN2 4GJ, Uk
e-mail: m.petridis@brighton.ac.uk

M. Bramer and M. Petridis (eds.), *Research and Development in Intelligent Systems XXX*, DOI: 10.1007/978-3-319-02621-3_14, © Springer International Publishing Switzerland 2013

1 Introduction

Representing and reasoning about causal relationships between events and their effects is essential in modeling the dynamic aspects of the world. Over the past half century, a multitude of alternative formalisms have been proposed in this area, including McCarthy and Hayes' framework of the situation calculus [1, 2], McDermott's temporal logic [3], Allen's interval based theory [4, 5], Kowalski and Sergot's event calculus [6], Shoham's point-based reified logic and theory [7, 8], and Terenziani and Torasso's theory of causation [9]. In particular, noticing that temporal reasoning plays an important role in reasoning about actions/events and change, a series of revised formalisms have been introduced to characterize richer temporal features in the situation calculus or the event calculus, such that of Lifschitz [10], of Sandewall [11], of Schubert [12], of Gelfond et al. [13], of Lin and Shoham [14], of Pinto and Reither [15, 16], of Miller and Shanahan [17, 18], and of Baral et al. [19–21].

In most existing formalisms for representing causal relationships between events and their effects, such as the situation calculus and the event calculus, the result of an event is represented by assuming that the effect takes place immediately after the occurrence of the event. However, as noted by Allen and Ferguson [22], temporal relationships between events and their effects can in fact be quite complicated. In some cases, the effects of an event take place immediately after the end of the event and remain true until some further events occur. E.g., in the block-world, as soon as the action "moving a block from the top of another block onto the table" is completed, the block being moved should be on the table (immediately). However, sometimes there may be a time delay between an event and its effect(s). E.g., 30 s after you press the button at the crosswalk, the pedestrian light turns to green [13]. Also, in some other cases, the effects of an event might start to hold while the event is in progress, and stop holding before or after the end of the event. Examples can be found later in the paper.

The objective of this paper is to introduce a formalism, which allows expression of versatile temporal causal relationships between events and their effects. As the temporal basis for the formalism, a general time theory taking both points and intervals as primitive on the same footing is presented in Sect. 2, which allows expression of both absolute time values and relative temporal relations. In Sect. 3, fluents and states are associated with times by temporal reification [7, 23]. Section 4 deals with representation of event/action and change, as well as temporal constraints on the causal relationships between events and their effects. Finally, Sect. 5 provides the summary and conclusions.

2 The Time Basis

In this paper, we shall adopt the general time theory proposed in [24] as the temporal basis, which takes both points and interval as primitive. On one hand, it provides a satisfactory representation of relative temporal knowledge, and hence retains the appealing characteristics of interval-based theory [4]. Specially, it can successfully by-pass puzzles like the so-called Dividing Instant Problem [4, 25–28]. On the other hand, it provides a means of making temporal reference to instantaneous phenomena with zero duration, such as "The database was updated at 00:00 a.m.", "The light was automatically switched on at 8:00 p.m.", and so on, which occur at time points rather than intervals (no matter how small they are).

The time theory consists of a nonempty set, *T*, of primitive time elements, with an *immediate predecessor* relation, Meets, over time elements, and a *duration assignment* function, Dur, from time elements to non-negative real numbers. If Dur(t) = 0, then t is called a point; otherwise, that is Dur(t) >0, t is called an interval. The basic set of axioms concerning the triad (*T*, Meets, Dur) is given as below [24]:

$$\text{(T1)}\ \forall t_1, t_2, t_3, t_4(\text{Meets}(t_1, t_2) \wedge \text{Meets}(t_1, t_3) \wedge \text{Meets}(t_4, t_2) \Rightarrow \text{Meets}(t_4, t_3))$$

That is, if a time element meets two other time elements, then any time element that meets one of these two must also meet the other. This axiom is actually based on the intuition that the "place" where two time elements meet is unique and closely associated with the time elements [29].

$$\text{(T2)}\ \forall t \exists t_1, t_2(\text{Meets}(t_1, t) \wedge \text{Meets}(t, t_2))$$

That is, each time element has at least one immediate predecessor, as well as at least one immediate successor.

$$\text{(T3)}\forall t_1, t_2, t_3, t_4(\text{Meets}(t_1, t_2) \wedge \text{Meets}(t_3, t_4) \Rightarrow \\ \text{Meets}(t_1, t_4) \nabla\ \exists t'(\text{Meets}(t_1, t') \wedge \text{Meets}(t', t_4)) \nabla \exists t''(\text{Meets}(t_3, t'') \wedge \text{Meets}(t'', t_2)))$$

where ∇ stands for "exclusive or". That is, any two meeting places are either identical or there is at least a time element standing between the two meeting places if they are not identical.

$$\text{(T4)}\ \forall t_1, t_2, t_3, t_4(\text{Meets}(t_3, t_1) \wedge \text{Meets}(t_1, t_4) \wedge \text{Meets}(t_3, t_2) \wedge \text{Meets}(t_2, t_4)) \Rightarrow t_1 = t_2)$$

That is, the time element between any two meeting places is unique.
N.B. For any two adjacent time elements, that is time elements t_1 and t_2 such that $\text{Meets}(t_1, t_2)$, we shall use $t_1 \oplus t_2$ to denote their ordered union. The existence of such an ordered union of any two adjacent time elements is guaranteed by axioms (T2) and (T3), while its uniqueness is guaranteed by axiom (T4).

$$\text{(T5)}\ \forall t_1, t_2(\text{Meets}(t_1, t_2) \Rightarrow \text{Dur}(t_1) > 0 \vee \text{Dur}(t_2) > 0)$$

That is, time elements with zero duration cannot meet each other.

$$(T6)\ \forall t_1, t_2(\text{Meets}(t_1, t_2) \Rightarrow \quad \text{Dur}(t_1 \oplus t_2) = \text{Dur}(t_1) + \text{Dur}(t_2))$$

That is, the "ordered union" operation, $\oplus$, over time elements is consistent with the conventional "addition" operation over the duration assignment function, i.e., Dur.

Analogous to the 13 relations introduced by Allen for intervals [Allen 1983, 1984], there are 30 exclusive temporal order relations over time elements including both time points and time intervals, which can be derived from the single Meets relation and classified into the following 4 groups:

- Order relations relating a point to a point:
  {Equal, Before, After}
- Order relations relating an interval to an interval:
  {Equal, Before, After, Meets, Met_by, Overlaps, Overlapped_by, Starts, Started_by, During, Contains, Finishes, Finished_by}
- Order relations relating a point to an interval:
  {Before, After, Meets, Met_by, Starts, During, Finishes}
- Order relations relating an interval to a point:
  {Before, After, Meets, Met_by, Started_by, Contains, Finished_by}

The definition of the derived temporal order relations in terms of the single relation Meets is straightforward [Ma and Knight 1994]. For instance,

$$\text{Before}(t_1, t_2) \Leftrightarrow \exists t \in \mathbf{T}(\text{Meets}(t_1, t) \wedge \text{Meets}(t, t_2))$$

For the convenience of expression, we define two non-exclusive temporal relations as below:

$$\text{In}(t_1, t_2) \Leftrightarrow \text{Starts}(t_1, t_2) \vee \text{During}(t_1, t_2) \vee \text{Finishes}(t_1, t_2)$$

$$\text{Sub}(t_1, t_2) \Leftrightarrow \text{Equal}(t_1, t_2) \vee \text{In}(t_1, t_2)$$

An important fact needs to be pointed out is that the distinction between the assertion that "point p Meets interval t" and the assertion that "point p Starts interval t" is critical: while Starts(p, t) states that point p is the starting part of interval t, Meets(p, t) states that point p is one of the immediate predecessors of interval t but p is not a part of t at all. In other words, Starts(p, t) implies interval t is left-closed at point p, and Meets(p, t) implies interval t is left-open at point p. Similarly, this applies to the distinction between the assertion that "interval t is Finished-by point p" and the assertion that "interval t is Met-by point p", i.e., Finished-by(t, p) implies interval t is right-closed at point p, and Met-by(t, p) implies interval t is right-open at point p.

3 Fluents and States

Representing the dynamic aspects of the world usually involves reasoning about various states of the world under consideration. In this paper, we shall define a state (denoted by, possibly scripted, s) of the world in the discourse as a collection of fluents (denoted by, possibly scripted, f), where a fluent is simply a Boolean valued proposition whose truth-value is dependent on the time.

The set of fluents, **F**, is defined as the minimal set closed under the following two rules:

$$\text{(F1)}\, f_1,\ f_2 \in \mathbf{F} \Rightarrow f_1 \vee f_2 \in \mathbf{F}$$

$$\text{(F2)}\, f \in \mathbf{F} \Rightarrow \text{not}(f) \in \mathbf{F}$$

In order to associate a fluent with a time element, we shall use Holds(f, t) to denote that fluent f holds true over time t.

As pointed out by Allen and Ferguson [22], as well as by Shoham [7], there are two ways we might interpret the negative sentence. In what follows, the sentence-negation will be symbolized by "¬", e.g., ¬Holds(t, f), distinguished from the negation of fluents, e.g., not(f) [Galton 1990]. In the weak interpretation, ¬Holds(t, f) is true if and only if it is not the case that f is true throughout t, and hence ¬Holds(t, f) is true if f changes truth-values over time t. In the strong interpretation of negation, ¬Holds(t, f) is true if and only if f holds false throughout t, so neither Holds(t, f) nor ¬Holds(t,f) would be true in the case that fluent f is true over some sub-interval of t and also false over some other sub-interval of t.

In this paper, we shall take the weak interpretation of negation as the basic construct:

$$\text{(F3)}\, \text{Holds}(f, t) \Rightarrow \forall t'(\text{Sub}(t', t) \Rightarrow \text{Holds}(f, t'))$$

That is, if fluent f holds true over time element t, then f Holds true over any part of t.

$$\text{(F4)}\, \text{Holds}(f_1, t) \vee \text{Holds}(f_2, t) \Rightarrow \text{Holds}(f_1 \vee f_2, t)$$

That is, the disjunction of fluent f_1 and fluent f_2 Holds true over time t if at least one of them holds true over time t.

$$\text{(F5)}\, \text{Holds}(f, t_1) \wedge \text{Holds}(f, t_2) \wedge \text{Meets}(t_1, t_2) \Rightarrow \text{Holds}(f, t_1 \oplus t_2)$$

That is, if fluent f Holds true over two time elements t_1 and t_2 that meets each other, then f holds over the ordered-union of t_1 and t_2.

Following the approach proposed in [18], we use Belongs(f, s) to denote that fluent f belongs to the collection of fluents representing state s:

$$\text{(F6)}\, s_1 = s_2 \Leftrightarrow \forall f(\text{Belongs}(f, s_1) \Leftrightarrow \text{Belongs}(f, s_2))$$

That is, two states are equal if and only if they contain the same fluents.

$$(F7)\ \exists s \forall f(\neg Belongs(f, s))$$

That is, there exists a state that is an empty set.

$$(F8)\ \forall s_1 f_1 \exists s_2 (\forall f_2 (Belongs(f_2, s_2) \Leftrightarrow Belongs(f_2, s_1) \vee f_1 = f_2))$$

That is, any fluent can be added to an existing state to form a new state.
Without confusion, we shall also use Holds(s, t) to denote that state s holds true over time t, provided:

$$(F9)\ Holds(s, t) \Leftrightarrow \forall f(Belongs(f, s) \Rightarrow Holds(f, t))$$

4 Causal Relationships Between Events and Their Results

The concepts of change and time are deeply related since changes are caused by events occurring over the time. In order to express the occurrence of events (denoted by e, possibly scripted), following Allen's approach [5], we use Occurs(e, t) to denote that event e occurs over time t, and impose the following axiom:

$$(C1)\ Occurs(e, t) \Rightarrow \forall t'(In(t', t) \Rightarrow \neg Occurs(e, t'))$$

We shall use formula $Changes(t_1, t, t_2, s_1, e, s_2)$ to denote a causal law, which intuitively states that, under the precondition that state s_1 holds over time t_1, the occurrence of event e over time t will change the world from state s_1 into state s_2, which holds over time t_2. Formally, we impose the following axiom about causality to ensure that if the precondition of a causal law holds and the event happens, then the effect expected to be caused must appear:

$$(C2)\ Changes(t_1, t, t_2, s_1, e, s_2) \wedge Holds(s_1, t_1) \wedge Occurs(e, t) \Rightarrow Holds(s_2, t_2)$$

In order to characterize temporal relationships between events and their effects, we impose the following temporal constraints:

$$(C3)\ Changes(t_1, t, t_2, s_1, e, s_2) \Rightarrow Meets(t_1, t) \wedge (Meets(t_1, t_2) \vee Before(t_1, t_2))$$

It is important to note that axiom (C3) presented above actually specifies the so-called (most) General Temporal Constraint (GTC) (see [3, 4, 8, 9]). Such a GTC guarantees the common-sense assertion that "the beginning of the effect cannot precede the beginning of the cause".

There are actually in total 8 possible temporal order relations between times t_1, t and t_2 which satisfy (C3), which can be illustrated in Fig. 1 as below:

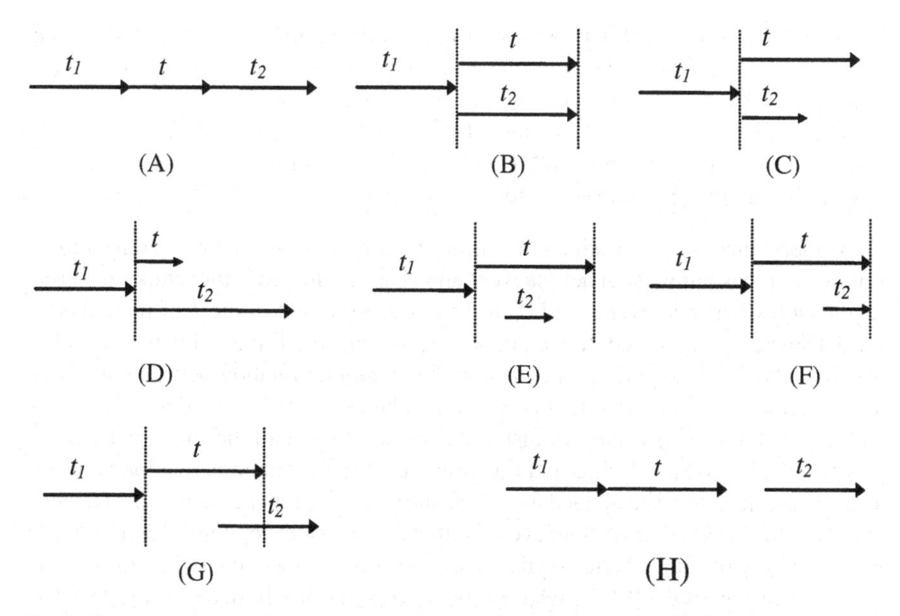

Fig. 1 Temporal order relations between events and effects satisfying the GTC

- Case (A) where the effect becomes true immediately after the end of the event and remains true for some time after the event. E.g., the event of putting a book on the table has the effect that the book is on the table immediately after the event is completed.
- Case (B) where the effect holds only over the same time over which the event is in progress. E.g., as the effect of the event of pressing the horn of a car, the horn makes sounds only when the horn is being pressed.
- Case (C) where the beginning of the effect coincides with the beginning of the event, and the effect ends before the event completes. E.g., consider the case where a man takes a course of antibiotics. He is getting better from the start of the course, but not after he is cured which happens before the end of the course.
- Case (D) where the beginning of the effect coincides with the beginning of the event, and the effect remains true for some time after the event. E.g., as the effect of the event of pressing the button of the bell on the door (say, for 1 s), the bell sounds a tune for 15 s.
- Case (E) where the effect only holds over some time during the progress of the event. E.g., a runner goes through a wall of tiredness over the 15 min of the event of running for 4 h.
- Case (F) where the effect becomes true during the progress of the event and remains true until the event completes. E.g., consider the event of discharging some water from a basket by means of lifting one side of the basket. In the case where the basket is not full, the effect that the water flows out takes place only after it has been lifted to the edge of the basket, and will keep flowing out until the event ends.

- Case (G) where the effect becomes true during the progress of the event and remains true for some time after the event. E.g., a runner becomes tired for days after the 13 min of the event of running along the track for 3 h.
- Case (H) where there is a time delay between the event and its effect. E.G., 25 s after the button at the crosswalk is pressed, the pedestrian light turns to yellow; and after another 5 s, it turns to green.

As mentioned in the introduction, various theories have been proposed for representing and reasoning about action/event and change. However, the temporal causal relationships between events and their effects as specified in most of the existing formalism are quite limited. In fact, in both McCarthy and Hayes' situation calculus and Kowalski and Sergot's event calculus, the temporal relation between an event and its effect is simply unique, that is "immediately before" (or "Meets" in Allen and Hayes' terminology [29]); whilst in the revised version of the situation calculus proposed by Gelfond, Lifschitz and Rabinov, etc. [13], the temporal relation between an event and its effect is extended to include the case of "proper before", that is, there may be a time delay between an event and its effect. An exception is the relatively general theory of Time, Action-Types, and Causation, introduced by Terenziani and Torasso's in the 1990s [9]. In what follows, we shall briefly demonstrate that the causal relationships characterized in this paper are more general than those of Terenziani and Torasso, and therefore versatile enough to subsume those representatives in the literature. In fact:

If t and t_2 are specified as: a point and a point, a point and an interval, an interval and a point, and an interval and an interval, respectively, by applying the temporal order relations as classified in Sect. 2, we can reach the following four theorems straightforwardly:

$$\begin{aligned}&(\text{Th1})\text{Changes}(t_1, t, t_2, s_1, e, s_2) \wedge \text{Dur}(t) = 0 \wedge \text{Dur}(t_2) = 0\\ &\quad \Rightarrow \text{Equal}(t, t_2) \vee \text{Before}(t, t_2)\end{aligned}$$

That is, if the event and the effect are both punctual, then either the event precedes (strictly) the effect, or they coincide with each other (i.e., they happen simultaneously at the same time point).

$$\begin{aligned}&(\text{Th2})\text{Changes}(t_1, t, t_2, s_1, e, s_2) \wedge \text{Dur}(t) = 0 \wedge \text{Dur}(t_2) > 0\\ &\quad \Rightarrow \text{Starts}(t, t_2) \vee \text{Meets}(t, t_2) \vee \text{Before}(t, t_2)\end{aligned}$$

That is, if the event is punctual and the effect is durative, then either the event precedes (immediately or strictly) the effect, or the event coincides with the beginning of the effect.

$$\begin{aligned}&(\text{Th3})\text{Changes}(t_1, t, t_2, s_1, e, s_2) \wedge \text{Dur}(t) > 0 \wedge \text{Dur}(t_2) = 0\\ &\quad \Rightarrow \text{Started} - \text{by}(t, t_2) \vee \text{Contains}(t, t_2)\\ &\quad \vee \text{Finished} - \text{by}(t, t_2) \vee \text{Meets}(t, t_2) \vee \text{Before}(t, t_2)\end{aligned}$$

That is, if the event is durative and the effect is punctual, then either the event precedes (immediately or strictly) the effect, or the effect coincides with either the beginning or the end of the event, or the effect happens during the event's occurrence.

$$\text{(Th4)}\mathrm{Changes}(t_1, t, t_2, s_1, e, s_2) \wedge \mathrm{Dur}(t) > 0 \wedge \mathrm{Dur}(t_2) > 0$$
$$\Rightarrow \mathrm{Started-by}(t, t_2) \vee \mathrm{Contains}(t, t_2) \vee \mathrm{Finished-by}(t, t_2) \vee \mathrm{Equal}(t, t_2)$$
$$\vee \mathrm{Starts}(t, t_2) \vee \mathrm{Overlaps}(t, t_2) \vee \mathrm{Meets}(t, t_2) \vee \mathrm{Before}(t, t_2)$$

That is, if both the event and the effect are durative, then the beginning of the event either precedes (immediately or strictly) or coincides with the beginning of the effect, where the end of the event can either precede (immediately or strictly), coincides with, or succeed (immediately or strictly) the end of the effect.

It is easy to see that (Th1) and (Th4) are equivalent to Terenziani and Torasso's Theorem 4 and Theorem 1, respectively while (Th2) and (Th3) can be seen as the extension to Terenziani and Torasso's Theorem 3 and Theorem 2 respectively This is due to the fact that, while the "Meets" relation between a punctual event and a durative effect, and between a durative event and a punctual effect, is accommodated in (Th2) and (Th3) respectively, Terenziani and Torasso's Theorem 3 and Theorem 2 do not allow such relations.

In fact, follow Terenziani and Torasso's Theorem 3, either there must be a gap between the punctual cause and its durative effect, or the punctual cause must coincide with the beginning part of its durative effect. In other words, the interval over which the effect happens must be either "After" or "closed at" the point at which the cause happens. Therefore, the case where a punctual cause "Meets" its durative effect (that is, the interval over which the effect happens is "open" at the point at which the cause happens) is not allowable. However, consider the following example:

Immediately after the power was switched on, a robot that had been stationary started moving.

If we use $s_{\mathrm{Stationary}}$ to represent the state that "the robot was stationary, e_{SwitchOn} to represent the event that "the power was switched on", and s_{Moving} to represent the corresponding effect that "the robot was moving", then

$$\mathrm{Changes}(t_{\mathrm{Stationary}}, t_{\mathrm{SwitchOn}}, t_{\mathrm{Moving}}, s_{\mathrm{Stationary}}, e_{\mathrm{SwitchOn}}, s_{\mathrm{Moving}})$$

should be consistent with:

$$\mathrm{Meets}(t_{\mathrm{SwitchOn}}, t_{\mathrm{Moving}})$$

That is, the "Switching" point t_{SwitchOn} is immediately followed by the "Moving" interval, but not included in the "Moving" interval itself. In other word, the robot was moving immediately after the "Switching" point t_{SwitchOn}, but at the time point when the power was switching on, the robot was not moving. Obviously, such a scenario cannot be expressed in Terenziani and Torasso's Theorem 3.

Similarly, in Terenziani and Torasso's Theorem 2, the case where a durative event "Meets" its punctual effect (that is, the interval over which the cause happens is "open" at the point at which the effect happens) is not allowable. Then again, consider the following example:

Immediately after the ball being falling down from the air, it touched the ground.

If we use s_{InAir} to represent the precondition that "the ball was at a certain position in the air", $e_{FallingDown}$ to represent the event that "the ball was falling down", and $s_{TouchGround}$ to represent effect that "the ball touched the ground", respectively, then

$$\text{Changes}(t_{InAir}, t_{FallingDown}, t_{TouchGround}, s_{InAir}, e_{FallingDown}, s_{TouchGround})$$

should be consistent with:

$$\text{Meets}(t_{FallingDown}, t_{TouchGround})$$

That is, the interval over which the ball was falling down is immediately followed by the point when the ball touched the ground, but does not include the point itself. In other word, the ball was falling down immediately before the instant when it touched the group, but at the time point when the ball touched the ground, the ball was no longer falling down. Again, such a scenario is not allowed in Terenziani and Torasso's Theorem 2.

5 Conclusions

By means of adopting a general time theory based on both points and intervals as primitive on the same footing, we have introduced in this paper a formalism for representing flexible temporal causal relationships between events and their effects. Such a formalism formally specifies the (most) General Temporal Constraint (GTC), ensuring the common-sense assertion that "the beginning of the effect cannot precede the beginning of the causal event". It is shown that the temporal causal relationships characterized here are versatile enough to subsume those representatives in the literature. An interesting topic for further research is to extend this framework to include representing and reasoning about concurrent actions/events and their effects. Ideally, any handy theory about action/event and change has to be able to handle the frame problem adequately. However, due to the length of this paper, we didn't tackle such a problem here.

References

1. McCarthy, J., Situations, actions and causal laws, *Stanford Artificial Intelligence Project: Memo 2*, 1963.

2. McCarthy, J., Hayes, P.: Some philosophical problems from the standpoint of artificial intelligence, in *Machine Intelligence*, 4, Eds. Meltzer B. and Michie D., Edinburgh University Press, pages 463–502, 1969.
3. McDermott, D.: A Temporal Logic for Reasoning about Processes and Plans, *Cognitive Science*, 6: 101–155, 1982.
4. Allen, J.: Maintaining Knowledge about Temporal Intervals, *Communication of ACM*, 26: 832–843, 1983.
5. Allen, J.: Towards a General Theory of Action and Time, *Artificial Intelligence*, 23: 123–154, 1984.
6. Kowalski, R., Sergot, M.: A Logic-based Calculus of Events, *New Generation Computing*, 4: 67–95, 1986.
7. Shoham, Y.: Temporal logics in AI: Semantical and Ontological Considerations, *Artificial Intelligence*, 33: 89–104, 1987.
8. Shoham, Y.: *Reasoning about Change: Time and Causation from the Standpoint of Artificial Intelligence*, MIT Press, 1988.
9. Terenziani, P., Torasso, P.: Time, Action-Types, and Causation: an Integrated Analysis, *Computational Intelligence*, 11(3): 529–552, 1995.
10. Lifschitz, V.: Formal theories of action, *In Proceedings of the Tenth International Joint Conference on, Artificial Intelligence*, pages 966–972, 1987.
11. Sandewall, E.: Filter preferential entailment for the logic of action in almost continuous worlds. In *Proceedings of the 12th International Joint Conference on, Artificial Intelligence*, pages 894–899, 1989].
12. Schubert, L.: Monotonic Solution of the Frame Problem in the Situation Calculus: an Efficient Method for Worlds with Fully Specified Actions, in: H.E. Kyburg, R. Loui and G. Carlson, eds., *Knowledge Representation and Defeasible Reasoning*, pages 23–67, Kluwer Academic Press, 1990.
13. Gelfond, M., Lifschitz, V., Rabinov, A.: What are the Limitations of the Situation Calculus? In Working Notes of AAAI Spring Symposium Series. Symposium: Logical Formalization of Commonsense Reasoning, pages 59–69, 1991.
14. Lin, F., Shoham, Y.: Concurrent Actions in the Situation Calculus, In *Proceedings of AAAI*-92, pages 590–595, 1992.
15. Pinto, J., Reiter, R.: Temporal Reasoning in Logic Programming: A Case for the Situation Calculus, In *Proceedings of 10th Int. Conf. on Logic Programming*, Budapest, Hungary, pages 21–24, 1993.
16. Pinto, J., Reiter, R.: Reasoning about Time in the Situation Calculus, *Annals of Mathematics and Artificial Intelligence*, 14(2–4): 251–268, 1995.
17. Miller, R., Shanahan, M.: Narratives in the Situation Calculus, *the Journal of Logic and Computation*, 4(5): 513–530, 1994.
18. Shanahan, M.: A Circumscriptive Calculus of Events, *Artificial Intelligence*, 77: 29–384, 1995.
19. Baral, C.: Reasoning about actions: non-deterministic effects, constraints, and qualification, In *Proceedings of IJCAI'95*, pages 2017–2023, 1995.
20. Baral, C., Gelfond, M.: Reasoning about Effects of Concurrent Actions, *Journal of Logic Programming*, 31(1–3): 85–117, 1997.
21. Baral, C., Son, T., Tuan, L.: A transition function based characterization of actions with delayed and continuous effects, In *Proceedings of KR'02*, pages 291–302, 2002.
22. Allen J., Ferguson, G.: Actions and Events in Interval Temporal Logic, *the Journal of Logic and Computation*, 4(5): 531–579, 1994.
23. Ma, J., Knight, B.: A Reified Temporal Logic, *the Computer Journal*, 39(9): 800–807, 1996.
24. Ma, J., Knight, B.: A General Temporal Theory, *the Computer Journal*, 37(2): 114–123, 1994.
25. van Benthem, J.: *The Logic of Time, Kluwer Academic*, Dordrech, 1983.
26. Galton, A.: A Critical Examination of Allen's Theory of Action and Time, *Artificial Intelligence*, 42: 159–188, 1990.
27. Vila, L.: A survey on temporal Reasoning in Artificial Intelligence. *AI Communications*, 7: 4–28, 1994.

28. Ma, J., Hayes, P.: Primitive Intervals Vs Point-Based Intervals: Rivals Or Allies? *the Computer Journal*, 49: 32–41, 2006.
29. Allen J., Hayes, P.: Moments and Points in an Interval-based Temporal-based Logic, *Computational Intelligence*, 5: 225–238, 1989.

Machine Learning and Constraint Programming

The Importance of Topology Evolution in NeuroEvolution: A Case Study Using Cartesian Genetic Programming of Artificial Neural Networks

Andrew James Turner and Julian Francis Miller

Abstract NeuroEvolution (NE) is the application of evolutionary algorithms to Artificial Neural Networks (ANN). This paper reports on an investigation into the relative importance of weight evolution and topology evolution when training ANN using NE. This investigation used the NE technique Cartesian Genetic Programming of Artificial Neural Networks (CGPANN). The results presented show that the choice of topology has a dramatic impact on the effectiveness of NE when only evolving weights; an issue not faced when manipulating both weights and topology. This paper also presents the surprising result that topology evolution alone is far more effective when training ANN than weight evolution alone. This is a significant result as many methods which train ANN manipulate only weights.

1 Introduction

NeuroEvolution (NE) is the application of evolutionary algorithms to artificial neural networks (ANNs). NE has many advantages over traditional gradient based training methods [19]. NE does not require the neuron transfer functions to be differentiable in order to find a fitness gradient; a process which can be computationally expensive. NE is resilient to becoming trapped in local optima. Gradient based methods are restricted to certain topologies and struggle to train deep ANNs [3, 6]; whereas NE does not. NE also does not require a target behaviour to be known in advance, which is required to produce the error value for gradient methods. This allows NE to be applied to open ended problems.

A. J. Turner (✉) · J. F. Miller
The University of York, York, United Kingdom
e-mail: at568@york.ac.uk

J. F. Miller
e-mail: julian.miller@york.ac.uk

M. Bramer and M. Petridis (eds.), *Research and Development in Intelligent Systems XXX*,
DOI: 10.1007/978-3-319-02621-3_15,

NE methods can be split into two groups [19], those which adjust only the connection weights of a user defined network topology and those which adjust both the connection weights and topology. There are many possible advantages to evolving topology as well as weights, including: access to topologies which would otherwise not be considered, searching in a topology space as well as a weight space and not requiring the user to choose a topology before it is know which topologies would be suitable.

Although it is thought that topology and weight evolution offers a significant advantage over weight evolution alone [2, 19], to the authors knowledge there are no publications which truly compare the two approaches. It is true that comparisons can be made between different NE methods which do and do not evolve topologies, but differences in implementation are likely to influence the result, weakening any comparison. It is also not known whether weight evolution is more beneficial than topology evolution, or vice versa, or if they are more effective when used together.

This paper investigates the relative importance of only connection weight evolution, only topology evolution, and both connection weight and topology evolution in NE. This is achieved using the NE method Cartesian Genetic Programming of Artificial Neural Networks (CGPANN) [5, 15] which is the application of the genetic programming technique Cartesian Genetic Programming (CGP) [8, 10] to ANNs. CGPANN is a topology and weight evolving NE method which can easily be adapted to evolve only connection weights or only topology. This makes CGPANN very suitable for investigating the relative importance of topology and weight evolution. CGPANN can also be seeded with random or user-defined topologies meaning that the influence of different topologies for weight only evolution can also be studied.

The remainder of the paper is structured as follows: Sect. 2 discusses NE in terms of weight and topology evolution, Sect. 3 describes CGP and its application to ANN, Sect. 4 describes the experiments presented in this paper with the results given in Sect. 5 and finally Sects. 6 and 7 give an evaluation of the results and closing conclusions.

2 NeuroEvolution - Weights and Topologies

In the NE literature it is thought that topologies, as well as weights, are highly important in the training of ANNs [2, 19]. Theoretically an ANN can be thought of in terms of a topology and weight search space; or as weight spaces associated with any given topology. Using only fixed topologies therefore limits the search to one subset weight space within the topology space; a possible disadvantage of methods which only manipulate weights such as traditional back propagation. Clearly, if a suitable topology is known in advance it could be used to decrease the dimensionality of the search. However, often this is not known in advance. Interestingly, the strategy of just increasing or decreasing the number of neurons during the search, in order to navigate the topology landscape, could be thought of as structural hill climbing [1].

However, this approach is likely to get trapped in topology local optima just as back propagation does with weights.

In the growing number of NE techniques found in the literature, some methods manipulate only weights [4, 11] and others manipulate both weights and topology [1, 5, 12, 13]. It is difficult however to draw empirical conclusions about the benefit of evolving topology as each of these techniques use different encodings to describe the ANN during evolution. When, for example, a weight and topology evolving NE technique outperforms a weight only technique, we may assume that the increase in performance was due to its ability to evolve topologies but it could equally be due to the differences in implementation (or both). For instance, the weight only evolving method Symbiotic Adaptive Neuro-Evolution (SANE) [11] evolves ANNs at a neuron level, with the complete networks assembled using neurons selected from the population. In contrast, the weight and topology evolving method NeuroEvolution of Augmenting Topologies (NEAT) [13] evolves ANNs at a network level and employs strategies to track when ancestral changes took place in order to more effectively make use of the crossover operator. So when it is shown that NEAT out performs SANE on a given benchmark [13] it is not clear if this is due to the ability to evolve topologies or due to other differences between the two methods, or both. Additionally, to the authors knowledge there are no NE techniques which solely rely on the evolution of topology with no alterations to the weights. For these reasons a direct comparison between the importance of weight evolution and topology evolution has not been able to be made.

3 Cartesian Genetic Programming Artificial Neural Networks

CGPANN [5] is a highly competitive [15] NE strategy which uses CGP [8, 10] to evolve both the weights and topology of ANNs. CGPANN chromosomes are usually initialised as random networks, but can be seeded with any starting topology. CGPANN can also be configured to only evolve weights, only evolve topology or evolve weights and topology. For these reasons CGPANN is a suitable tool for empirically investigating the effect topology has when evolving weights and the relative importance of evolving weight and topology for NE strategies.

3.1 CGP

CGP [8, 10] is a form of Genetic Programming which represents computational structures as directed, usually acyclic graphs indexed by their Cartesian coordinates. Each node may take its inputs from any previous node or program input. The program outputs are taken from the output of any internal node or program input. This structure leads to many of the nodes described by the CGP chromosome not contributing to the final operation of the phenotype, these inactive, or "junk", nodes have been shown to greatly aid the evolutionary search [7, 17, 20]. The existence of inactive genes in CGP are also useful in that they suppress a phenomenon in GP known as bloat. This is the uncontrolled growth in size of evolved computational structures over evolutionary time [9].

The nodes described by CGP chromosomes are arranged in a rectangular $r \times c$ grid of nodes, where r and c respectively denote the user-defined number of rows and columns. In CGP, nodes in the same column are not allowed to be connected together (as in multi-layer perceptrons). CGP also has a connectivity parameter l called "levels-back" which determines whether a node in a particular column can connect to a node in a previous column. For instance if $l = 1$ all nodes in a column can only connect to nodes in the previous column. Note that levels-back only restricts the connectivity of nodes; it does not restrict whether nodes can be connected to program inputs (terminals). It is important to note that any architecture (limited by the number of nodes) can be constructed by arranging the nodes in a $1 \times n$ format where the n represents the maximum number of nodes (columns) and choosing $l = n$. Using this representation the user does not need to specify the topology, which is then automatically evolved along with the program.

Figure 1 gives the general form of a CGP showing that a CGP chromosome can describe multiple input multiple output (MIMO) programs with a range of node transfer functions and arities. In the chromosome string, also given in Fig. 1, F_i denote the function operation at each node, C_i index where the node gathers its inputs and each O_i denote which nodes provide the outputs of the program. It should be noted that CGP is not limited to only one data type, it may be used for Boolean values, floats, images, audio files, videos etc. CGP generally uses the evolutionary strategy (ES) algorithm $(1 + \lambda) - ES$. In this algorithm in each generation there are $1 + \lambda$ candidates and the fittest is chosen as the parent. The next generation is formed by this parent and λ offspring obtained through mutation of the parent. It is important to note that if no offspring are fitter than the parent, but at least one has the same fitness as the parent, then the offspring is chosen as the new parent. In

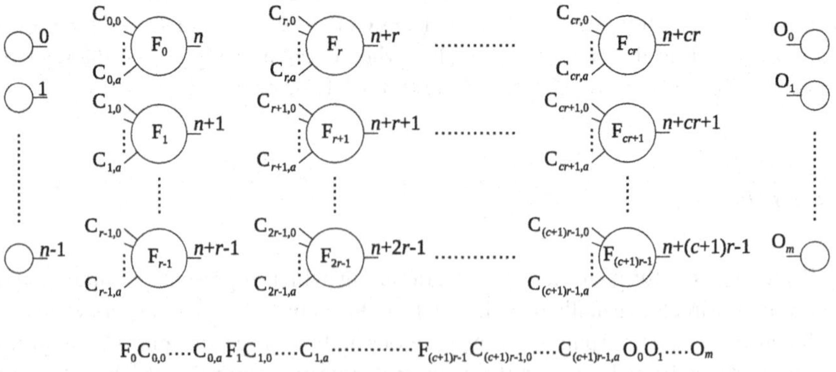

Fig. 1 Depiction of a cartesian genetic programs structure with chromosome encoding below, taken from [10]

CGP, the λ value is commonly set as four. The connection genes in the chromosomes are initialised with random values that obey the constraints imposed by the three CGP structural parameters, r, c, l. The function genes are randomly chosen from the allowed values in the function lookup table. The output genes O_i are randomly initialised to refer to any node or input in the graph. The standard mutation operator used in CGP works by randomly choosing a valid allele at a randomly chosen gene location. The reason why both a simple operator and a simple evolutionary algorithm are so effective is related to the presence of non-coding genes. Simple mutations can connect or disconnect whole sub-programs. For a more detailed description of CGP see [10].

3.2 CGP Encoded Neural Networks

ANNs are a natural application for CGP as both ANNs and CGPs are structured as directed acyclic[1] graphs. CGP is directly compatible with ANNs by using node transfer functions suited to ANNs (radial bias, sigmoidal etc) and encoding extra chromosome values for the weights assigned to each nodes/neurons input. These extra weight values are evolved with the rest of the CGP chromosome and are subject to mutation operations like other genes. The range of values that these weight genes can take is specified by the user; for example $[-10, 10]$.

In order to independently study the effect of weight evolution and topology evolution, the mutation percentage used by the CGPANN was split into two values; weight mutation percentage and topology mutation percentage. The weight mutation percentage is the percentage of weight genes which are changed in the chromosome and the topology mutation percentage is the percentage of connection genes and output genes which are changed. Weight only evolution can then be studied by setting the topology mutation percentage to zero and vice versa. If CGPANN is used with the topology mutation percentage set to zero, the CGPANN method becomes equivalent to a simple genetic algorithm manipulating the weights of a fixed topology. The reason for using CGPANN in this way is so that the same code could be used for all experiments; it also an example of how flexible the CGPANN approach is to different requirements.

In CGPANNs it is possible for there to be multiple connections between the same two nodes [15]. This is because when a connection gene is mutated, no effort is made to ensure that multiple connections between nodes do not occur; although this could easily be implemented if desired. In contrast, the approach taken in this paper is to allow only the first of any connections betweens two nodes into the phenotype. This is done for two reasons. Firstly, it gives CGPANN the ability to indirectly evolve each nodes arity. For instance, if a node had arity six but three inputs were from one node and three from another, then the node effectively has arity two as the majority of the connections would not be decoded into the phenotype. If a mutation operation then

[1] Both CGP and ANNs can also be structured in a recurrent form.

changed one of these inputs to a new different node, then the effective arity would increase to three; it changes during evolution. Secondly, if multiple connections between the same two nodes were encoded into the phenotype, then the maximum weight range set by the user could be exceeded. This results from two connections between the same two nodes being equivalent to one connection with the sum of their weights. It is unknown if this offers an advantage during evolution or not. However as this paper is concerned with a comparison with weight only evolving methods which cannot exceed the user defined maximum weight range, it is required that this ability is removed for a fair comparison.

Since this paper investigates the effect of evolving topology compared to weight only evolution, the implementation of CGPANN used here is restricted so that it only evolves acyclic networks with the node transfer functions fixed. In its unrestricted form CGPANN can evolve weights, topology, node arity and node transfer functions for both recurrent and non-recurrent ANNs. CGPANN also has all of the advantages of regular CGP, including: natural resilience to bloat [9] and redundancy in the chromosomes aiding evolution [7].

4 Experiments

Two experiments are presented in this paper. The first investigates the effect of topology when only evolving connection weights. The second investigates the relative importance of connection weight evolution and topology evolution. Both of these experiments are carried out using CGPANN. The evolutionary algorithm is $(1 + 4) - ES$. Probabilistic mutation was used where each gene is altered with a given probability. The weights between nodes are limited to $[-10, 10]$. When evolving topologies the CGP levels back parameter is set so that each node can connect to any previous node or the inputs, the outputs can also be taken from any node or input. All of the experiments were repeated fifty times in order to produce averages which are presented in this paper.

4.1 Effect of Topology on Weight Evolution

The first experiment investigates the effect of topology when using weight evolving NE. This is achieved by comparing the results from a range of topologies when only evolving connection weights. This experiment was designed to highlight the impact of the chosen topology on the effectiveness of the search. The results of this experiment are compared with an experiment in which both topology and connection weights evolve (i.e. using CGPANN in its regular form).

When using fixed topologies the initial chromosomes are seeded with an ANN that is fully connected[2]; each with a given number of hidden layers and nodes per layer.

When evolving both topology and weights, the number of hidden layers and nodes per layer are also varied, but here they represent the upper limits of the network topology; the maximum number of rows and columns in the CGPANN chromosome, see Sect. 3. In this case, the CGPANN chromosomes are not seeded with any predefined topology but initialised with randomly created networks. When using fixed topologies the arity of each node is determined by the number of nodes in the previous layer. However, when evolving the topology, the user must specify a node arity. For this experiment the node arity used was ten. Each node can however effectively lower its arity by connecting to the same previous node multiple times; as the implementation of CGPANN used here only decodes the first connection between the same two nodes into the phenotype. All cases were run for five thousand generations (20001 evaluations) and the final fitnesses reported. The connection weight and topology mutation rates were five and zero percent respectively for the fixed topology case and both five percent when evolving weights and topology.

4.2 Weight Evolution Versus Topology Evolution

The second experiment investigates the relative importance of weight evolution and topology evolution when using NE. This is investigated by comparing NE performance in three situations: (a) only evolving weights, (b) only evolving topology and (c) evolving both weights and topology. In all experiments initial populations were seeded with random weights and topologies. We achieved these comparisons by comparing the results obtained using the following three methods:

1. Fixing the topology mutation rate as zero and varying the weight mutation rate.
2. Fixing the weight mutation rate as zero and varying the topology mutation rate.
3. Varying both the weight and topology mutation rate (both with the same value).

For the first method the randomly generated topologies remained fixed and only the connection weights mutated with a number of mutation rates. For the second method the randomly generated weights remained fixed and only the topology mutated with a number of mutation rates. For the third method both the randomly generated weights and topologies were mutated with a number of mutation rates. All of these methods were run for one thousand generations (4001 evaluations). The maximum number of layers/rows was set as thirty with one node-per-layer/column. These CGPANN row and column limits were used as they allow the highest possible number of topologies for the amount of nodes available; for instance all topologies possible when rows = 15 and columns = 2 are also possible when rows = 30 and

[2] Fully connected between layers i.e. a node in hidden layer two has an input from every node in hidden layer one.

columns = 1. This structure can not encode all topologies however, due to the arity limit set for each node.[3] Here the arity of each node was set as ten. Again only the first connection between the same two nodes was decoded into the phenotype.

4.3 Test Problems

In order to investigate the described experiments, benchmark problems are required. Such test problems should be typical of the types of problems ANN are often applied to and should be challenging enough to draw out any differences between weight and topology evolution. The following two test problems were used for the experiments presented in this paper; the first is a control problem and the second is a classification problem (two common applications for ANN).

4.3.1 Double Pole

The Double Pole benchmark [18] is a classic control problem in the artificial intelligence literature and is often used for comparing NE methods [15]. The task is to balance two poles hinged, at their base, on a controllable cart. All movements are limited to one dimension. The cart is placed on a 4.8 m track, with the cart always starting in the centre. The longer of the two poles starts at 1 deg from vertical and the shorter starts at vertical. The cart is controlled by applying a force in the range [−10, 10] N thus moving the cart from side to side. See [18] for diagrams and further details.

The equations which govern the dynamics of the poles and cart are given in Eqs. 1 and 2 with the parameter definitions and limits given in Table 1. The simulations

Table 1 Parameters used for the Double Pole balancing benchmark

Symbol	Description	Values
x	Cart position	$[-2.4, 2.4]$ m
θ_i	ith pole angle	$[-36, 36]$ deg
F	Force applied to cart	$[-10, 10]$ N
$\tilde{F}_i$	Force on cart due to ith pole	-
l_i	Half length of ith pole	$l_1, l_2 = 0.5, 0.05$ m
M	Cart mass	1.0kg
m_i	Mass of ith pole	$m_1, m_2 = 0.1, 0.01$ kg
μ_c	Friction coefficient between cart and track	0.0005
μ_{pi}	Friction coefficient between ith pole and cart	0.000002

[3] If the arity is set high enough however all topologies are possible as each node can lower its own arity by only utilizing the first of multiple connections between two nodes.

are run using Euler integration with a time step of 0.01 s. The outputs of the ANN are updated every 0.02 s. The simulations are then run for 100,000 time steps using the ANN being tested as the controller. The assigned fitness is the number of time steps (out of a maximum 100,000) over which the ANN keeps both poles within [−36, 36] deg from vertical whilst keeping the cart within the track range.

The inputs to the ANN are cart position, cart velocity, both poles' angles from vertical and both poles' angular velocity. Each of these inputs are scaled to a [−1, 1] range by assuming the following ranges: cart position [−2.4, 2.4] m, cart velocity [−1.5, 1.5] m/s, pole positions [−36, 36] deg and pole velocities [−115, 115] deg/s.

The output from the ANN, also in the range [−1, 1], is also scaled to a force in the range [−10, 10] N. As an extra restraint the magnitude of the applied force is constrained to be always greater than $\frac{1}{256} \times 10$ N. This slightly increases the difficulty of the task. All the node transfer functions use the bipolar sigmoid which produces an output in the range [−1, 1]; mapping easily to [−10, 10].

$$\ddot{x} = \frac{F - \mu_c sgn(\dot{x}) + \sum_{i=1}^{N} \tilde{F}_i}{M + \sum_{i=1}^{N} \tilde{m}_i}, \qquad \ddot{\theta}_i = -\frac{3}{4l_i}\left(\ddot{x}\cos\theta_i + g\sin\theta_i + \frac{\mu_{pi}\dot{\theta}_i}{m_i l_i}\right) \tag{1}$$

$$\tilde{F}_i = m_i l_i \dot{\theta}_i^2 \sin\theta_i + \frac{3}{4} m_i \cos\theta_i \left(\frac{\mu_{pi}\dot{\theta}_i}{m_i l_i} + g\sin\theta_i\right), \qquad \tilde{m}_i = m_i\left(1 - \frac{3}{4}\cos^2\theta_i\right) \tag{2}$$

4.3.2 Monks Problems

The Monks Problems [14] are a set of classification benchmarks intended for comparing learning algorithms. The classification tasks are based on the appearance of robots which are described by six attributes each with a range of values: head_shape (round, square, octagon), body_shape (round, square, octagon), is_smiling (yes, no), holding (sword, balloon, flag), jacket_color (red, yellow green, blue), has_tie (yes, no). There are three classification tasks described in [14] but only the first is used here. Where the robot belongs to the class if (head_shape = body_shape) or (jacket_color = red) else it does not.

The original document used a randomly selected 124 (of the possible 432) combinations to be used for the training set. As this paper is not concerned with generalisation the ANNs were both trained and tested on the complete training set. All results quoted in this paper are the ANNs final classification error on the training set. The fitness is the classification error (which was to be minimised).

The implementation used here assigns each possible attribute value its own input to the ANN; totalling seventeen inputs. Each of these inputs is set as one if the particular attributes value is present and as zero otherwise. The ANN classifies each sample as belonging to the class if the single ANN output is greater or equal to 0.5. The transfer function used by all the nodes was the unipolar sigmoid.

5 Results

5.1 Effect of Topology on Weight Evolution

The results of the first experiment, investigating the effect of topology when evolving connection weights, are given in Figs. 2 and 3 for the Double Pole and Monks Problem test problems respectively. The left images show the average fitness for a range of fixed topologies when only evolving the weights. The right images, for comparison, show the average fitnesses when evolving both weights and topology for a range of CGPANN topology limits.

Figures 2 and 3 show that when weights are evolved with a fixed topology, the search effectiveness is highly dependent on the topology used. This result has been

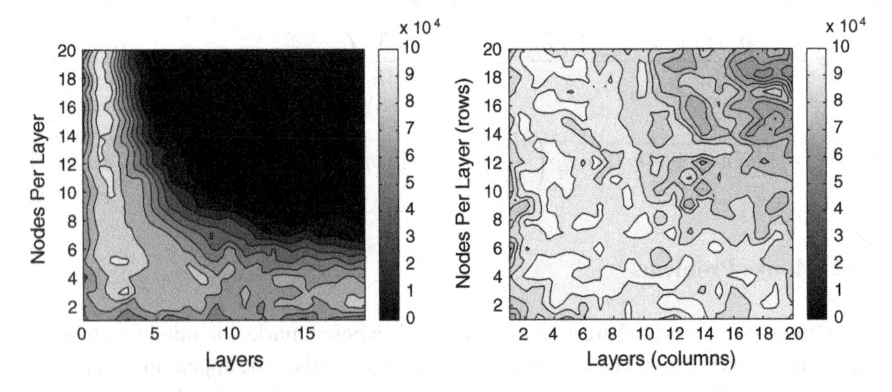

Fig. 2 Double Pole test problem results. *Left*: Fixed topology only evolving weights. *Right*: Random initial topology evolving weights and topology. The higher values (lighter) represent a better fitness. The same fitness scale (time steps balanced) is used for both figures for comparison

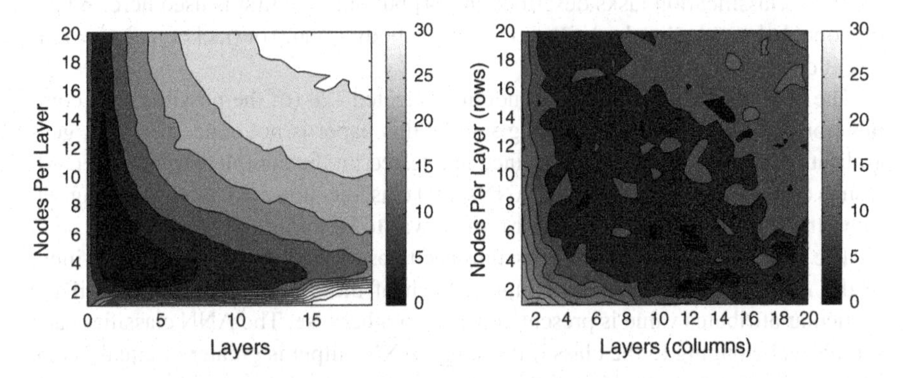

Fig. 3 Monks Problem test problem results. *Left*: Fixed topology only evolving weights. *Right*: Random initial topology evolving weights and topology. The lower values (darker) represent a better fitness. The same fitness scale (percentage error) is used for both figures for comparison

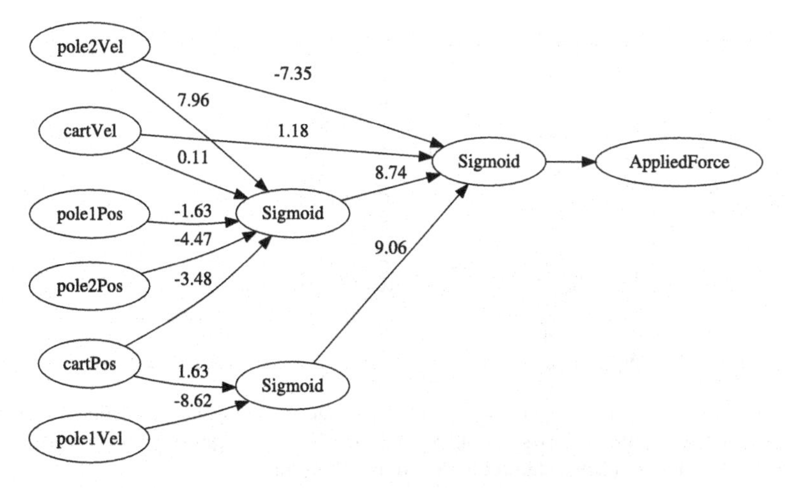

Fig. 4 Example of ANN which scored maximum fitness on the Double Pole test problem. Created using standard CGPANN with both row and column limits set as three

shown previously [4] but, to the authors knowledge, never in so much detail and so explicitly. Similar results are also found when training ANN using back propagation [6]; highlighting the importance of topology across training methods. It can also be seen that when only weights are evolved, the most effective topologies use a small number of layers with many nodes per layer; topologies which are often used when using traditional weight only methods, as other topologies are harder to train [6]. When using CGPANN to evolve both weights and topology, it can be seen that the effectiveness of the search is much more uniform across a range of topology constraints; the issue of selecting a suitable topology prior to training the ANN is vastly reduced. An example of the type of topology created when using CGPANN is given in Fig. 4. Clearly the generated topology is very unconventional, with no clear layers and variable node arity.

5.2 *Weight Evolution Versus Topology Evolution*

The results of the second experiment, comparing the relative effects of weight evolution and topology evolution, are shown in Fig. 5 for a range of mutation rates.

The results are also analysed using non-parametric statistical tests; as it is likely the distributions are non-normal. The Mann-Whitney U-test and the Kolmogorov-Smirnoff test (KS) are used to identify if the differences between approaches are statistically significant; with $\alpha = 0.05$ in both cases. The effect-size (A), as defined by Vargha et al. [16], is also used to indicate the importance of any statistical difference; with values >0.56 indicating a small effect size, >0.64 a medium and >0.71 a large. The results of these statistical tests are given for both the Double Pole and the

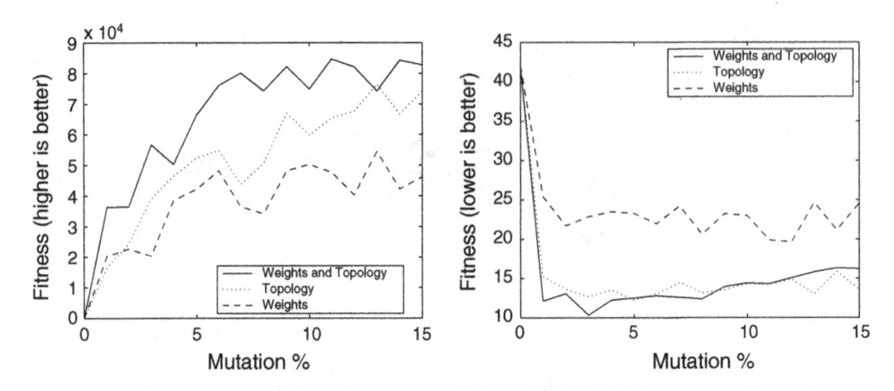

Fig. 5 Comparison of only weight, only topology and weight and topology evolution for Double Pole (*left*) and Monks Problem (*right*) test problems. Higher values represent a better fitness a for the Double Pole test problem and lower for the Monks Problem

Table 2 Statistical analysis of Double Pole and Monks Problem test problem

Problem	Comparison	Mutation %	U-test	KS	A
Double Pole	W & T | T	3	0.176	0.358	0.574
Double Pole	W & T | T	5	0.141	0.508	0.576
Double Pole	W & T | W	3	2.31×10^{-6}	2.76×10^{-6}	0.766
Double Pole	W & T | W	5	2.59×10^{-3}	1.71×10^{-2}	0.661
Double Pole	W | T	3	4.43×10^{-5}	1.78×10^{-4}	0.734
Double Pole	W | T	5	8.11×10^{-2}	3.12×10^{-2}	0.596
Monks Problem	W & T | T	3	0.124	0.241	0.589
Monks Problem	W & T | T	5	0.593	0.358	0.531
Monks Problem	W & T | W	3	9.46×10^{-6}	1.78×10^{-4}	0.757
Monks Problem	W & T | W	5	2.67×10^{-5}	2.76×10^{-5}	0.743
Monks Problem	W | T	3	8.35×10^{-4}	4.43×10^{-3}	0.694
Monks Problem	W | T	5	2.48×10^{-5}	4.23×10^{-4}	0.744

| separates the methods under comparison. *W* weight evolving, *T* topology evolving, *W & T* weight and topology evolving

Monks Problem test problems at three and five percent mutation rates (commonly used mutation values); see Table 2.

It can be seen from Fig. 5 that topology evolution is providing a better results than weight evolution when used independently. This is confirmed to be statistically significant in all cases examined except for the rank-sum test on the Double Pole test problem using a five percent mutation rate. The effect-size values for the weight only and topology only comparison also indicate that the statistical difference is important. This same statistical difference is also present when comparing the use of topology and weight evolution with weight evolution alone. However there is no statistical difference between the use of weight and topology evolution with topology evolution alone. This confirms the unexpected result that topology evolution is significantly more important to the search for effective neural networks than weight evolution.

6 Evaluation

Figures 2 and 3 show that only evolving connection weights leads a to search which is highly dependent upon the topology used. However, by allowing both topology and weight evolution this issue is drastically reduced. It is true that when only evolving connection weights using standard topologies, a few layers and many nodes per layer, good results are produced. However, it can also be seen in Fig. 2 that topologies which have many layers and few nodes per layer also produced good results. It is clear that simply using standard topologies ignores large areas of potentially beneficial search space. In addition, for more challenging problems a suitable network size would not be known in advance.

In Fig. 3, the Monks Problem test problem, it is seen that selecting a suitable topology and then evolving only the weights produces better results than evolving both topology and weights. This result is unsurprising as evaluating the topologies in this way is likely to find suitable locations in the topology space. As with any search, if domain knowledge is known in advance then this can be used to restrict the search space to suitable areas. Interestingly, this was not the case in Fig. 2, where evolving topology produced a better result in nearly all cases.

In the second experiment, we obtained the surprising result that topology evolution alone is significantly superior to weight evolution alone when training ANNs; when using random initial weights and topologies. It has long been thought that topology manipulation offers an advantage when training ANNs. This paper has not only empirically demonstrated that topology manipulation is advantageous, but that it offers benefits in the search more than weight manipulation.

To confirm these results future experiments should include additional test problems and the use of other topology manipulating training methods, such as developmental methods [19]. At the very least however it can be said that CGPANN appears to benefit highly from topology evolution; far more in fact, than weight evolution.

7 Conclusion

This paper has shown the significant benefits of using NE methods, which evolve both weights and topologies. These methods are not dependent upon selecting the correct topology before the search begins and are also strongly aided by the presence of topology manipulation. The surprising result that topology manipulation is more important than weight mutation has also been presented. This result may be one of the reasons some NE methods are so successful at training ANNs. CGPANN has been used throughout this investigation and has been shown to be very versatile; capable of using fixed user chosen topologies, randomly generated topologies, only evolving weights, only evolving topology and evolving both weights and topology. CGPANN is also capable of evolving the activation function used within the neurons; giving CGPANN complete control of the evolved ANN.

References

1. P. Angeline, G. Saunders, and J. Pollack. An Evolutionary Algorithm that Constructs Recurrent Neural Networks. *IEEE Transactions on Neural Networks*, 5(1):54–65, 1994.
2. D. Floreano, P. Dürr, and C. Mattiussi. Neuroevolution: from Architectures to Learning. *Evolutionary Intelligence*, 1(1):47–62, 2008.
3. X. Glorot and Y. Bengio. Understanding the Difficulty of Training Deep Feedforward Neural Networks. In *Proceedings of the International Conference on Artificial Intelligence and Statistics (AISTATS10). Society for Artificial Intelligence and, Statistics*, 2010.
4. C. Igel. Neuroevolution for Reinforcement Learning using Evolution Strategies. In *Evolutionary Computation*, volume 4, pages 2588–2595. IEEE, 2003.
5. M. M. Khan, G. M. Khan, and J. F. Miller. Evolution of Neural Networks using Cartesian Genetic Programming. In *Proceedings of IEEE World Congress on Computational Intelligence*, 2010.
6. H. Larochelle, Y. Bengio, J. Louradour, and P. Lamblin. Exploring Strategies for Training Deep Neural Networks. *The Journal of Machine Learning Research*, 10:1–40, 2009.
7. J. Miller and S. Smith. Redundancy and Computational Efficiency in Cartesian Genetic Programming. *IEEE Transactions on Evolutionary Computation*, 10(2):167–174, 2006.
8. J. Miller and P. Thomson. Cartesian Genetic Programming. In *Proceedings of the Third European Conference on Genetic Programming (EuroGP2000)*, volume 1802, pages 121–132. Springer-Verlag, 2000.
9. J. F. Miller. What bloat? Cartesian Genetic Programming on Boolean Problems. In *2001 Genetic and Evolutionary Computation Conference Late Breaking Papers*, pages 295–302, 2001.
10. J. F. Miller, editor. *Cartesian Genetic Programming*. Springer, 2011.
11. D. Moriarty and R. Mikkulainen. Efficient Reinforcement Learning through Symbiotic Evolution. *Machine learning*, 22(1):11–32, 1996.
12. R. Poli. Some Steps Towards a Form of Parallel Distributed Genetic Programming. In *Proceedings of the First On-line Workshop on, Soft Computing*, pages 290–295, 1996.
13. K. Stanley and R. Miikkulainen. Evolving Neural Networks through Augmenting Topologies. *Evolutionary computation*, 10(2):99–127, 2002.
14. S. Thrun, J. Bala, E. Bloedorn, I. Bratko, B. Cestnik, J. Cheng, K. De Jong, S. Dzeroski, S. Fahlman, D. Fisher, et al. The Monk's Problems a Performance Comparison of Different Learning Algorithms. Technical report, Carnegie Mellon University, 1991.
15. A. J. Turner and J. F. Miller. Cartesian Genetic Programming encoded Artificial Neural Networks: A Comparison using Three Benchmarks. In *Proceedings of the Conference on Genetic and Evolutionary Computation (GECCO-13)*, pages 1005–1012. ACM, 2013.
16. A. Vargha and H. D. Delaney. A Critique and Improvement of the CL Common Language Effect Size Statistics of McGraw and Wong. *Journal of Educational and Behavioral Statistics*, 25(2):101–132, 2000.
17. V. K. Vassilev and J. F. Miller. The Advantages of Landscape Neutrality in Digital Circuit Evolution. In *Proc. International Conference on Evolvable Systems*, volume 1801 of LNCS, pages 252–263. Springer Verlag, 2000.
18. A. Wieland. Evolving Neural Network Controllers for Unstable Systems. In *Neural Networks, 1991, IJCNN-91-Seattle International Joint Conference on*, volume 2, pages 667–673. IEEE, 1991.
19. X. Yao. Evolving Artificial Neural Networks. *Proceedings of the IEEE*, 87(9):1423–1447, 1999.
20. T. Yu and J. F. Miller. Neutrality and the Evolvability of a Boolean Function Landscape. *Genetic programming*, pages 204–217, 2001.

Inferring Context from Users' Reviews for Context Aware Recommendation

F. Z. Lahlou, H. Benbrahim, A. Mountassir and I. Kassou

Abstract Context Aware Recommendation Systems are Recommender Systems that provide recommendations based not only on users and items, but also on other information related to the context. A first challenge in building these systems is to obtain the contextual information. In this paper, we explore how accurate it is possible to infer contextual information from users' reviews. For this purpose, we use Text Classification techniques and conduct several experiments to identify the appropriate Text Representation settings and classification algorithm to the context inference problem. We carry out our experiments on two datasets containing reviews related to hotels and cars, and aim to infer the contextual information 'intent of purchase' from these reviews. To infer context from reviews, we recommend removing terms that occur once in the data set, combining unigrams, bigrams and trigrams, adopting a TFIDF weighting schema and using the Multinomial algorithm rather Naïve Bayes than Support Vector Machines.

1 Introduction

Traditional Recommender Systems (RS) focus on users and items when computing predictions. However, there is other contextual information (such as time, weather or accompanying persons) that may influence user decisions. Indeed, the same item

F. Z. Lahlou (✉) · H. Benbrahim · A. Mountassir · I. Kassou
ALBIRONI Research Team, ENSIAS, Mohamed V University, Souissi, Rabat, Morocco
e-mail: fatimazahra.lahlou@um5s.net.ma

H. Benbrahim
e-mail: benbrahim@ensias.ma

A. Mountassir
e-mail: asmaa.mountassir@gmail.com

I. Kassou
e-mail: kassou@ensias.ma

M. Bramer and M. Petridis (eds.), *Research and Development in Intelligent Systems XXX*, 227
DOI: 10.1007/978-3-319-02621-3_16, © Springer International Publishing Switzerland 2013

can be of interest to a user in a given context, and absolutely uninteresting in another one. For example, a user may book at the same time two different hotels, one for his business trip, with some specific accommodations (such as Wi-Fi, conference center or meeting area), and another for his vacation with his family with other accommodations (such as family rooms, swimming pool or even childcare services). Consequently, even if the user is the same, the RS should not suggest the same hotel to him in both situations. Hence, contextual information should be taken into account in the recommendation process. Moreover, research proved that including contextual information when computing recommendation improves its accuracy [1]. Such systems are called Context Aware Recommender Systems.

Context is defined by Dey et al. as "any information that can be used to characterize the situation of an entity" [2]. Getting contextual information is a first step for building any context aware recommender system. In e-commerce websites, such information is hardly ever explicitly provided by users. Moreover, most of current e-commerce web sites allow users, in addition to expressing ratings, to write reviews on the evaluated items. While the primary goal of these reviews is to express sentiments and preferences about items, they often carry some contextual information. Figures 1 and 2 show examples of reviews containing some contextual information. As one can see, the review in Fig. 1 reveals who was accompanying the reviewer, the purpose of his trip, the duration and the date of his stay, while the review in Fig. 2 indicates the date of car purchase, the number of vehicles the user owned before, and the purpose of using the car.

I stayed at this hotel with two colleagues on a business trip for 7 nights in December of 2007. I have stayed in many hotels across the world and hotel Maamoura ranks among the top of my hotel experiences. The staff is incredibly friendly and helpful. The rooms are very spacious, clean, well furnished and quiet. The bathrooms are great as well. The hotel location is excellent in the middle of the city but on a quiet side street. There are tons of bank machines, pay phones and food options nearby. The hotel has 2 elevators. After long days of work I often found myself staying up to talk to the hotel manager simply because he is a really great guy. You will also be hard pressed to find a hotel manager who is more proud of Morocco and more dedicated to ensuring that everything goes well for you on your trip. I will definitely stay there again upon a return trip to Casablanca.

Fig. 1 A hotel review containing some contextual information

"Purhased 4runner LTD in October my 56th vehicle and my favorite.Use for business travel. 20mpg within the first week. Drives effortlessy,very quiet,comfortable,roomy, blue tooth GREAT! Rated reliability on past performance and future expectation."

Fig. 2 A car review containing some contextual information

In this paper, we start from this observation, and explore how accurate it is possible to infer contextual information from users' reviews. We focus on the contextual information 'intent of purchase', that we aim to infer from reviews. As illustrated by aforementioned examples, users' decisions may be affected by different intents of purchase. We consider the problem of inferring context as a Text Classification one, where categories are the set of possible values that can take the 'intent of purchase' for each domain. We conduct a study on two datasets containing reviews from two different domains: hotels and cars. We encounter different challenges while inferring context from reviews. To deal with these challenges, we compare different text representation settings and classification algorithms in order to determine the ones that give the best results.

The present work makes three main contributions:

- It draws attention to a novel problem of inferring context from reviews.
- It lists the main challenges faced with this problem.
- It investigates the use of text classification techniques to address this issue.

The rest of the paper is organized as follows. In the second section, we discuss the context inference problem and present some related works. Used data collection is described in the third section. The fourth section lists the encountered challenges. The adopted classification process is described in the fifth section. Details about our experiments and discussion on obtained results are presented in the sixth section. Finally, the last section concludes the paper and provides directions for future research.

2 The Context Inference Problem

In the literature [1], there are three ways to obtain contextual information from users' data, namely: (i) explicitly, by asking direct questions to relevant people; (ii) implicitly from the data or the environment. For example, temporal information can be extracted from the timestamp of a transaction; (iii) by inference using statistical or Machine Learning methods.

Inferring the contextual information 'intent of purchase' has been studied in [3], where researchers explored the possibility of inferring this information from the existing customers' demographic and transactional data. They demonstrated that it is possible to infer fairly accurately the context on condition that a proper segmentation level is identified, and good predictive model is selected. However this work and our share the same objective, they are different since the type of data from which we aim to infer context is different: [3] aim to get the contextual information from the demographic and transactional data, where we try to infer it from users' reviews.

Over the last years, research studying opinion mining has received a considerable attention [4]. Inferring context from reviews, even tackling the same type of data (reviews), is a different research issue for many reasons. First, the task in opinion mining is to get user's evaluation (positive, negative or neutral), while in a context inference problem, the goal is to obtain the contextual information that characterize user's experience with the reviewed item. Furthermore, users' opinions always exist in reviews as it is for this objective that they has been written, while contextual information may not exist in a review at all. Finally, user's evaluation is contained in used words, (as 'great', 'pleased', 'horrible', 'unsatisfied', 'love' ...), while in a context inference problem, we look for clues, as 'double bed', 'my children', 'work' ... . Note here that the context inference problem is not an Information Extraction problem, as for this last, the information to be extracted is explicit and needs only to be detected.

In the literature, to the best of our knowledge, very few works until now deal with contextual information inference from user reviews. Aciar [5] notices that sentences in reviews are either 'contextual' (that contain information about the context), or 'preferences' ones (that contain opinion about features evaluated by the user). She then tries to identify contextual sentences in user reviews using rule set defined by Text Mining tools. This work though is significant remains limited as it only identifies contextual sentences and does not provide the contextual information existing in these sentences.

The closest works to our study are those of Hariri et al. [6] and Li et al. [7] since they also investigate the use of Text Classification techniques to detect contextual information. Hariri et al. [6] consider the context inference as a multi-labeled text classification problem in which documents can be classified into one or more categories. For this purpose, they use, as a Text Categorization technique, the Labeled Latent Dirichlet Allocation (L-LDA). The recommendation is then computed using the contextual information in combination with user ratings history to compute a utility function over the set of items. Li et al. [7] work on restaurant review data and try to extract the contextual information 'companion'. For this task, they use and compare two methods: (i) a web semantic string matching based method; (ii) a text classification method, in which they build and compare three classifiers: a rule based classifier, a Maximum Entropy with bags of words features, and a hybrid classifier combining the two ones. As result, they find that the hybrid classifier is the best performing one with up to 83.8 % in terms of F1 measure. They also compare several approaches for integrating contextual information into recommendation process.

In this paper, we address the issue of inferring context from reviews using a Text Classification based approach. Unlike precedent works, we focus principally on the context inferring problem. We first present different challenges faced by this issue. We try to deal with these challenges by comparing different text representation settings. We also apply classification algorithms known by their effectiveness in standard text classification tasks and which are different from those used in the aforementioned works. Finally, we conduct a study on a dataset from cars domain, which, at the best of our knowledge, has not been studied before.

3 Data Collection

We conduct our study on two datasets: a hotel reviews dataset that we collect from 'TripAdvisor'[1] website, and a car reviews dataset that we collect from 'Cars'[2] website. For both datasets, the contextual information 'intent of purchase' is sometimes provided by the reviewers with their reviews. We use this provided information as labels in order to perform a supervised classification. In the rest of this section, we describe these two datasets.

3.1 'TripAdvisor' Dataset

A typical 'TripAdvisor' review, as illustrated by Fig. 3, is accompanied by some information, such as user name, hotel information, ratings, period of stay and trip type. In our study, we focus on the contextual information 'trip type', which represent the 'intent of purchase' when booking hotel.

Though some 'TripAdvisor' datasets are freely available on the web, these ones do not contain reviews' trip type information that we use as a label for our classification. For this reason, we collect our own dataset composed of 777 reviews, roughly uniformly distributed over each category, as illustrated by Table 1.

'TripAdvisor' website enables users to choose a trip type among these five values: "Business", "Couple", "Family", "Friends" and "Solo". We consider these five values as target categories for our classification.

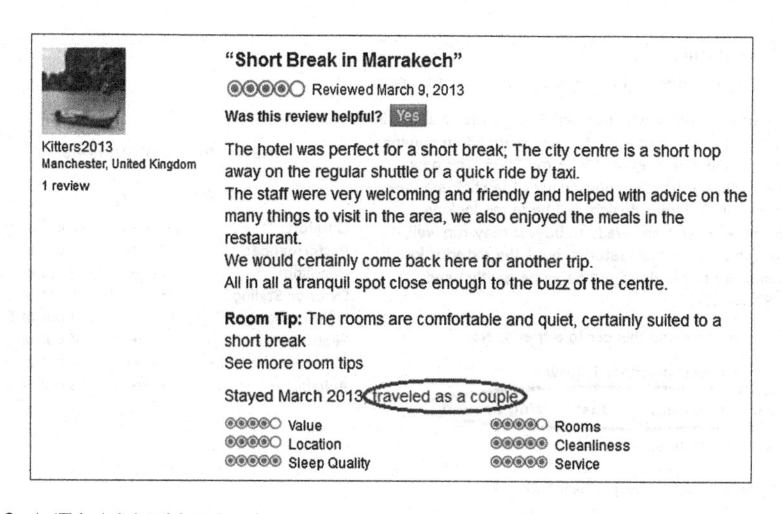

Fig. 3 A 'TripAdvisor' hotel review

[1] http://www.tripadvisor.com

[2] http://www.cars.com

Table 1 The distribution of 'TripAdvisor' dataset over categories

Category	Number of reviews
Business	157
Couple	158
Family	150
Friends	149
Solo	163

3.2 'Cars' Dataset

For the second dataset, we collect reviews from the car reviewing site 'Cars'. A typical 'Cars' review is illustrated in Fig. 4. As for 'TripAdvisor', 'Cars' enable reviewers to provide extra information in addition to their reviews. Among these information, we are interested on the 'primary use for this car', which we consider as the contextual information 'intent of purchase' for this domain. Users can fill the field 'primary use for this car' by choosing among these eight values: "Transporting Family", "Commuting", "Just getting around", "Towing", "Having fun", "Outdoor sports", "off roading" and "work". We consider the provided 'primary use for this car' as a label for our classification, and the eight values as target categories.

We collect 2136 labeled reviews from 'Cars' website. Unlike 'TripAdvisor' dataset, these reviews are not uniformly distributed over categories. The distribution of this dataset over categories is presented in Table 2. Note that "Outdoor sports" and "off roading" are minority categories, with a very small number of documents. We

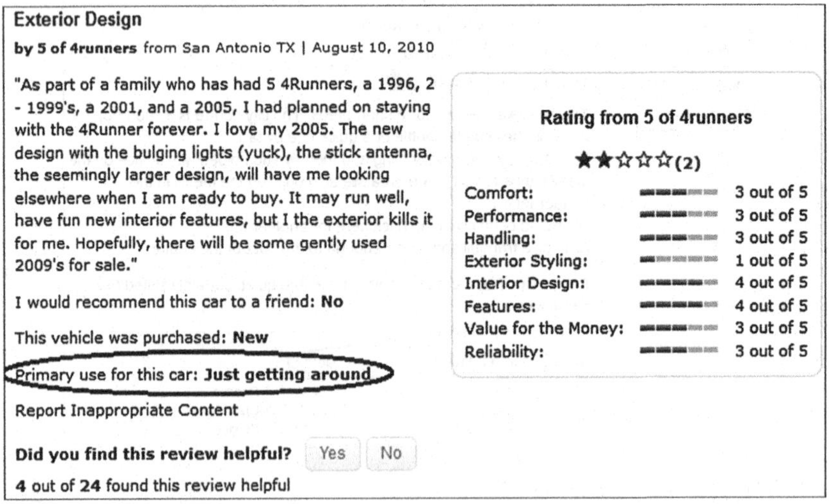

Exterior Design

by 5 of 4runners from San Antonio TX | August 10, 2010

"As part of a family who has had 5 4Runners, a 1996, 2 - 1999's, a 2001, and a 2005, I had planned on staying with the 4Runner forever. I love my 2005. The new design with the bulging lights (yuck), the stick antenna, the seemingly larger design, will have me looking elsewhere when I am ready to buy. It may run well, have fun new interior features, but I the exterior kills it for me. Hopefully, there will be some gently used 2009's for sale."

I would recommend this car to a friend: **No**

This vehicle was purchased: **New**

Primary use for this car: **Just getting around**

Report Inappropriate Content

Did you find this review helpful? Yes No

4 out of **24** found this review helpful

Rating from 5 of 4runners

★★☆☆☆(2)

Comfort:	3 out of 5
Performance:	3 out of 5
Handling:	3 out of 5
Exterior Styling:	1 out of 5
Interior Design:	4 out of 5
Features:	4 out of 5
Value for the Money:	3 out of 5
Reliability:	3 out of 5

Fig. 4 A 'Cars' car review

Table 2 The distribution of 'Cars' dataset over categories

Category	Number of reviews
Transporting family	297
Commuting	717
Just getting around	701
Towing	32
Having fun	165
Outdoor sports	18
Off roading	4
work	202

expect that classifying reviews belonging to these categories will affect the classification performance.

4 Encountered Challenges

Research in Text Classification is considered relatively mature as it was studied on well written corpora with coherent vocabulary like newspapers, and high classification accuracy was reached for some algorithms [8]. However, classifying reviews in order to infer context poses new research challenges for several reasons.

First, reviews are typically written by ordinary internet users; therefore, they are not always well written and often contain misspelling and typing errors. Moreover, used vocabulary is diverse due to the variety of authors. Yet, sometimes ambiguous words are used. Finally, reviews are typically short, with an average of 155 words per review for 'TripAdvisor' dataset, and 106 words per review for 'Cars' dataset.

In addition to these challenges related to any classification task that handles reviews, we face other problems specific to the context inference problem:

- First, and as the main purpose of writing reviews is to express opinions not to describe the context, only few words in the reviews report the context, while the majority of words express user evaluation on the item. Therefore, the data is considered very noisy for classification.
- Furthermore, some reviews do not contain, at a first look, any contextual information describing the 'intent of purchase' provided by the reviewer. Figures 5 and 6 illustrate examples of such review.
- Finally, we notice that for our data set, there is an overlap between some classes. For example, some reviews belong to Business and Solo categories at the same time, as a business travel is typically in solo. Likewise, other reviews can be classified as either Family or Couple. Indeed, reviews about family travels can be similar to those about couple travels since the reviewer can talk about his wife or her husband. The overlap between categories in the 'Cars' dataset is more important, as a car can have multiple use at the same time. For example, a user may buy a

This hotel is very nice, good position, even if is on a main street, so the traffic is very busy.
service is good, restorant is very good, the room is big and also the bed, especially the double bed. internet is available in the room only with cable connection.
taxi is available almost always, but better to ask reception to call one if you need...

Stayed October 2011, traveled on business

Fig. 5 Example of a 'TripAdvisor' review without clues on the trip type information

"The title of my post says it all. This car never failed me, got a lot of compliments and was fun to drive."

I would recommend this car to a friend: **Yes**

This vehicle was purchased: **Used**

Primary use for this car: **Just getting around**

Fig. 6 Example of a 'Cars' review without clues on the first use information

really good experience, the staff was very warm although they do not speak good english, they do the best they can. food was great and the room was very nice. i strongly recommend this hotel for a business trip or familly trip.

Stayed June 2011, traveled on business

Fig. 7 Example of a 'TripAdvisor' review that can belong to two categories (Business and Family)

car for family, going to work and commuting at the same time. Figure 7 shows an example of a 'TripAdvisor' review that belongs to the Business category and that can also be classified as family.

It is worth noting that, while reading reviews from 'Cars' dataset, we notice that this one is much more challenging compared to the 'TripAdvisor' for many reasons. First, there is a lack of clues indicating context in a significant number of reviews. Furthermore, overlapping in this dataset is more important, as we mentioned. Finally, the data is noisier as users tend to describe cars features rather than reporting context.

5 Document Preprocessing and Representation

Before we classify a set of text documents, it is necessary to preprocess these documents first and then map them into a vector space. The first step is called document preprocessing. The second step is known as document representation.

Document preprocessing consists of cleaning, normalizing and preparing text documents to classification step. We describe below some of the most common tasks of preprocessing phase.

- Tokenization: This process aims at transforming a text document into a sequence of tokens separated by white spaces. The output of this process is a text free of

punctuation marks and special characters. We can tokenize a given text into either single words or phrases depending on the adopted model. The most common models are the bag-of-words, which consists on splitting a given text into single words, and the n-gram word model [9]. This later is a sequence of n words in a given document. If n is equal to 1, we talk about unigrams, if it is equal to 2, they are called bigrams, and so on. For example, in the sentence "I traveled with my family", the unigrams that we can extract are "I", "traveled", "with", "my" and "family". While the bigrams that we can find are "I traveled", "traveled with", "with my", and "my family".
- Stemming: Word stemming is a crude pseudo-linguistic process which removes suffices to reduce words to their word stem [10]. For example, the stemming of the two words "teachers" and "teaching" will give the same stem "teach". Consequently, the dimensionality of the feature space can be reduced by mapping morphologically similar words onto their word stem.
- Term Frequency Thresholding: This process used to eliminate words whose frequencies are either above a pre-specified upper threshold or below a pre-specified lower threshold. This process helps to enhance classification performance since terms that rarely appear in a document collection will have little discriminative power and can be eliminated. Likewise, high frequency terms are assumed to be common and thus not to have discriminative power either. We specify that there is no theoretical rule to set the threshold; it is chosen empirically and hence its value depends on the experimental environment.

Once the preprocessing stage accomplished, each document is mapped onto a vector representation. We refer to the Vector Space Model (VSM) [11]. The retained terms (single words or phrases) after preprocessing are called features. The document vector is obtained by computing the weight of each term with respect to this document. There are several weighting schemes where the most common ones are presence, frequency and TFIDF-based weightings [12]. After weighting each feature with respect to a document d, this one is represented by a vector $d := (n_1(d), \ldots, n_m(d))$, where $n_i(d)$ denotes the weight of the *ith* feature regarding document d.

Once we represent text documents by their vectors in the feature space, these vectors can serve as input for classification algorithms so as to learn classifiers from training document vectors and, as a final step, to classify unseen documents.

6 Experiments

In our experiments, we perform a single-label multi-class classification in order to classify each review in the category corresponding to the 'intent of purchase' contextual information. For this task, we compare different text representation settings and classification algorithms in order to determine those that are appropriate for the context inference problem. In this section, we describe the experimental environment, present the obtained results and discuss the different findings.

6.1 Experimental Design

In our experiments, we study four text representation settings, namely term frequency thresholding (using native features, that we denote by Freq1, or removing terms that occur once in the data set, that we denote by Freq2), n-gram words (unigrams only (Uni), or the combination of unigrams and bigrams (UniBi), or combination of unigrams and bigrams and trigrams (UniBiTri)), stemming (application of stemming or not), and weighting schema (TFIDF or normalized frequency (NormFreq)).

Note that for all the tested settings, we convert all words to lower case. We remove punctuation marks, special characters and numbers. We also remove stop words that include just single words. The stop list that we have does not contain pronouns, as we think that pronouns can be relevant in our classification problem.

For the classification task, we use two classifiers known by their effectiveness in the literature of text classification, namely Naïve Bayes (NB) and Support Vector Machine (SVM) [12]. We specify that for NB, we use the Multinomial NB [13]. Concerning SVM classifier, we use the Sequential Minimum Optimization (SMO) [14]. As a validation method, we use the 10-fold cross-validation method [15]. We finally note that for our tasks of preprocessing and classification, we use the data mining package Weka [16].

6.2 Results

In Tables 3 and 4, we present for each dataset the obtained results in terms of the macro-averaged F1 following the application of each setting combination for each classifier. For each setting combination, we show the number of features so as to highlight the effect of the application of each setting.

Note that, for 'cars' dataset, when we performed experiments on the whole dataset we had poor results since documents of minority classes are always misclassified. This is why we do not include the two minority classes "Off roading" and "Outdoor sports" in the experiments that we report in this study.

The first setting we consider is related to the minimum term frequency thresholding. We are interested in comparing the use of native features (i.e. Freq1) with the removal of terms (features) that occur once in the data set (i.e. Freq2). Note that these terms are called hapaxes. By comparing results recorded following the application of (S1 and S2), and (S3 and S4), and also (S5 and S6), we can see that, for both datasets, the removal of hapaxes enhances the performance for the two classifiers NB and SVM. Furthermore, removing such terms reduces also significantly the dimensionality. For example, for 'TripAdvisor' dataset, the dimensionality decreases by 58.14 % when we move from S1 to S2.

By listing terms that occur once, we can see that these terms either correspond to errors made by users or reflect the rich vocabulary used by the different users. We can say that the improvement in classification performance is due to the fact that removing hapaxes reduces the noise in the data. Moreover, since we use as validation method k-fold cross-validation, data is divided into training and test data. Therefore,

Table 3 Classification results in terms of F1 and number of features for each setting for 'TripAdvisor' dataset

Settings	NB	SVM	#features
Stem + NormFreq + Freq1 + Uni (S1)	51.9	46.1	5400
Stem + NormFreq + Freq2 + Uni (S2)	55.6	47.1	2260
Stem + NormFreq + Freq1 + UniBi (S3)	51.0	39.0	55724
Stem + NormFreq + Freq2 + UniBi (S4)	62.8	50.5	10891
Stem + NormFreq + Freq1 + UniBiTri (S5)	51.2	31.3	149561
Stem + NormFreq + Freq2 + UniBiTri (S6)	61.6	**52.0**	17879
NoStem + NormFreq + Freq2 + UniBiTri (S7)	61.3	50.9	18163
Stem + TFIDF + Freq2 + UniBiTri (S8)	**72.58**	**52.0**	17879
NoStem + TFIDF + Freq2 + UniBiTri (S9)	72.52	50.9	18163

Table 4 Classification results in terms of F1 and number of features for each setting for 'Cars' dataset

Settings	NB	SVM	#features
Stem + NormFreq + Freq1 + Uni (S1)	35.2	33.3	6243
Stem + NormFreq + Freq2 + Uni (S2)	39.0	33.7	2839
Stem + NormFreq + Freq1 + UniBi (S3)	31.8	31.9	84223
Stem + NormFreq + Freq2 + UniBi (S4)	43.4	36.9	17171
Stem + NormFreq + Freq1 + UniBiTri (S5)	32.3	30.8	246292
Stem + NormFreq + Freq2 + UniBiTri (S6)	46.7	40.8	27802
NoStem + NormFreq + Freq2 + UniBiTri (S7)	47.2	39.9	28142
Stem + TFIDF + Freq2 + UniBiTri (S8)	59.7	**40.8**	27802
NoStem + TFIDF + Freq2 + UniBiTri (S9)	**60.0**	39.9	28142

a feature that occurs once in the data set appears either in a training or test document. In both cases, this feature will not contribute in classification. This is why we can say that these features do not have discriminate power and it is recommended to remove them.

The next test deals with the n-gram word model. By comparing S2 and S4, we can see that bigrams is of a great importance for the two classifiers. Moreover, the introduction of trigrams and bigrams enhances considerably results for all the classifiers, as we can see by comparing S6 to S2. These findings lead us to conclude that the task of context inference needs the identification of some specific phrases that can lose their meaning when split into single words. We give as examples these phrases that we extracted from 'TripAdvisor' data set: "double bed", "my company", "swimming pool", "wireless internet", "king size beds", "separate living room", and from 'Cars' dataset: "take the family", "race car", "sports car", etc.

The next setting that we consider is related to the weighting schema. According to S6 and S8, we can see that the most suitable scheme for NB is the TFIDF based weighting, while SVM performs similarly using the normalized frequency based weighting and TFIDF.

Finally, the last setting deals with the application of stemming. We compare the application of stemming with the use of raw text (that we denote by NoStem). By comparing results obtained following the application of respectively S6 and S7, and also S8 and S9, we can see that the results of the classifiers for both datasets are very close, with most of the time a very slight degradation for using the raw text. Furthermore, as the application of stemming reduces considerably the dimensionality, we can say that it is more suitable to use stemming instead of raw text.

As a comparison of the studied classifiers, we can see from Tables 3 and 4 that NB outperforms SVM in all cases for both datasets. Indeed, for 'TripAdvisor' dataset, the best result yielded by NB corresponds to 72.5 % in terms of F1, while the best result obtained by SVM corresponds to 52.9 %. Whereas, for 'Cars' dataset, the best results obtained by NB and SVM are respectively 60 % and 40.8 %. So, it is clear that the most effective classifier for a task of context inference is the multinomial NB. We explain this finding by the fact that SVM does not perform well with noisy data [17].

So far, the best combination setting identified by our experiments is the removal of terms that occur once in the data set, the combination of unigrams, bigrams and trigrams. We also recommend, for a context inferring problem, the application of stemming and the use of NB with a TFIDF based weighting.

We get as best result in terms of the macro-averaged F1-measure 72.5 % for the 'TripAdvisor' dataset, and 60 % for the 'Cars' dataset. We can say that the difference in the classification performance was expected, since the 'TripAdvisor' dataset is uniformly distributed unlike 'Cars' dataset. But also because the 'Cars' dataset is much more challenging as explained at the end of Sect. 4.

Furthermore, the bad performance of SVM with the 'Cars' dataset is due to its sensitivity with unbalanced datasets.

We consider the obtained results, (up to 72.5 % for 'TripAdvisor' dataset and 60 % for 'Cars' dataset) not very high but promising if we take into account challenges that we face while extracting context from reviews. Recall that all the used classifiers are shown as the most effective in many classical classification applications. We can summarize the encountered challenges as follows: (i) shortness of documents, (ii) richness of the used vocabulary, (iii) misspelling errors, (iv) the noisy data, (v) absence of clues indicating context in many reviews, (vi) and overlap of categories.

7 Conclusion

In this paper, we explore how accurate it is possible to infer the contextual information 'intent of purchase' from users' reviews. For this purpose, we follow a Text Classification based approach, and aim to classify each review in its corresponding value of 'intent of purchase'. We encounter many challenges and try to deal with it by investigating different text representation settings and classification algorithms known by their effectiveness in classical text classification problems. We conduct our experiments on datasets collected internally from the two reviewing website 'TripAdvisor' and 'Cars'. Our results show that it is recommended to remove terms that occur once in the data set, to combine unigrams, bigrams and trigrams, to adopt a TFIDF weighting schema and to use the Multinomial Naïve Bayes rather than Support Vector Machines. We achieve as best result 72.5 % in terms of the macro-averaged F1-measure for the 'TripAdvisor' dataset, and 60 % for the 'Cars' dataset. We consider that these results are promising given the different challenges that present the task of context inference. As future work, we intend to incorporate the inferred contextual information into the recommendation computations.

References

1. Adomavicius, G., Tuzhilin, A.: Context-aware recommender systems. Recommender Systems Handbook. 217–253 (2011).
2. Dey, A.K.: Understanding and using context. Personal and ubiquitous computing. 5, 4–7 (2001).
3. Palmisano, C., Tuzhilin, A., Gorgoglione, M.: Using context to improve predictive modeling of customers in personalization applications. Knowledge and Data Engineering, IEEE Transactions on. 20, 1535–1549 (2008).
4. Pang, B., Lee, L.: Opinion mining and sentiment analysis. Now Publishers Inc. (2008).
5. Aciar, S.: Mining context information from consumers reviews. Proceedings of Workshop on Context-Aware Recommender System. ACM (2010).
6. Hariri, N., Mobasher, B., Burke, R., Zheng, Y.: Context-Aware Recommendation Based On Review Mining. Proceedings of the 9th Workshop on Intelligent Techniques for Web Personalization and Recommender Systems (ITWP) (2011).
7. Li, Y., Nie, J., Zhang, Y., Wang, B., Yan, B., Weng, F.: Contextual recommendation based on text mining. Proceedings of the 23rd International Conference on Computational Linguistics: Posters. pp. 692–700 (2010).
8. Benbrahim, H., Bramer, M.: Neighbourhood exploitation in hypertext categorization, (2005).
9. Shannon, E.: A mathematical theory of communication. The Bell System Technical Journal. 27, 379–423 (1948).
10. Smeaton, A.: Information retrieval: Still butting heads with natural language processing? Information Extraction A Multidisciplinary Approach to an Emerging Information, Technology. 115–138 (1997).
11. Salton, G., Wong, A., Yang, C.-S.: A vector space model for automatic indexing. Communications of the ACM. 18, 613–620 (1975).
12. Sebastiani, F.: Machine learning in automated text categorization. ACM computing surveys (CSUR). 34, 1–47 (2002).
13. McCallum, A., Nigam, K.: Employing EM in pool-based active learning for text classification. Proceedings of ICML-98, 15th International Conference on, Machine Learning. pp. 350–358 (1998).
14. Hastie, T., Tibshirani, R.: Classification by pairwise coupling. The annals of statistics. 26, 451–471 (1998).
15. Mitchell, T.: Machine Learning. McCraw Hill. (1996).
16. Witten, I.H., Frank, E.: Data Mining: Practical machine learning tools and techniques. Morgan Kaufmann (2005).
17. Wu, Y., Liu, Y.: Robust truncated hinge loss support vector machines. Journal of the American Statistical Association. 102, 974–983 (2007).

Constraint Relationships for Soft Constraints

Alexander Schiendorfer, Jan-Philipp Steghöfer, Alexander Knapp, Florian Nafz and Wolfgang Reif

Abstract We introduce *constraint relationships* as a means to define qualitative preferences on the constraints of soft constraint problems. The approach is aimed at constraint satisfaction problems (CSPs) with a high number of constraints that make exact preference quantizations hard to maintain manually or hard to anticipate—especially if constraints or preferences change at runtime or are extracted from natural language text. Modelers express preferences over the satisfaction of constraints with a clear semantics regarding preferred tuples without assigning priorities to concrete domain values. We show how a CSP including a set of constraint relationships can linearly be transformed into a k-weighted CSP as a representative of c-semirings that is solved by widely available constraint solvers and compare it with existing techniques. We demonstrate the approach by using a typical example of a dynamic and interactive scheduling problem in AI.

A. Schiendorfer (✉) · J.-P. Steghöfer · A. Knapp · F. Nafz · W. Reif
Institute for Software and Systems Engineering, Augsburg University, Augsburg, Germany
e-mail: alexander.schiendorfer@student.uni-augsburg.de

J.-P. Steghöfer
e-mail: steghoefer@informatik.uni-augsburg.de

A. Knapp
e-mail: knapp@informatik.uni-augsburg.de

F. Nafz
e-mail: nafz@informatik.uni-augsburg.de

W. Reif
e-mail: reif@informatik.uni-augsburg.de

M. Bramer and M. Petridis (eds.), *Research and Development in Intelligent Systems XXX*,
DOI: 10.1007/978-3-319-02621-3_17,

1 Introduction

Numerous real world problems in AI including planning, scheduling, or resource allocation have been addressed successfully using the generic framework of constraint satisfaction. One has to identify the decision variables of a problem as well as constraints that regulate legal assignments.

While classical constraints can be considered "law-like" formulations that *must not* be violated, preferences constitute desired properties a solution *should* have. For instance, in a university timetabling problem, a constraint states that a professor can never give two lectures at the same time. We might *prefer* solutions that do not include Friday afternoon lectures.

Real-world problems tend to become too rigid as problems become over-constrained due to additional constraints representing preferences. Pioneering approaches to this problem either change the problem by relaxing existing constraints by adding domain values as in Partial CSP [11] or look for solutions that fulfill as many constraints as possible as in MaxCSP [13]. Usually, we are interested in assignments that satisfy all mandatory constraints, and enable preferences as well as possible. We present a qualitative formalism that enables to make statements such as "We prefer a solution that violates constraint X and satisfies Y to another one that violates Y but satisfies X".

Our contribution consists of two parts. First, we propose constraint relationships that provide a useful and time-saving modeling and elicitation tool to abstractly denote preferences. We illustrate their usage by analyzing scenarios for a typical example of the scheduling problem. Second, we give a transformation into a k-weighted CSP that respects the dominance properties we formalized and can be used with off-the-shelf constraint solvers.

1.1 Constraint Relationships

Constraint relationships constitute a qualitative "is more important than"-relation over constraints to denote which constraints should rather be satisfied and which ones can be dropped if necessary. For instance, stating "having a nice view" is more important than "spending less than three hours in a bus" models decisions where—other parameters being equal—plans satisfying the duration constraint but not the landscape requirement are considered worse than those offering interesting landscapes. We argue that this is a useful and realistic approach that enables an easy-to-use formalism for special application areas and positions itself among the AI formalisms presented in Sect. 1.2. Constraint relationships are targeted at problems with many constraints where preference levels are hard to maintain manually, or dynamic constraint satisfaction problems with frequently changing constraints and preferences. Typical applications facing these circumstances are multi-agent systems with incomplete preferences [17].

In some applications (e.g., [9] constructed a utility model of goals of their agents that is transferred into a numerical utility function), it is considered more natural to denote preferences in a qualitative rather than a quantitative way which results in an increasing interest in qualitative formalisms in AI [7]. For instance, vague preference statements in natural language can be easier formulated in a qualitative way.

On the quantitative side, existing soft constraint approaches work well if the CSP is known completely at design time. Designers can assign weights, categorize constraints into a hierarchy or assign coherent coefficients that yield acceptable solutions. However, assigning penalties that express preferences correctly is not straightforward in systems of several dozen constraints, in open-world scenarios, or dynamic CSPs [16]. Constraints may be added or removed at runtime to generate new CSPs [2, 8]. New constraints can change the system considerably and assigning penalties can become difficult, usually requiring manual intervention.

The decision whether a constraint is hard or soft and how important it is results from a "preference elicitation" process (see, e.g., [12]) and uses techniques from decision analysis. A fixed quantitative model performs poorly when autonomous agents face new situations and need to change their preferences [9]. In particular, numerical mappings lose underlying rationales.

Thus, we suggest *constraint relationships* as the elicited preference model being abstracted from numerical values to keep original motivations and facilitate changes.To use our preference formalism with existing solvers, we give a transformation that preserves the desired dominance property (see Sect. 2.1) into a k-weighted CSPs. The resulting problem is then solved by weighted CSP solvers or encoding weights as penalties in a constraint satisfaction optimization approach.

1.2 Preference Formalisms in AI

Several AI preference formalisms have been devised [15]. Among these are several kinds of soft constraints such as Fuzzy CSP [10, 12] or Weighted CSP (WCSP) [20] as well as other mechanisms like conditional preference networks (CP-nets) [5] or constraint hierarchies [4].

Generic frameworks for soft constraints such as c-semirings [3] or valued constraints [19] have been proposed to design generic algorithms and prove useful properties over a common structure. Assignments are labeled with preference levels or violation degrees that are combined using a specific operator to find preferred solutions. Formally, a *c-semiring* $\langle E, +_s, \times_s, \mathbf{0}, \mathbf{1}\rangle$ comprises a set of preference levels E; a binary operation $+_s$ closed in E used to compare assignments (defining $e \geq_s e' \leftrightarrow e +_s e' = e$) for which $\mathbf{0}$ is a neutral element (i.e., $e +_s \mathbf{0} = e$) and $\mathbf{1}$ is an annihilator (i.e., $e +_s \mathbf{1} = \mathbf{1}$); and a binary operation $\times_s$ also closed in E used to combine preference levels where $\mathbf{0}$ is an annihilator and $\mathbf{1}$ a neutral element.

In WCSPs, a cost function is provided for each constraint. The function assigns a weight from $\mathbb{R}_{\geq 0}$ to each tuple of values. Solving a WCSP then consists of minimizing the sum of weights of all violated constraints. This basic form has been refined to *k-weighted CSPs* that include a special weight k for hard constraints, meaning that assignments over k are unacceptable [14]. The resulting c-semiring is then defined as $\langle 0...k, \min, +^k, k, 0\rangle$ (with $0...k = \{0, \dots, k\}$ and $x +^k y = \min\{k, x + y\}$) which is the instance we will use to denote the preference semantics of constraint relationships.

Outside the class of soft constraint problems, in *constraint hierarchies* [4], constraints are categorized into strict levels, represented as sets $H_0, \dots, H_n$ in which constraints in H_i take precedence over constraints in H_{i+1} (that is, level i dominates level $i+1$). Valid solutions fulfill all constraints in H_0 and as many as possible in the dominated levels. Constraint hierarchies offer a similar design paradigm to constraint relationships. This formalism is suited well for problems that incorporate a "totalitarian" semantics [7]—i.e., it is better to satisfy a very important constraint rather than many less important ones. We argue that different semantics are required for other, more "egalitarian" problem classes. In constraint relationships, indifference is expressed by leaving out orderings—hierarchies only enable this for constraints on the same level. Furthermore, we show in Sect. 2.2 that constraint relationships can express a particular class of constraint hierarchies, viz. *locally-predicate-better* hierarchies, but model additional solution preferences.

CP-nets [6] are a qualitative tool to represent preferences over assignments by specifying orders over domain values. They intend to reduce the complexity of all possible preference orderings by structuring the variables according to their mutual impact. Concretely, it is assumed, that only some variable assignments affect the preference order on the domain values of others. Given this structural information about influences, a decision maker is asked to explicitly specify her preferences over the values of a variable X for *each* assignment to its *parent* variables, i.e., the ones that affect the preference on the domain of X. Hence, a set of total orders on the values of finite domains is kept for every variable in a conditional preference table. Lifting these orders to solutions generally results in a preorder [18]. While many real-world examples can be adequately modeled using CP-nets if user preferences on the actual values can be elicited [6], we argue that for complex constraint networks with infinite domains it is easier to make generalizing statements [7] that refer to a coarser level of granularity—preferences on constraints rather than on domain values for single variables. As the number of constraints is arguably lower than the number of variables and possible domain values for larger problems, the definition of preferences over constraints simplifies the preference elicitation problem considerably. Whereas CP-nets offer efficient solving algorithms due to the extensional structure of variables and values, our approach works with intensional constraints as well. A comparison to CP-nets can be found in Sect. 2.3.

Preliminaries

We follow and extend the definitions for classical constraint networks and soft constraints used in [15].

A *constraint network* $\langle X, D, C\rangle$ is given by a finite set $X = \{x_1, \dots, x_n\}$ of n *variables*; a set $D = \{D_1, \dots, D_n\}$ of corresponding *domains* such that x_i takes values from D_i; and a finite set C of *constraints*. Each $c \in C$ is given by a pair (r, v) where $v \subseteq X = sc(c)$ represents the scope of the constraint and $r \subseteq \prod_{x_i \in v} D_i = rl(c)$ is a relation containing the allowed combinations for c. We distinguish between *hard constraints* $C_h \subseteq C$ and *soft constraints* $C_s \subseteq C$, such that $C_h \cup C_s = C$ and $C_h \cap C_s = \emptyset$.

Assignments are tuples $t_V \in \prod_{x_i \in V} D_i$ with $V \subseteq X$. If $W \subseteq V$, $t_V[W]$ returns the *projection* of a tuple containing only elements in W. An assignment is *complete* if $V = X$, and *partial* otherwise. A constraint $c \in C$ is *fully assigned* by t_V if $sc(c) \subseteq V$. An assignment is *consistent with* c if c is fully assigned by t_V and $t_V[sc(c)] \in rl(c)$; we then write $t_V \models c$. An assignment is a *solution* if $\forall c \in C_h : t_V \models c$. We often omit the variables from assignments and write t instead of t_V.

A *weighted constraint network* $\langle X, D, C, w \rangle$ is given by a constraint network $\langle X, D, C \rangle$ and a *weighting function* and $w : C \to \mathbb{R}_{\geq 0}$. The weight of an assignment t_V is the combined weight of all unsatisfied constraints: $w(t_V) = \sum_{c \in C : t_V \not\models c} w(c)$. The purpose of a weighted CSP is to find assignments having minimal weight.

We write $C'_1 \uplus C'_2$ to denote the *disjoint union* of C'_1 and C'_2, where we require that $C'_1 \cap C'_2 = \emptyset$.

A binary relation $Q \subseteq M \times M$ on a set M is *asymmetric* if $(m, m') \in Q$ implies that $(m', m) \notin Q$; it is *transitive* if $(m, m') \in Q$ and $(m', m'') \in Q$ implies $(m, m'') \in Q$; it is a *partial order relation* if it is asymmetric and transitive. The *transitive closure* of Q, denoted by Q^+, is inductively defined by the rules that (a) if $(m, m') \in Q$, then $(m, m') \in Q^+$ and (b) if $(m, m') \in Q$ and $(m', m'') \in Q^+$, then $(m, m'') \in Q^+$.

2 Constraint Relationships

A set of *constraint relationships* for the soft constraints C_s of a constraint network $\langle X, D, C \rangle$ is given by a binary asymmetric relation $R \subseteq C_s \times C_s$ whose transitive closure R^+ is a partial order relation. We write $c' \prec_R c$ or $c \succ_R c'$ iff $(c, c') \in R$ to define c to be *more important* than c', analogously for R^+. If $c' \prec_R c$ we call c' a *direct predecessor*, if $c' \prec_{R^+} c$ a *transitive predecessor* of c. Moreover, we refer to the *constraint relationship graph* as the directed graph spanned by $\langle C_s, R \rangle$.

2.1 Semantics of Dominance Properties

We have defined constraint relationships syntactically as a relation that denotes some constraint being "more important" than another one. However, we need to express *how* much more important a constraint is than another one in order to address questions such as "Is it better to satisfy a more important constraint than *all* its less important predecessors". Concretely, we examine ways to lift the binary relation over soft constraints to sets of soft constraints that are violated by an assignment.

Such a *violation set* is denoted by capitalizing the letter used for the assignment; i.e., for some assignment t its violation set is $T = \{c \in C_s \mid t \not\models c\}$.

We consider several possibilities for worsening a violation set and will use $T \longrightarrow_R^p U$ to express that T is worsened to U by using the strategy or *dominance property* p. All strategies share two generic rules that we present first. On the one hand, a set of violated constraints T gets worse if some additional constraint c is violated:

$$T \longrightarrow_R^p T \uplus \{c\} \tag{W1}$$

On the other hand, worsening two independent parts of a violation set leads to a worsening of the whole violation set: If T_1 is worsened to U_1 and T_2 is worsened to U_2, then $T_1 \uplus T_2$ is also worsened to $U_1 \uplus U_2$:

$$\frac{T_1 \longrightarrow_R^p U_1 \quad T_2 \longrightarrow_R^p U_2}{T_1 \uplus T_2 \longrightarrow_R^p U_1 \uplus U_2} \tag{W2}$$

In fact, we can derive

$$T \longrightarrow_R^p T \uplus \{c_1, \ldots, c_k\} \tag{W1'}$$

We have $T \longrightarrow_R^p T \uplus \{c_1\}$ by (W1) and $\emptyset \longrightarrow_R^p \{c_2\}$ again by (W1), and thus $T \longrightarrow_R^p T \uplus \{c_1, c_2\}$ by (W2), from which the result follows by induction.

We now introduce three particular dominance properties. In the first approach, violating a less important constraint rather than an important one should be considered *better—ceteris paribus*. We call this criterion *single predecessor dominance* (SPD):

$$T \uplus \{c\} \longrightarrow_R^{\text{SPD}} T \uplus \{c'\} \quad \text{if } c \prec_R c' \tag{SPD}$$

For instance, if $C_s = \{\mathsf{a}, \mathsf{b}\}$ and $\mathsf{a} \succ_R \mathsf{b}$ (constraint a is considered more valuable than constraint b), $\{\mathsf{a}\} \longrightarrow_R^{\text{SPD}} \{\mathsf{a}, \mathsf{b}\}$ by (W1) (with $p = \text{SPD}$), $\{\mathsf{b}\} \longrightarrow_R^{\text{SPD}} \{\mathsf{a}\}$ by (SPD). Since SPD does not have a single constraint dominate a set of others it is well suited for "egalitarian" problems. It is therefore similar to a MaxCSP instance where we are interested in satisfying a large number of constraints rather than discriminating strongly by their individual importance.

However, a stronger notion is needed when some constraints contribute more to the quality of a solution than a whole set of others—in particular the constraints that are explicitly denoted less important. This property is called *direct predecessors dominance* (DPD) and is motivated by the fact that human preference decisions can be intransitive [1]:

$$T \uplus \{c_1, \ldots, c_k\} \longrightarrow_R^{\text{DPD}} T \uplus \{c'\} \quad \text{if } \forall c \in \{c_1, \ldots, c_k\} : c \prec_R c' \tag{DPD}$$

For a minimal example, consider $C_s = \{\mathsf{a}, \mathsf{b}, \mathsf{c}\}$ and $\mathsf{a} \succ_R \mathsf{b}$, $\mathsf{a} \succ_R \mathsf{c}$. Then violating a is more detrimental to a solution than any combination of b and c, e.g., $\{\mathsf{b}, \mathsf{c}\} \longrightarrow_R^{\text{DPD}} \{\mathsf{a}\}$. It follows by definition that $T \longrightarrow_R^{\text{SPD}} U$ implies $T \longrightarrow_R^{\text{DPD}} U$.

The most "hierarchical" notion, *transitive predecessors dominance*, consists of extending DPD to include transitive predecessors as well. It is motivated by the natural extension of constraint relationships R to its transitive closure R^+ to obtain a partial order and the ability to express a subset of constraint hierarchies with constraint relationships. Note that this property could also be achieved by using DPD rules and R^+ explicitly in the model:

$$T \uplus \{c_1, \dots, c_k\} \longrightarrow_R^{\mathrm{TPD}} T \uplus \{c'\} \quad \text{if } \forall c \in \{c_1, \dots, c_k\} : c \prec_{R^+} c' \qquad \text{(TPD)}$$

If $C_s = \{\mathsf{a}, \mathsf{b}, \mathsf{c}\}$ and $\mathsf{a} \succ_R \mathsf{b}, \mathsf{b} \succ_R \mathsf{c}$, then $\{\mathsf{b}, \mathsf{c}\} \longrightarrow_R^{\mathrm{TPD}} \{\mathsf{a}\}$, but also $\{\mathsf{c}\} \longrightarrow_R^{\mathrm{TPD}} \{\mathsf{b}\}$. Again, by definition $T \longrightarrow_R^{\mathrm{DPD}} U$ implies $T \longrightarrow_R^{\mathrm{TPD}} U$. As mentioned before, $T \longrightarrow_R^{\mathrm{TPD}} U$ iff $T \longrightarrow_{R^+}^{\mathrm{DPD}} U$.

Each relation $\longrightarrow_R^p$ over assignments induced by the used semantics describes *how* an assignment is worsened. It only is a partial order if the generating relation (R or R^+) already is a partial order. In general, the relation does not need to be transitive as this property is not required for R. Consider for example $C_s = \{\mathsf{a}, \mathsf{b}, \mathsf{c}\}$, $\mathsf{a} \succ_R \mathsf{b}$, $\mathsf{b} \succ_R \mathsf{c}$, and an SPD semantics; then $\{\mathsf{c}\} \longrightarrow_R^{\mathrm{SPD}} \{\mathsf{b}\}$ and $\{\mathsf{b}\} \longrightarrow_R^{\mathrm{SPD}} \{\mathsf{a}\}$, but not $\{\mathsf{c}\} \longrightarrow_R^{\mathrm{SPD}} \{\mathsf{a}\}$ since $\mathsf{a} \not\succ_R \mathsf{c}$.

We can enforce partial orders on assignments for each dominance property $p \in \{\mathrm{SPD}, \mathrm{DPD}, \mathrm{TPD}\}$, denoted by $t >_R^p u$ and to be read as "t is better than u", using $T\ (\longrightarrow_R^p)^+\ U$ (meaning repeated sequential application of the rules); we will prove the asymmetry of these transitive closures in Sect. 3 using weights.

2.2 Connection to Constraint Hierarchies

Constraint hierarchies [4] offer different *comparators* to discriminate solutions based on their satisfaction degree of constraints at different levels given by an error function and additional weights for constraints. Comparators are divided into *locally better* and *globally better*. Locally better compares based on the error functions only, whereas globally better predicates also take constraint-specific weights into account.

Error functions are either *predicate* functions, i.e., $e(c, t) = 0$ if $t \models c$, and 1 otherwise; or *metric* functions that give a continuous degree of violation (e.g., for $c \triangleq (X = Y)$, $e(c, t)$ could return the difference between the valuations of the variables X and Y in t). Since our approach is not concerned with metric error functions or user-defined weights, we restrict ourselves to comparison with *locally predicate better* (LPB) and show that these hierarchies can be encoded in constraint relationships. We then show that constraint relationships generalize LPB-hierarchies by providing an example that cannot be expressed by them.

First, consider the definition of LPB given by a constraint hierarchy $H = \{H_0, \dots, H_n\}$. The operator $>_{LPB}$ compares two solutions t and u (the constraints in H_0 are taken to be the hard constraints) and $t >_{LPB} u$ should be read as "t is better than u"; it is defined by

$$t >_{LPB} u \leftrightarrow \exists k > 0 : (\forall i \in 1...k-1 : \forall c \in H_i : e(c,t) = e(c,u)) \land$$
$$(\forall c \in H_k : e(c,t) \leq e(c,u)) \land (\exists c \in H_k : e(c,t) < e(c,u))$$

Our encoding of the constraint hierarchy H in constraint relationships R_H is defined as follows: $C_h = H_0$, $C_s = \bigcup_{i \in 1...n} H_i$, and

$$c \succ_{R_H} c' \leftrightarrow c \in H_i \land c' \in H_{i+1}.$$

We write T_i for all constraints in hierarchy level i that are violated by an assignment t, i.e., $T_i = \{c \in H_i \mid t \not\models c\}$ and, analogously, U_i for an assignment u; we abbreviate $\bigcup_{k \leq i \leq l} T_i$ by $T_{k...l}$.

Theorem 1. *If* $t >_{LPB} u$, *then* $T \longrightarrow^{\mathrm{TPD}}_{R_H} U$.

Proof Observe that $e(c,t) < e(c,u)$ iff $t \models c \land u \not\models c$; and $e(c,t) \leq e(c,u)$ iff $u \models c \rightarrow t \models c$.

Let $t >_{LPB} u$; we have to show that T is worsened to U by application of TPD-rules. Let $k > 0$ be such that (*) $\forall i \in 1...k-1 : \forall c_i \in H_i : t \models c_i \leftrightarrow u \models c_i$, (**) $\forall c_k \in H_k : u \models c_k \rightarrow t \models c_k$, and let $c \in H_k$ such that $t \models c \land u \not\models c$. By (*) we have that $T_{1...k-1} = U_{1...k-1}$. Furthermore, $T_{1...k} \subseteq U_{1...k} \setminus \{c\}$ since $T_{1...k-1} = U_{1...k-1}$, $\forall c_k \in H_k : t \not\models c_k \rightarrow u \not\models c_k$ by (**), and $c \notin T_{1...k}$. In particular, $T_{1...k} \subseteq U_{1...k} \setminus \{c\} \subseteq U_{1...n} \setminus \{c\}$. If $T_{1...k} = U_{1...n} \setminus \{c\}$, then $T = T_{1...k} \uplus T_{k+1...n} \longrightarrow^{\mathrm{TPD}}_{R_H} T_{1...k} \uplus \{c\} = U$ by (TPD), since all constraints in $T_{k+1...n}$ are transitively dominated by $c \in H_k$. If $T_{1...k} \subsetneq U_{1...n} \setminus \{c\}$, then $T_{1...k} \longrightarrow^{\mathrm{TPD}}_{R_H} U_{1...n} \setminus \{c\}$ by (W1′); applying rule (TPD) in (W2), we again have $T = T_{1...k} \uplus T_{k+1...n} \longrightarrow^{\mathrm{TPD}}_{R_H} (U_{1...n} \setminus \{c\}) \uplus \{c\} = U$. □

Conversely, Fig. 1a shows a constraint relationship problem that is not expressible in LPB hierarchies. Let $H : \{\mathsf{a}, \mathsf{b}, \mathsf{c}, \mathsf{d}, \mathsf{e}\} \rightarrow \mathbb{N} \setminus \{0\}$ be a mapping from the constraints to their respective hierarchy levels. We consider solutions that only satisfy *one* constraint and violate all others and write a for "a solution satisfying only a". We show that every admissible choice of H introduces too much ordering: The constraint relationships require a to be better than b which in turn should be better than c, thus we have to have $H(\mathsf{a}) < H(\mathsf{b}) < H(\mathsf{c})$. Since we expect a to be better than d as well, but require b and d to be incomparable, $H(\mathsf{d})$ has to be equal to $H(\mathsf{b})$. Similarly, $H(\mathsf{e})$ has to be $H(\mathsf{c})$ as e and c should be incomparable. But then b would be better than e, a relation that is explicitly not modeled in the underlying constraint relationships.

2.3 Connection to CP-Nets

CP-nets [6] specify total orders over the domain of a variable depending on an assignment to other variables in a conditional preference table. Concretely, a preference statement for a variable y is written as $x_1 = d_1, \ldots, x_n = d_n : y = w_1 \succ$

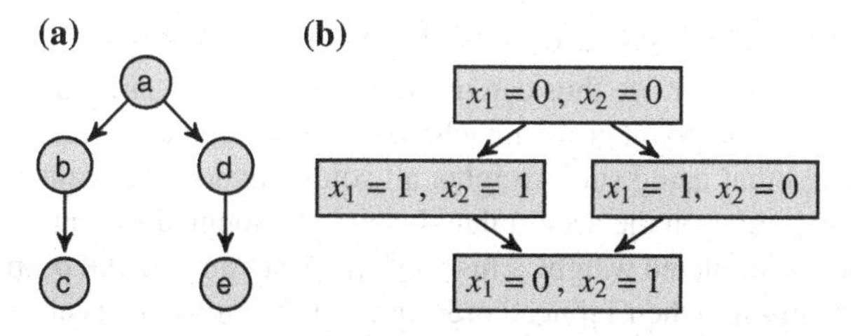

Fig. 1 Constraint Relationships compared to other formalisms **a** A set of constraint relationships not expressible in LPB-hierarchies. **b** Desired solution order $x_1 = 0,\ x_2 = 0$ should be the best

$\cdots \succ y = w_k$ where $x_1, \ldots, x_n$ are the *parent* variables of y and $w_1, \ldots, w_k$ are all domain values of y given in a total order $\succ$. Such an order needs to be specified for all assignments to $x_1, \ldots, x_n$. A preference statement should be interpreted as "Given that $x_1 = d_1, \ldots, x_n = d_n$, all other variables being equally assigned, prefer a solution that assigns w_i to y over one that assigns w_j to y iff $i > j$" which is the *ceteris paribus* assumption. The change of value for y from w_i to w_j is then called a "worsening flip". A complete assignment t to the variables of a CP-net is preferred to another one, say t', if t' can be obtained from t via a sequence of worsening flips [15].

On the one hand, the induced "better-as" relation on assignments need not be a partial order since cycles may arise [5]. By contrast, constraint relationships always lead to a partial order $>_R^p$ on assignments. On the other hand, CP-nets cannot express all partial orders on assignments [18]. Consider the minimal example depicted in Fig. 1b: $X = \{x_1, x_2\}$, $D_1 = D_2 = \{0, 1\}$. The proposed solution order cannot be expressed in CP-nets since $x_1 = 1,\ x_2 = 0$ and $x_1 = 1,\ x_2 = 1$ differ only by the assignment of x_2 and have to be comparable because of the total order requirement and *ceteris paribus* semantics in CP-nets. But this solution ordering is easily expressible in constraint relationships defining a constraint for each possible assignment.

Thus, the two frameworks are incomparable regarding the solution order. An extension, however, of the constraint relationship approach with conditional statements as in CP-nets might turn out to be fruitful.

3 Transforming Constraint Relationships into Weighted CSPs

Once the constraint problem and its constraint relationships are defined and the dominance property is chosen, we want to apply this information by transforming it into a k-weighted CSP that we can solve with standard solvers.

For a given CSP $\langle X, D, C\rangle$, a set of constraint relationships R, and a dominance property $p \in$ {SPD, DPD, TPD, we devise a weighting function $w_R^p : C \to \mathbb{N} \setminus \{0\}$ which we prove to be strictly monotonic w.r.t. to the dominance property p for soft constraints in the sense that $\sum_{c\in T} w_R^p(c) < \sum_{c\in U} w_R^p(c)$ if $t >_R^p u$. For each hard constraint $c \in C_h$, we set $w_R^p(c) = k_R^p$ with $k_R^p = 1 + \sum_{c\in C_s} w_R^p(c)$; therefore a solution of $\langle X, D, C\rangle$ satisfying only hard constraints still has a combined weight less than k^p_R. In particular, we thus can use the c-semiring $\langle 0...k^p_R, \min, +^{k^p_R}, k^p_R, 0\rangle$ for solving the soft-constraint problem, where solvers are readily available. However, it has to be noted that using such weights all solutions become comparable due to the totality of the order on the accumulated weights, though different solutions may show the same accumulated weight. This "defect" is not an inherent property of constraint relationships and their induced orderings on solutions, but is an artifact of the transformation into k-weighted CSPs. It would be interesting to find a c-semiring for representing constraint relationships that does not introduce such artificial orderings.

Each w^p_R will be defined recursively relying on the weights of predecessors w.r.t. R to have been already calculated. This is well defined since C and hence R are finite and R^+ is a partial order (i.e., the graph of R^+ is acyclic). Algorithmically, the weights can be determined in a bottom-up fashion or using a depth-first strategy.

For establishing the respective strict monotonicity properties, let $W^p_R(T) = \sum_{c \in T} w^p_R(c)$; we then have to prove $W^p_R(T) < W^p_R(U)$ if $t >^p_R u$. Since $>^p_R$ is defined via $(\longrightarrow^p_R)^+$ and all dominance properties are defined inductively by rules, it suffices to show $W^p_R(T) < W^p_R(U)$ for each rule application $T \longrightarrow^p_R U$.

Rule (W1) says that $T \longrightarrow^p_R T \uplus \{c\}$; and indeed, $W^p_R(T) < W^p_R(T \uplus \{c\})$, since all weights are in $\mathbb{N} \setminus \{0\}$. Rule (W2) has the premises $T_i \longrightarrow^p_R U_i$ for $i \in \{1, 2\}$; these amount to the assumptions $W^p_R(T_i) < W^p_R(U_i)$, from which we can conclude that $W^p_R(T_1 \uplus T_2) = W^p_R(T_1) + W^p_R(T_2) < W^p_R(U_1) + W^p_R(U_2) = W^p_R(U_1 \uplus U_2)$. It thus remains to prove the strict monotonicity of (SPD), (DPD), and (TPD).

Single predecessor dominance. For SPD, we propose the function that takes the maximum weight of its predecessors and adds 1 (we take $\max(\emptyset)$ to be 0):

$$w^{\text{SPD}}_R(c) = 1 + \max\{w^{\text{SPD}}_R(c') \mid c' \in C_s : c \succ_R c'\} \quad \text{for } c \in C_s .$$

This mapping is indeed strictly monotonic for applications of rule (SPD): Let $T \uplus \{c\} \longrightarrow^{\text{SPD}}_R T \uplus \{c'\}$ with $c \prec_R c'$. Then $w^{\text{SPD}}_R(c) < w^{\text{SPD}}_R(c')$ and hence $W^{\text{SPD}}_R(T \uplus \{c\}) = W^{\text{SPD}}_R(T) + w^{\text{SPD}}_R(c) < W^{\text{SPD}}_R(T) + w^{\text{SPD}}_R(c') = W^{\text{SPD}}_R(T \uplus \{c'\})$.

Direct predecessors dominance. For DPD we take the sum of weights of all predecessors and add 1 (summation over an empty index set is taken to be 0):

$$w^{\text{DPD}}_R(c) = 1 + \textstyle\sum_{c' \in C_s : c \succ_R c'} w^{\text{DPD}}_R(c') \quad \text{for } c \in C_s .$$

Rule (DPD) requires that violating a single constraint is worse than violating all its direct predecessors and hence this weight assignment assures that the weight of a constraint is strictly greater than the sum of the set of *all* its direct predecessors. In fact, for strict monotonicity, let $T \uplus \{c_1, \dots, c_k\} \longrightarrow^{\text{DPD}}_R T \uplus \{c'\}$ with $c_i \prec_R c'$ for all $1 \le i \le k$. Then $W^{\text{DPD}}_R(\{c_1, \dots, c_k\}) = \sum_{1 \le i \le k} w^{\text{DPD}}_R(c_i) < w^{\text{DPD}}_R(c) = W^{\text{DPD}}_R(\{c'\})$, by definition of w^{DPD}_R, and $W^{\text{DPD}}_R(T \uplus \{c_1, \dots, c_k\}) < W^{\text{DPD}}_R(T \uplus \{c'\})$.

Transitive predecessors dominance. Analogous to the DPD case, a TPD preserving weight assignment function w_R^{TPD} needs to make sure that $w_R^{\text{TPD}}(c) > \sum_{c' \in C_s : c \succ_{R^+} c'} w_R^{\text{TPD}}(c')$. Since, as mentioned in Sect. 2.1, TPD is DPD for R^+, the function $w_{R^+}^{\text{DPD}}$ would suffice. However, we can avoid computing the transitive closure of R by using the following function that also only depends on the direct predecessors:

$$w_R^{\text{TPD}}(c) = 1 + \sum_{c' \in C_s : c \succ_R c'} (2 \cdot w_R^{\text{TPD}}(c') - 1) \quad \text{for } c \in C_s .$$

We can establish that $w_R^{\text{TPD}}(c) \geq 1 + \sum_{c' \in C_s : c \succ_{R^+} c'} w_R^{\text{TPD}}(c')$ for all $c \in C_s$ with this definition by induction over the number of transitive predecessors of c: If $\{c' \in C_s \mid c \succ_{R^+} c\} = \emptyset$, then $w_R^{\text{TPD}}(c) = 1$ and thus the claim holds. Let $\{c' \in C_s \mid c \succ_{R^+} c\} \neq \emptyset$ and assume that $w_R^{\text{TPD}}(c') \geq 1 + \sum_{c'' \in C_s : c' \succ_{R^+} c''} w_R^{\text{TPD}}(c'')$ holds for all $c' \in C_s$ such that $c \succ_R c'$. Then $w_R^{\text{TPD}}(c) \geq 1 + \sum_{c' \in C_s : c \succ_{R^+} c'} w_R^{\text{TPD}}(c')$.

In summary, we have

Theorem 2. *If $t >_R^p u$, then $W_R^p(T) < W_R^p(U)$ for $p \in \{\text{SPD, DPD, TPD}\}$.* □

In particular, $>_R^p$ is asymmetric for $p \in \{\text{SPD, DPD, TPD}\}$ since the order $<$ on the weights is asymmetric: If we would have $t >_R^p u$ and $u >_R^p t$, then $W_R^p(T) < W_R^p(U)$ and $W_R^p(U) < W_R^p(T)$ by the theorem, which is impossible.

4 Example Scenario–Ski Day Planner

We demonstrate an application of constraint relationships using a simplified fictional scenario. First, we show how varying sets of relationships lead to different preferred assignments. Then we discuss changing constraints and preferences.

Consider an application that guides travelers exploring a new ski area by offering a plan for a ski day. Each skier has different priorities that can be set interactively.

Assume the following soft constraints are defined on the set of possible tours (that need to respect hard constraints such as weather induced blockages or daylight time):

- Avoid black slopes (ABS): Beginners avoid difficult (marked “black”) slopes.
- Variety (VT): Different slopes should be explored.
- Fun-park (FP): A feature for Freestyle fans
- Little Wait (LW): Impatient visitors prefer not to wait too long at a lift.
- Only Easy Slopes (OE): People restrict their tours to easy (“blue”) slopes.
- Lunch Included (LI): Some travelers enjoy a good mountain dish.

For clarity, we abstract from details such as concrete tuple representations and leave hard constraints aside by assuming the following three assignments are solutions but differ in their performance on soft constraints.

- $t_X^{(1)} \models \{\text{LW, OE, LI, FP}\} \wedge t_X^{(1)} \not\models \{\text{VT, ABS}\}$

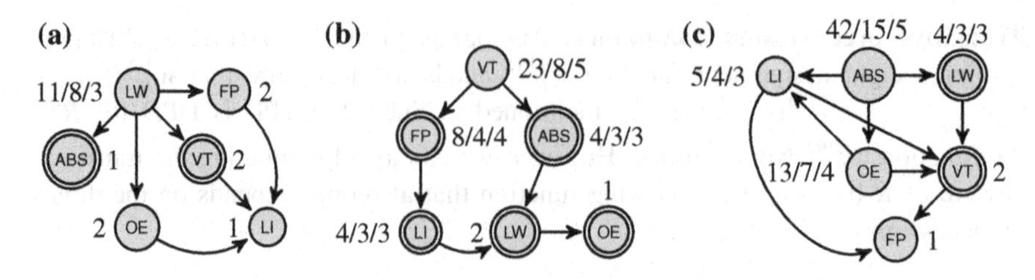

Fig. 2 The constraint relationship graphs for each persona. Double borders indicate that this constraint was violated in the TPD-preferred assignment according to Table 1(c). Weights are printed for TPD/DPD/SPD, only one number indicates that the weights are equal for all dominance semantics. **a** Graph of skier A **b** Graph for boarder B **c** Graph for rookie C

- $t_X^{(2)} \models \{\mathsf{VT}\} \wedge t_X^{(2)} \not\models \{\mathsf{FP}, \mathsf{ABS}, \mathsf{LW}, \mathsf{LI}, \mathsf{OE}\}$
- $t_X^{(3)} \models \{\mathsf{OE}, \mathsf{FP}, \mathsf{ABS}, \mathsf{LI}\} \wedge t_X^{(3)} \not\models \{\mathsf{VT}, \mathsf{LW}\}$

Assume three personas as prototypical customers: Skier A is impatient, skilled in skiing, wants to explore a fun-park but is not afraid of difficult slopes or needs lunch since he wants his workout. Boarder B is an explorer, she wants a large number of different slopes (except for black ones) but accepts to wait. Rookie C started skiing and wants to avoid black slopes. He appreciates a tour of easy slopes and lunch.

Figure 2 depicts corresponding constraint relationship graphs extracted from this description and Table 1 shows how the assignments are evaluated using these relationships. Every user favors a different assignment. A indicated little interest in variety, avoiding black slopes or easy tracks and got the only assignment that does not require waiting. Similarly, we get a match for B's requirements and do not force C to take difficult slopes. The calculated assignment winners thus make sense and show how the different graphs influence the decision process to a strong degree. Interestingly for B, the selected dominance property affects the preferred solution. Since $t_X^{(2)}$ only satisfies VT and violates the 5 other constraints, it is considered worse than $t_X^{(3)}$ in SPD and DPD semantics. However, as VT is the most important constraint for B and is only satisfied by $t_X^{(2)}$, in a TPD semantics this solution is still preferred over the others satisfying more constraints.

Table 1 Different tours rated by different relationship graphs

	A	B	C
$t_X^{(1)}$	**3**	8	7
$t_X^{(2)}$	9	13	16
$t_X^{(3)}$	5	**7**	**5**

(a) SPD semantics

	A	B	C
$t_X^{(1)}$	**3**	11	17
$t_X^{(2)}$	14	13	30
$t_X^{(3)}$	10	**10**	**5**

(b) DPD semantics

	A	B	C
$t_X^{(1)}$	**3**	27	44
$t_X^{(2)}$	17	**19**	65
$t_X^{(3)}$	13	25	**6**

(c) TPD semantics

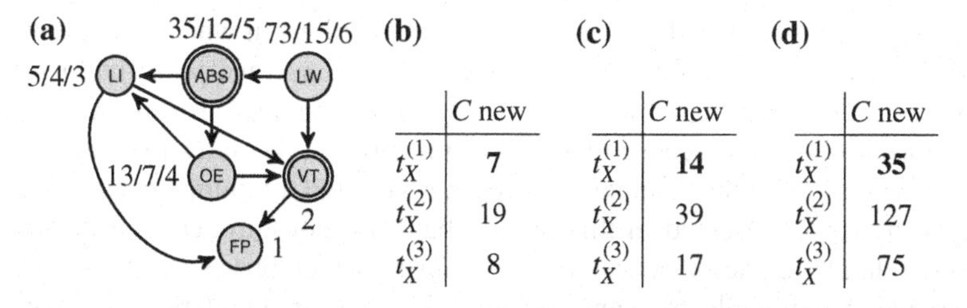

(b)

	C new
$t_X^{(1)}$	**7**
$t_X^{(2)}$	19
$t_X^{(3)}$	8

(c)

	C new
$t_X^{(1)}$	**14**
$t_X^{(2)}$	39
$t_X^{(3)}$	17

(d)

	C new
$t_X^{(1)}$	**35**
$t_X^{(2)}$	127
$t_X^{(3)}$	75

Fig. 3 After adaption, $t_X^{(1)}$ is now preferred by rookie C. **a** Graph for rookie C new **b** SPD semantics **c** DPD semantics **d** TPD semantics

Changing Preferences

In multi-agent-systems, agents change their goals depending on their perceived environmental situation [9]. Similarly, our personas may change their constraint relationships given new circumstances.

Assume C has gotten enough practice such that avoiding black slopes is not as important as before—but he refuses to wait. Hence, the edge $\mathsf{ABS} \succ_R \mathsf{LW}$ gets inverted, making LW the most important constraint. Assignment $t_X^{(1)}$ is the only one that has a route without much waiting—and it is now favored by C (see Fig. 3).

Changing Constraints

In open-world scenarios and dynamic CSPs, the set of constraints frequently changes. Assume that slopes have been evaluated for beautiful landscapes (BL) and foggy slopes (FS) can be avoided.

Given that A does not care too much about landscapes it is safe to assume that those constraints would be ranked even less important than LI. Boarder B, however, does care about BL and marks them more important than FS, FP, and ABS. Assume further that $t_X^{(1)} \models \{\mathsf{BL}, \mathsf{FS}\}$, $t_X^{(2)} \models \{\mathsf{FS}\}$ and $t_X^{(3)} \models \{\mathsf{FS}\}$.

It is easy to calculate that A still ranks $t_X^{(1)} >_R^{\mathrm{TPD}} t_X^{(3)} >_R^{\mathrm{TPD}} t_X^{(2)}$ even though different numeric values are placed. For B the situation is different, as BL is only satisfied by $t_X^{(1)}$ which is why it would then be the preferred solution of B.

5 Conclusions and Future Work

We have presented constraint relationships, an approach to flexibly and intuitively express preferences over constraints in a qualitative way. Constraint relationships establish a partial preference order on the constraints that can be transformed into a k-weighted CSP while preserving dominance properties. The process is tool-supported and can be applied to static and dynamic problems. Classical solution algorithms for weighted CSPs or off-the-shelf constraint optimizers can be used to solve the resulting problem. We showed a typical example that benefits from constraint relationships.

The approach is especially well-suited for applications in which constraints are added and changed at runtime. As we are mainly concerned with the engineering of self-adaptive systems, we are going to apply the approach to the dynamic adaptation of CSPs that occur in these systems. From a theoretical standpoint, we want to explore different dominance properties and find a minimal c-semiring that respects the rules presented in this paper—but does not introduce any additional ordering relation unlike the k-weighted c-semiring which makes any two solutions comparable. Additionally, we want to examine how constraint relationships can contribute to preference elicitation or preference learning. In particular, elicitation of constraint relationships by comparing solutions satisfying different soft constraints using abductive reasoning is planned.

Currently, we are looking into the use of constraint relationships to synthesize sub-models into a global CSP in a multi-agent-system. Such a process allows the combination of characteristics of individual agents that are not known at design time into a common model that can then be solved centrally. One of the applications of this technique are decentralized energy management systems [21] in which power plants are combined in a self-organizing fashion to satisfy a changing power load.

Acknowledgments This work has been partially funded by the German Research Foundation (DFG) in the research unit FOR 1085 "OC Trust–Trustworthy Organic Computing Systems". We would like to thank María Victoria Cengarle for fruitful discussions and the anonymous reviewers for the constructive feedback that led to an improvement of the manuscript.

References

1. Andréka, H., Ryan, M., Schobbens, P.Y.: Operators and Laws for Combining Preference Relations. J. Log. Comput. **12**(1), 13–53 (2002).
2. Bessière, C.: Arc-Consistency in Dynamic Constraint Satisfaction Problems. In: T.L. Dean, K. McKeown (eds.) Proc. 9th Nat. Conf. Artificial Intelligence (AAAI'91), pp. 221–226. AAAI Press (1991).
3. Bistarelli, S., Montanari, U., Rossi, F.: Constraint Solving over Semirings. In: Proc. 14th Int. Joint Conf. Artificial Intelligence (IJCAI'95), vol. 1, pp. 624–630. Morgan Kaufmann (1995).
4. Borning, A., Freeman-Benson, B., Wilson, M.: Constraint Hierarchies. LISP Symb. Comp. **5**, 223–270 (1992).
5. Boutilier, C., Brafman, R.I., Domshlak, C., Hoos, H.H., Poole, D.: CP-nets: A Tool for Representing and Reasoning with Conditional Ceteris Paribus Preference Statements. J. Artif. Intell. Res. **21**, 135–191 (2004).
6. Boutilier, C., Brafman, R.I., Geib, C., Poole, D.: A Constraint-based Approach to Preference Elicitation and Decision Making. In: AAAI Spring Symp. Qualitative Decision Theory, pp. 19–28 (1997).
7. Brafman, R., Domshlak, C.: Preference Handling - An Introductory Tutorial. AI Magazine **30**(1), 58–86 (2009).
8. Dechter, R., Dechter, A.: Belief Maintenance in Dynamic Constraint Networks. In: H.E. Shrobe, T.M. Mitchell, R.G. Smith (eds.) Proc. 7th Nat. Conf. Artificial Intelligence (AAAI'88), pp. 37–42. AAAI Press (1988).
9. Doyle, J., McGeachie, M.: Exercising Qualitative Control in Autonomous Adaptive Survivable Systems. In: Proc. 2nd Int. Conf. Self-adaptive software: Applications (IWSAS'01), *Lect. Notes Comp. Sci.*, vol. 2614, pp. 158–170. Springer (2003).

10. Dubois, D., Fargier, H., Prade, H.: The Calculus of Fuzzy Restrictions as a Basis for Flexible Constraint Satisfaction. In: Proc. 2nd IEEE Int. Conf. Fuzzy Systems, vol. 2, pp. 1131–1136 (1993).
11. Freuder, E.C., Wallace, R.J.: Partial Constraint Satisfaction. Artif. Intell. **58**(1–3), 21–70 (1992).
12. Gelain, M., Pini, M.S., Rossi, F., Venable, K.B.: Dealing with Incomplete Preferences in Soft Constraint Problems. In: C. Bessière (ed.) Proc. 13th Int. Conf. Principles and Practice of Constraint Programming (CP'07), *Lect. Notes Comp. Sci.*, vol. 4741, pp. 286–300. Springer (2007).
13. Jampel, M.: A Brief Overview of Over-constrained Systems. In: M. Jampel, E. Freuder, M. Maher (eds.) Over-Constrained Systems, *Lect. Notes Comp. Sci.*, vol. 1106, pp. 1–22. Springer (1996).
14. Larrosa, J.: Node and Arc Consistency in Weighted CSP. In: R. Dechter, R.S. Sutton (eds.) Proc. 18th Nat. Conf. Artificial Intelligence (AAAI'02), pp. 48–53. AAAI Press (2002).
15. Meseguer, P., Rossi, F., Schiex, T.: Soft Constraints. In: F. Rossi, P. van Beek, T. Walsh (eds.) Handbook of Constraint Programming, chap. 9. Elsevier (2006).
16. Mittal, S., Falkenhainer, B.: Dynamic Constraint Satisfaction. In: H.E. Shrobe, T.G. Dietterich, W.R. Swartout (eds.) Proc. 8th Nat. Conf. Artificial Intelligence (AAAI'90), vol. 1, pp. 25–32. AAAI Press (1990).
17. Rossi, F.: Preferences, Constraints, Uncertainty, and Multi-Agent Scenarios. In: Proc. Int. Symp. Artificial Intelligence and Mathematics (ISAIM'08) (2008).
18. Rossi, F., Venable, K.B., Walsh, T.: Preferences in Constraint Satisfaction and Optimization. AI Mag. **29**(4), 58–68 (2008).
19. Schiex, T., Fargier, H., Verfaillie, G.: Valued Constraint Satisfaction Problems: Hard and Easy Problems. In: Proc. 14th Int. Joint Conf. Artificial Intelligence (IJCAI'95), vol. 1, pp. 631–639. Morgan Kaufmann (1995).
20. Shapiro, L.G., Haralick, R.M.: Structural descriptions and inexact matching. IEEE Trans. Pattern Analysis and, Machine Intelligence **PAMI-3**(5), 504–519 (1981).
21. Steghöfer, J.P., Anders, G., Siefert, F., Reif, W.: A System of Systems Approach to the Evolutionary Transformation of Power Management Systems. In: Proc. INFORMATIK 2013 Wsh. "Smart Grids", Lect. Notes Inform. Bonner Köllen Verlag (2013).

Short Papers

A Fuzzy Logic-Based Decision Support System for the Diagnosis of Arthritis Pain for Rheumatic Fever Patients

Sanjib Raj Pandey, Jixin Ma, Choi-Hong Lai and Chiyaba Njovu

Abstract This paper describes our conceptual ideas and future development framework of Decision support System to diagnose of arthritis pain and determine whether the pain is associated with rheumatic fever or not. Also, it would be used for diagnosing arthritis pain (only for rheumatic fever patients), in four different stages, namely: Fairly Mild, Mild, Moderate and Severe. Our diagnostic tool will allow doctors to register symptoms describing arthritis pain using numerical values that are estimates of the severity of pain that a patient feels. These values are used as input parameters to the system, which invokes rules to determine a value of severity for the arthritis pain.

1 Introduction

Rheumatic Fever and Rheumatic Heart Diseases are considered to be two of the biggest health risks among children compares to other heart diseases in Nepal. Cases of RHD emanate from an untreated RF in children. In Nepal, there is no appropriate decision support system to capture and diagnose all the cases of RF. This situation could be improved with the use of Computer-based diagnostic tools that doctors could use to register all the symptoms observed in children across the country [1, 2].

S. R. Pandey (✉) · J. Ma · C.-H. Lai · C. Njovu
School of Computing & Mathematical Science, University of Greenwich,
Park Row SE10 9LS, London
e-mail: S.R.Pandey@gre.ac.uk

J. Ma
e-mail: J.Ma@gre.ac.uk

C.-H. Lai
e-mail: C.H.Lai@gre.ac.uk

C. Njovu
e-mail: C.Njovu@gre.ac.uk

M. Bramer and M. Petridis (eds.), *Research and Development in Intelligent Systems XXX*, DOI: 10.1007/978-3-319-02621-3_18,

In Nepal, doctors and other rural health workers rely on their personal professional knowledge, experiences and beliefs to diagnose cases of RF by recording observed symptoms in a suspected RF patient. This is by no means accurate as a judgment or a belief and varies from one individual to another. During our field visit to Nepal Heart Foundation in Nepal to understand the situation of RF, country's local practice and procedure to diagnose of RF, data collection of RF case etc. I have discussed with experts and concluded to divide the diagnosis of arthritic pain into four different stages (fairly mild, mild, moderate and severe) and identify weather the arthritis pain is related with RF or not. As a starting point, we described a conceptual framework in this paper, how to apply fuzzy logic in order to categories the arthritis pain in the four different categories and also identify the arthritis pain weather associated with RF or not.

2 Conceptual Model and Method

Our proposed conceptual model architecture that employs a fuzzy logic component is shown in Fig. 1. The fuzzy logic system maps a crisp input value (real- valued) into a crisp output value. The Fuzzy logic inference model consists of four components, namely: fuzzification, fuzzy inference engine, fuzzy rules and defuzzification. The fuzzification process takes an input crisp value representing doctors' beliefs of the severity in an observed patient. These values are then used in a fuzzy membership function that converts them into a value representing a degree of belief. Fuzzy inference engine is a reasoning process that will map the degree of belief into an output using a chosen membership function. These values are then used in conjunction with suitable logical operators and fuzzy IF-THEN rules. The result is the subject of defuzzification that is the process of translating fuzzy logic results into crisp values used as final output from the system. This process is described in the fuzzy knowledge and Rule base model architecture as shown in Fig. 1.

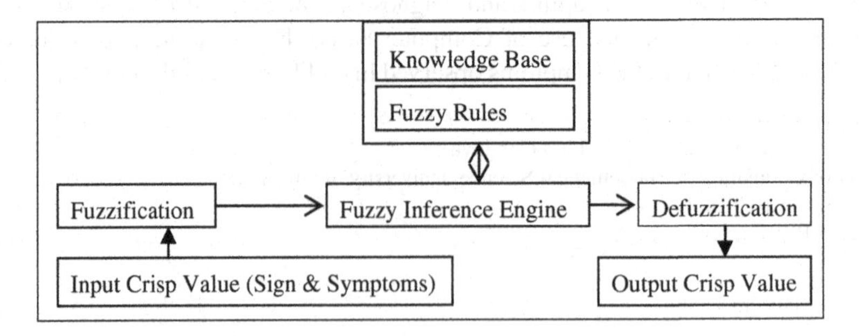

Fig. 1 Fuzzy knowledge and rule-based model

2.1 Fuzzy Inferences System for Arthritis Pain

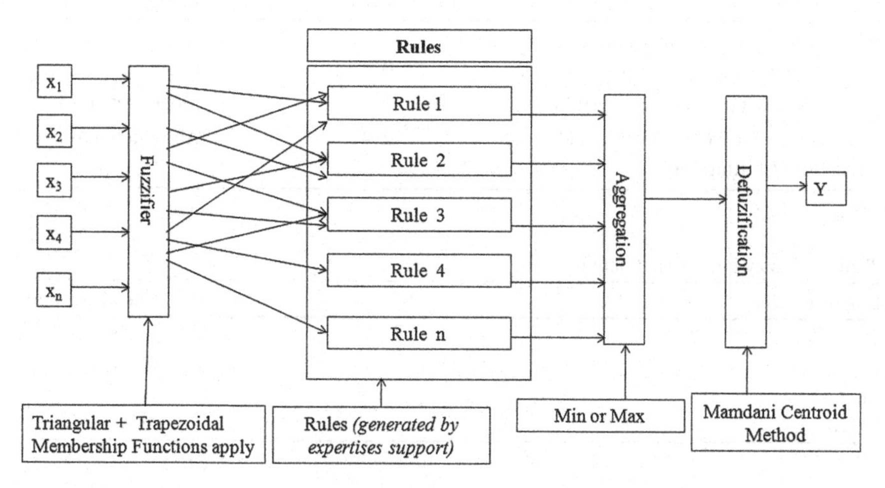

Fuzzy triangular and trapezoidal membership functions will be applied to determine the degree of symptoms to ascertain the doctor's belief of symptom's severity. The crisp value (0–4) will be assigned by a doctor based on his/her experiences and beliefs. These values will be used in the fuzzification process by applying the membership functions to obtain the degree (membership degree). The fuzzy inference mechanism will map the entire set of rules with membership degrees and fuzzy logical operators will be used to evaluate the strength of firing rules. Then the output of each rule is aggregated and Mamdani's Centre of Gravity (CoG) methods are applied for the defuzzification process.

The signs and symptoms of arthritis are determined by linguistic variables (e.g. pain, hotness, swelling, movement restriction etc) with fuzzy interval and linguistic label (none, fairly positive, moderately positive, and absolutely positive). Three input variables and one output variable are designed to diagnose the stage of arthritis pain for RF patients. The following signs and symptoms are considered of arthritis pain for RF patients are given below.

1. Severe pain in the joint:
    a. Ankle
    b. Wrists
    c. Knee
    d. Elbows
2. Pain associated with:
    a. Hotness
    b. Redness
    c. Swelling
    d. Movement Restriction

Table 1 Input variable, ranges and fuzzy set

Input variables	Ranges	Fuzzy sets
All arthritis signs and symptoms: pain (ankles, knees, elbows, wrists), pain associated with (hot-ness, swelling, movement restriction, redness) and migratory/shifting pain	0	None
	0–2	Fairly positive
	1–3	Moderately positive
	2–4	Absolutely positive

Table 2 Fuzzy interval and output

Linguistic label	Fuzzy interval
Fairly mild	<0.1
Mild	0.5–2
Moderate	1–3
Severe	2–4

3. Migratory or Shifting Pain

The input variables, value ranges and associated fuzzy sets for each signs and symptoms are given in Table 1. Fuzzy interval and linguistic label of output are given in Table 2.

2.2 Fuzzy Rule for Arthritis Pain

Fuzzy rules are formations to control the output variables. A fuzzy rule consists of IF <Condition> AND/OR/NOT <Consequences>. AND, OR, NOT logical operators can be applied if rule has multiple parts, choosing the logical operator depend upon the nature of the rule and its associated output factor. The format and structures of rules for diagnosis of arthritis pain for RF patients are given below in Table 3. In the rules AP, MP, FP stands for absolutely positive, moderately positive and fairly positive respectively. In the rule no 36, the output label of "Not RF" means these types of arthritis pain is not associated with RF, laboratory and other test is required to diagnose the problems.

3 Defuzzification

The defuzzification method translates the output from the fuzzy inference engine into crisp output. The output of the defuzzification process is in single number form. Various methods are available Mean of Maxima, Weighted Average etc but the most used method is Center–of–Gravity (COG).

Table 3 Sample of rules for arthritis pain diagnosis for rheumatic fever patients

Rule number	Rule descriptions
Rule 1:	IF arthritis pain is AP AND pain associated with is AP AND migratory pain is AP THEN diagnosis arthritis pain="Severe"
Rule 7:	IF arthritis pain is AP AND pain associated with is MP AND migratory pain is FP THEN diagnosis arthritis pain="Moderate"
Rule 24:	IF arthritis pain is FP AND pain associated with is MP AND migratory pain is FP THEN diagnosis arthritis pain="Mild"
Rule 36:	IF arthritis pain is FP AND pain associated with is AP AND migratory pain is NONE THEN diagnosis arthritis pain="Not RF"

4 Conclusion

The main aim of this paper is to design a conceptual framework to diagnosis the stages of arthritis pain into different stages by observing the patients' pain using fuzzy logic. Fuzzy logic has good capabilities in dealing with imprecise, uncertain data, fuzzy logic and fuzzy theory already been proven as a potential tool for developing and employing a medical decision support system [3–9]. We hope this framework will assist the DSS system to measure the level of arthritis pain in different stages that will support to determine whether the exiting arthritis pain is associated with RF or not. We are developing a model using Matlab Fuzzy Toolbox and will apply the aforementioned framework and its rules for diagnosis of arthritis pain for RF patients. Once the model has been completed then it will be evaluated and tested with experts. In further research, we will apply fuzzy logic to develop a complete model for diagnosis of RF based on the Nepal local's and expertise's guidelines.

References

1. Michael Edwards, PhD, MPH, Theo Lippeveld, MD, "Decision Support Systems for Improving the Analytic Capacity of Decentralized Routine Health Information Systems in Developing Countries" http://csdl2.computer.org/comp/proceedings/hicss/2004/2056/06/205660152b.pdf
2. Jonathan R Carapetis and Liesl J Zühlke, "Global research priorities in rheumatic fever and rheumatic heart disease", Ann Pediatr Cardiol. 2011 Jan-Jun; 4(1): 4–12. http://www.ncbi.nlm.nih.gov/pmc/articles/PMC3104531/?tool=pubmed
3. Carmen De Maio, Vincenzo Loia, Giuseppe Fenza, Mariacristina, Roberto Linciano, Aldo Morrone, "Fuzzy Knowledge Approach to Automatic Disease Diagnosis", 2011 IEEE International Conference on Fuzzy Systems, June 27–30, 2011, Taipei, Taiwan.
4. Hepu Deng and SantosoWibowo, "A Rule-Based Decision Support System for Evaluating and Selecting IS Projects", Proceedings of the International Multi Conference of Engineers and Computer Scientists 2008 Vol II, IMECS 2008, 19–21 March, 2008, Hong Kong.
5. Obi J.C., Imainvan A.A "Decision Support System for the Intelligent Identification of Alzheimer using Neuro Fuzzy Logic", International Journal on, Soft Computing, 2(2), May 2011.

6. X.Y. Djam1,*, G. M. Wajiga2, Y. H. Kimbi3 and N.V. Blamah4 "A Fuzzy Expert System for the Management of Malaria" Int. J. Pure Appl. Sci. Technol., 5(2) (2011), pp. 84–108, International Journal of Pure and Applied Sciences and Technology, ISSN 2229–6107.
7. William Siler and James Buckley, "Fuzzy Expert System and Fuzzy Reasoning" Wiley & Sons, Inc pp, 49–50 2005.
8. Stephen Yurkovich, Kevin M. Passino, Department of Electrical Engineering, The Ohio State University, "Fuzzy Control", Copyright 1998, Addison Wesley Longman, Inc. ISBN 0-201-18074-X.
9. Stephen Yurkovich, Kevin M. Passino, Department of Electrical Engineering, The Ohio State University, "Fuzzy Control", Copyright 1998, Addison Wesley Longman, Inc. ISBN 0-201-18074-X.

A Viewpoint Approach to Structured Argumentation

Nouredine Tamani, Madalina Croitoru and Patrice Buche

Abstract We introduce in this paper a viewpoint-based argumentation approach in the context of the EcoBioCap project, which requirements are different actor arguments expressed over several criteria, describing the objects of a domain, to support/oppose contradictory goals. A viewpoint is an ASPIC+ argumentation system defined over a subset of rules corresponding to a single criterion. Concepts of conflict between viewpoints, independent viewpoints, and collection of independent viewpoints are the basis of our argumentation approach.

1 Introduction and Motivation

Within the framework of the European project EcoBioCap (www.ecobiocap.eu) about the design of next generation packagings using advanced composite structures, we aim at developing a Decision Support System (DSS). The DSS is made of two part: (i) a flexible querying process [1], which is based on a bipolar approach dealing with imprecise data corresponding to the packaging characteristics such as the optimal permeance, its dimensions, shape, etc., and (ii) an argumentation process which aims at aggregating several stakeholders requirements and preferences expressed as simple textual arguments, to enrich the querying process. The different kinds of stakeholders (sanitary, commercial, ecology oriented, etc.) express the constraints

N. Tamani (✉)
IATE, INRA-Supagro, INRIA, 2 place Pierre Viala, Montpellier 34060, France
e-mail: nouredine.tamani@supargo.inra.fr; ntamani@gmail.com

M. Croitoru
INRIA GraphIK, LIRMM, 161 rue Ada, Montpellier 34095, France
e-mail: croitoru@lirmm.fr

P. Buche
IATE, CIRAD-INRA-Supagro-UM2, 2 place Pierre Viala, Montpellier 34060, France
e-mail: patrice.buche@supargo.inra.fr

M. Bramer and M. Petridis (eds.), *Research and Development in Intelligent Systems XXX*, 265
DOI: 10.1007/978-3-319-02621-3_19,

about packagings reusing their domain ontologies expressed as simple relational description logics such as the DLR-Lite [2] language. Stakeholder's specification i, made of concepts and facts, is extracted from a database enriched with the ontological rules of each domain, to build a partial knowledge base $\mathcal{K}_{\mathcal{S}_i}$. Unfortunately the union of the stakeholder knowledge bases $\mathcal{K} = \bigcup_{i=1,\ldots,n} \mathcal{K}_{\mathcal{S}_i}$ is inconsistent. Maximal consistent subsets of this inconsistent knowledge base should be identified for decision making. Having the domain expert rules already exist in different ontological modules, a structured argumentation system such as ASPIC+ [3, 4] can be used on top of the union of stakeholders ontologies.

ASPIC+ is based on a generic logical language equipped with a contrariness function and certain assumptions in inference rules form. The attack and defeat relations are then defined based on this logical language, and rationality postulates are used to ensure the quality of the output. However, when we reason about objects according to different aspects characterizing them, we can face a situation in which ASPIC+ returns empty extensions, under any considered semantics. This situation happens when an object is accepted according to some criteria and rejected according other. More precisely, the closure under transposition assumption on the language makes the system infer conclusions which could be counter intuitive.

Argumentation theory in general [5–7] is actively pursued in the literature, some approaches even combining argumentation and multi criteria decision making [8]. The argumentation process will allow to identify maximally consistent sets but also allow the justification of why an argument (a certain packaging is chosen or not) is accepted. The problem at hand does not simply consist in addressing a multi-criteria optimisation problem [9]: the domain experts would need to be able to justify why a certain packaging (or set of possible packagings) is chosen.

There are several ways of fixing this problem. A simple and natural solution consists of modeling the rules within the DeLP (combining logic programming and defeasible argumentation) approach [10]. However, the rules come from domain ontologies in each field and should be treated as strict ontological rules, even when put together with other strict ontological rules contradicting them. In addition, it has been proven in [11] that the DeLP approach can deliver incomplete and inconsistent conclusions. Another solution (and the one presented in this paper) is to separate the criteria possibly conflicting. In such way the rationality postulates are not violated. The decision would then be made based on viewpoints (corresponding to such criteria). Our approach is similar in spirit to the intuitions behind Value based Argumentation Frameworks [12], where the strength of an argument corresponds to the values it promotes. While this approach is very close in nature to our intuitions, it cannot be applied since in our case a value can be "split" into several audiences: there could be contradictory goals even from the same viewpoint. The notion of viewpoint and goals introduced in this setting also remind those proposed by [13].

In this paper, we introduce the notion of a "viewpoint" as an instance of ASPIC+ built on a single criterion, which allows stakeholders to argue about goals solely from the criterion's standpoint. We also define the concepts of conflicting viewpoints, independent viewpoints, and collections of viewpoints which forms groups of pros or cons. It is easy to show that collection output satisfies the rationality postulates (closure under the strict rules and consistency). Collections are finally used to express a consensual query addressed to the considered database.

2 Viewpoint-Based Argumentation Approach

Definition 1 (Viewpoint AS). A viewpoint argumentation system is a tuple VAS $= (\mathcal{L}, \mathcal{R}, \mathcal{G}, \mathbb{C}, \mathcal{V}, -, n)$ such that:

- $\mathcal{L}$: is the logical language of the system,
- $\mathcal{R} = \mathcal{R}_s \cup \mathcal{R}_d$: is the set of rules defined over formulas of $\mathcal{L}$, s.t. $\mathcal{R}_s$ is the set of strict rules (not closed under transposition), and $\mathcal{R}_d$ is the set of defeasible rules,
- $\mathcal{G}$: is the set of unitary goals to reach containing pairs of formulas of form $(g, -g)$, defined by the user.
- $\mathbb{C}$: is the set of criteria upon which we defined the function $\tau : \mathbb{C} \rightarrow 2^{\mathcal{R}}$ with $\nexists r \in \mathcal{R}_s$ and $\nexists c_1, c_2 \in \mathbb{C}$, such that $c_1 \neq c_2$ and $r \in \tau(c_1)$ and $r \in \tau(c_2)$,
- $\mathcal{V}$: is the set of views expressed over criteria $\mathbb{C}$ such that:
  - $\mathcal{V} = \bigcup_{i=1,\dots,m} v_i$, where v_i is a view, and $m = |\mathbb{C}|$,
  - $\forall i \in \{1, \dots, m\}, v_i = (\mathcal{L}, \mathcal{R}_{v_i}, c, -, n)$ is an ASPIC+ instance defined on criterion c, where $\mathcal{R}_{v_i} = \mathcal{R}_{sv_i} \cup \mathcal{R}_{dv_i}$ is the set of rules of v_i, such that:
    - $\mathcal{R}_{sv_i} = Cl_{tp}(\mathcal{R}_{scv_i})$, where $\mathcal{R}_{scv_i} \subseteq \tau(c)$ is the set of strict rules expressed on criterion c, and $Cl_{tp}(\mathcal{R}_{scv_i})$ stands for the closure under transposition of the set of strict rules $\mathcal{R}_{scv_i}$,
    - $\mathcal{R}_{dv_i} = \mathcal{R}_{dcv_i}$, where $\mathcal{R}_{dcv_i} \subseteq \tau(c)$ is the set of defeasible rules expressed on criterion c.
- $-$ is a contrariness function (see [3, 4] for further details),
- n is a naming convention of defeasible rules [3], not used.

A view defined by $(\mathcal{L}, \mathcal{R}, c, -, n)$ is an instance of the ASPIC+ AS.

Definition 2 (View output). Let VAS $= (\mathcal{L}, \mathcal{R}, \mathcal{G}, \mathbb{C}, \mathcal{V}, -, n)$ be a viewpoint AS. The output of view $v_i \in \mathcal{V}$ is $Output(v_i) = \bigcap_{j=1,\dots,n} E_{ij}$,[1] such that $E_{ij, j=1,\dots,n}$ are extensions under a given semantics of v_i. A view v_i is said tenable iff $Output(v_i) \neq \emptyset$.

Definition 3 (Conflict relation $\mathcal{C}_v$). Let VAS $= (\mathcal{L}, \mathcal{R}, \mathcal{G}, \mathbb{C}, \mathcal{V}, -, n)$ be a viewpoints AS. $\mathcal{C}_v \subseteq \mathcal{V} \times \mathcal{V}$ is a conflict relation such that v_i is in conflict with a view v_j, denoted $(v_i, v_j) \in \mathcal{C}_v$, iff $\exists \varphi \in Output(v_i), \exists \psi \in Output(v_j) : \varphi \in -\psi$.

Example 1 We consider researchers and cheese producers arguing about cheese packagings. Researchers focus on the atmosphere properties (denoted atm) and suggest that "wheat gluten based packagings are suitable because they provide a good

[1] This follows the line of work of [11], because credulous attitude can lead to inconsistencies.

atmosphere control", and cheese producers considering the *interaction* with the product (denoted int) retort with "gluten cannot be put in contact with cheese because the bacteria in the crust would eat the gluten, degrading the packaging". We define the VAS $= (\mathcal{L}, \mathcal{R}, \mathcal{G}, \mathbb{C}, \mathcal{V}, -, n)$ as:

- Strict rules $\mathcal{R}_s = \{ACP \to SP,\ DP \to -SP\}$ and defeasible rules $\mathcal{R}_d = \{\emptyset \Rightarrow GP\}$,
- $\mathcal{V} = \{v_{atm}, v_{int}\}$, s.t. v_{atm}, v_{int} are viewpoints defined on criteria $\mathbb{C} = \{atm, int\}$,
- The goals of the system are to get (non)suitable packagings: $\mathcal{G} = \{SP, -SP\}$,
- The contrariness function $-$ corresponds to the classical negation,
- The knowledge base is then $\mathcal{K}_a = \{Packaging, SP, ACP, DP\}$, and $\mathcal{K}_p = \{GP\}$.

The researchers' view $v_1 = v_{atm}$ contains the rules $\mathcal{R}_{v_1} = (\mathcal{R}_{sv_1}, \mathcal{R}_{dv_1})$ with:

- $\mathcal{R}_{sv_1} = \{1.\,ACP \to SP,\ 2. - SP \to -ACP,\ 3.\,GP \to ACP,\ 4. - ACP \to -GP\}$
- $\mathcal{R}_{dv_1} = \{\emptyset \Rightarrow GP\}$

Researchers' arguments under $\mathcal{R}_{sv_1}$ and $\mathcal{R}_{dv_1}$ are then:
• $A_1 : \emptyset \Rightarrow GP$ • $A_2 : A_1 \to ACP$ (rule 3) • $A_3 : A_2 \to SP$ (rule 1)
The set of arguments $a = \{A_1, A_2, A_3\}$ forms the only preferred extension in the defined view v_{atm}. $Output(v_{atm}) = Concs(a) = \{GP, ACP, SP\}$.
Cheese makers' view $v_2 = v_{int}$ defined by the rules $\mathcal{R}_{sv_2} = (\mathcal{R}_{sv_2}, \mathcal{R}_{dv_2})$ with:

- $\mathcal{R}_{sv_2} = \{\ 5.\,DP \to -SP,\ 6.\,SP \to -DP,\ 7.\,GP \to DP,\ 8. - DP \to -GP\}$
- $\mathcal{R}_{dv_2} = \{\emptyset \Rightarrow GP\}$.

The consumers' arguments are then:
• $B_1 : \emptyset \Rightarrow GP$ • $B_2 : B_1 \to DP$ (rule 7) • $B_3 : B_2 \to -SP$ (rule 5)
The set of arguments $b = \{B_1, B_2, B_3\}$ forms the only preferred extension in view v_{san}, and $Output(v_{san}) = Concs(b) = \{GP, DP, -SP\}$. View v_{atm} is in conflict with v_{int} and vice-versa. Please see [14] for more detailed examples.

The views of a *VAS* have to be pairwise independent to avoid anomalies in the reasoning, as discussed below. Let r be a rule of the form $\phi_1, \ldots, \phi_m \to / \Rightarrow \psi$. We denote $\{\phi_1, \ldots, \phi_m\}$ as $H(r)$ corresponding to hypotheses of rule r, and ψ as $C(r)$, corresponding to its consequence. A rule r can be then expressed $H(r) \to / \Rightarrow C(r)$.

Definition 4 (Dependent/independent views). Let VAS $= (\mathcal{L}, \mathcal{R}, \mathcal{G}, \mathbb{C}, \mathcal{V}, -, n)$ be a viewpoints AS. Let $\mathcal{D} \subseteq \mathcal{V} \times \mathcal{V}$ be a dependence relation. We say that a view v_i depends on v_j, denoted $(v_i, v_j) \in \mathcal{D}$, iff $\exists r \in \mathcal{R}_{v_i}, \exists r' \in \mathcal{R}_{v_j}$ s.t. $C(r') \notin \mathcal{G}$ and $C(r') \in H(r)$. Views v_i and v_j are said independent iff $(v_i, v_j) \notin \mathcal{D} \wedge (v_j, v_i) \notin \mathcal{D}$.

Proposition 1 *(Independency). If there exists dependent views in a VAS* $= (\mathcal{L}, \mathcal{R}, \mathcal{G}, \mathbb{C}, \mathcal{V}, -, n)$*, then the output of its collections can be incomplete and inconsistent.*

The proof of the proposition 1 is detailed in the technical report [14].

Definition 5 (Collection C_g of views on goal g). Let VAS $= (\mathscr{L}, \mathscr{R}, \mathscr{G}, \mathbb{C}, \mathscr{V}, -, n)$ be a viewpoints AS. A collection of views $C \subseteq \mathscr{V}$ on goal $g \in \mathscr{G}$, denoted by C_g, is the maximal set (w.r.t. inclusion) of non-conflicting and tenable views that share the same goal g. More formally:
$C_g = \{v \in \mathscr{V}, \nexists v' \in C_g : (v, v') \in \mathscr{C}_v \wedge ((v, v') \in \mathscr{D} \vee (v', v) \in \mathscr{D}) \wedge g \in Output(v)\}$.

Example 2 In Example 1, we can form collections $C_{SP} = \{v_{atm}\}$ and $C_{-SP} = \{v_{int}\}$ which indicate the (in)compatible criteria for packagings selection.

Definition 6 (Collection output). The output of C_g, a collection on goal g, is the set of justified conclusions of its views: $Output(C_g) = \bigcup_{i=1,\ldots,k} Output(v_i), v_i \in C_g$.

Using the collections in database querying. We can write a consensual query from the obtained collections by carrying out an analogical reasoning, which generalizes results delivered by the argumentation process applied upon instances. An instance can help to understand the involved stakeholders' needs and then to express, based on arguments pros and cons, a query reflecting the way objects should be selected from a database. In above example, the desired packagings are those that ensure a good atmosphere control and have no risky interaction with the packed product.

3 Rationality Postulates Inside a Collection

Each viewpoint in a VAS satisfies the rationality postulates. It is also important that all its delivered collections are consistent and closed under the strict rules involved in them to ensure the reliability and the completeness of the reasoning process.

Let $C_g = \{v_1, \ldots, v_{m'}\}$ be a collection of m' independent views on goal g. The strict rules of C_g is $\mathscr{R}_{sC_g} = \bigcup_{i=1,\ldots,m'} \mathscr{R}_{sv_i}$. C_g is closed under $\mathscr{R}_{sC_g}$ iff $Output(C_g) = Cl_{\mathscr{R}_{sC}}(Output(C_g))$.

Proposition 2 *C_g is closed under $\mathscr{R}_{sC_g}$.*

Proposition 3 *If $\mathscr{R}_{sC_g}$ is consistent then its output is also consistent.*

The proof of the propositions 2 and 3 are detailed in the technical report [14].

We notice that if $\mathscr{R}_{sC_g}$ is inconsistent then $\mathscr{R}_s$ of the related *VAS* is also inconsistent. Finally, a viewpoint argumentation system *VAS* $= (\mathscr{L}, \mathscr{R}, \mathscr{G}, \mathbb{C}, \mathscr{V}, -, n)$ limits the effect of contaminating formulas [15], since ill-formed rules cannot contaminate other syntactically disjoint views. Moreover, the closure under transposition does not propagate ill-formed rules to all formulas, and it is possible to discard an ill-formed rule from a view as in [16] to prevent the crash of the view.

4 Discussion

In the proposed approach, a view can be attached to an agent, as in [17–22], and a collection can form a coalition. However, we deal with criteria and we group non-conflicting views according to goals and not to achieve a task, and no utility function is needed to compute the action cost (attribute values [22]) or preordering relations between coalitions (as in [17, 18]). The approach can also be considered from merging Dung's based argumentation system standpoint [23], but our work is different since we do not add or delete any attacks between arguments.

Since we build a collection satisfying a goal and another satisfying its contradiction, then this approach can be seen as a bipolar approach for coalition building as in [24, 25]. The main difference lies in the fact that we define only one relation between arguments (attack relation) and not two relations: attack and support.

Acknowledgments The research leading to these results has received funding from the European Community's Seventh Framework Programme (FP7/ 2007-2013) under the grant agreement n°FP7-265669-EcoBioCAP project.

References

1. S. Destercke, P. Buche, and V. Guillard. A flexible bipolar querying approach with imprecise data and guaranteed results. *Fuzzy sets and Systems*, 169:51–64, 2011.
2. S. Colucci, T. D. Noia, A. Ragone, M. Ruta, U. Straccia, and E. Tinelli. *Semantic Web Information Management*, chapter 19 : Informative Top-k retrieval for advanced skill management, pages 449–476. Springer-Verlag Belin Heidelberg, 2010.
3. H. Prakken. An abstract framework for argumentation with structured arguments. Technical report, Department of Information and Computing Sciences. Utrecht University., 2009.
4. S. Modgil and H. Prakken. A general account of argumentation with preferences. *Artificial Intelligence*, 195:361–397, 2013.
5. P. Besnard and A. Hunter. *Elements of Argumentation*. The MIT Press, 2008.
6. P. M. Dung. On the acceptability of arguments and its fundamental role in nonmonotonic reasoning, logic programming and n-person games. *Artificial Intelligence Journal*, 77:321–357, 1995.
7. I. Rahwan and G. Simari. *Argumentation in Artificial Intelligence*. Springer, 2009.
8. L. Amgoud and H. Prade. Using arguments for making and explaining decisions. *Artificial Intelligence*, 173(3–4):413–436, 2009.
9. D. Bouyssou, D. Dubois, M. Pirlot, and H. Prade. *Decision-making process — Concepts and Methods*. Wiley, 2009.
10. A. J. García and G. R. Simari. Defeasible logic programming: An argumentative approach. *Theory and practice of logic programming*, 4:95–138, 2004.
11. M. Caminada and L. Amgoud. On the evaluation of argumentation formalisms. *Artificial Intelligence*, 171:286–310, 2007.
12. T. J. Bench-Capon. Persuasion in practical argument using value-based argumentation frameworks. *Journal of Logic and Computation*, 13(3):429–448, 2003.
13. Z. Assaghir, A. Napoli, M. Kaytoue, D. Dubois, and H. Prade. Numerical information fusion: Lattice of answers with supporting arguments. In *ICTAI*, pages 621–628, 2011.
14. N. Tamani, M. Croitoru, P. Buche. A viewpoint approach to structured argumentation. https://docs.google.com/file/d/0B0DPgJDRNwbLRmlqUVh4cGFrSVk/edit?usp=sharing. Technical report, INRA-SupAgro, 2013

15. M. W. A. Caminada, W. A. Carnielli, and P. E. Dunne. Semi-stable semantics. *Journal of Logic and Computation*, pages 1–45, 2011.
16. Y. Wu. *Between argument and conclusion. Argument-based approaches to discussion. Inference and Uncertainty*. PhD thesis, UniversitT du Luxembourg, 2012.
17. L. Amgoud. An argumentation-based model for reasoning about coalition structures. In *ArgMas*, pages 217–228, 2005.
18. L. Amgoud. Towards a formal model for task allocation via coalition formation. In *AAMAS*, pages 1185–1186, 2005.
19. G. Boella, L. van der Torre, and S. Villata. Social viewpoints for arguing about coalitions. In *PRIMA*, pages 66–77, 2008.
20. V. D. Dang and N. R. Jennings. Generating coalition structures with finite bound from the optimal guarantees. In *AAMAS*, pages 564–571, 2004.
21. S. Heras, J. Jordan, V. Botti, and V. Julian. Argue to agree: a case-based argumentation approach. *International Journal of Approximate reasoning*, 54:82–108, 2013.
22. T. L. van der Weide, F. Dignum, J.-J. C. Meyer, H. Prakken, and G. Vreeswijk. Arguing about preferences and decisions. In *ArgMAS*, pages 68–85, 2010.
23. S. Coste-Marquis, C. Devred, S. Konieczny, M.-C. Lagasquie-Schiex, and P. Marquis. On the merging of dung's argumentation systems. *Artificial Intelligence*, pages 730–753, 2007.
24. L. Amgoud, J.-F. Bonnefon, and H. Prade. An argumentation-based approach to multiple criteria decision. In *ECSQARU*, pages 269–280, 2005.
25. C. Cayrol and M.-C. Lagasquie-Schiex. Coalition of arguments in bipolar argumentation frameworks. In *7th CMNA*, pages 14–20, 2007.

Rule Type Identification Using TRCM for Trend Analysis in Twitter

João Bártolo Gomes, Mariam Adedoyin-Olowe, Mohamed Medhat Gaber and Frederic Stahl

Abstract This paper considers the use of Association Rule Mining (*ARM*) and our proposed Transaction based Rule Change Mining (*TRCM*) to identify the rule types present in tweet's hashtags over a specific consecutive period of time and their linkage to real life occurrences. Our novel algorithm was termed TRCM-RTI in reference to Rule Type Identification. We created Time Frame Windows (*TFWs*) to detect evolvement statuses and calculate the lifespan of hashtags in online tweets. We link *RTI* to real life events by monitoring and recording rule evolvement patterns in TFWs on the Twitter network.

1 Introduction

Twitter has been reported to have the highest number of users in the rank of microblogging applications [3]. Traditional media closely monitor the activities on the network to enhance the contents of their news broadcast [5]. Hashtag symbols (#) are often used as prefix to tweets' keywords to describe their contents [7].

J. B. Gomes
Institute for Infocomm Research (I2R), A*STAR, 1 Fusionopolis Way Connexis, Singapore 138632, Singapore
e-mail: bartologjp@i2r.a-star.edu.sg

M. Adedoyin-Olowe · M. M. Gaber
School of Computing, University of Portsmouth, Hampshire, England PO1 3HE, UK
e-mail: mariam.adedoyin-olowe@myport.ac.uk

M. M. Gaber
e-mail: Mohamed.Gaber@port.ac.uk

F. Stahl (✉)
School of Systems Engineering, University of Reading, PO Box 225, Whiteknights, Reading RG6 6AY, UK
e-mail: F.T.Stahl@reading.ac.uk

M. Bramer and M. Petridis (eds.), *Research and Development in Intelligent Systems XXX*, 273
DOI: 10.1007/978-3-319-02621-3_20,

Our previous work [1] proposed *TRCM* methodology by identifying four association rules pattern present in tweets' hashtags. In this paper, we propose a new technique, based on Transaction based Rule Change Mining (*TRCM*), for Rule Type Identification (RTI), which is termed *TRCM-RTI*. We apply the technique to discover the rule trend of tweets' hashtags over a consecutive period of time. We create Time Frame Windows (*TFWs*) showing different rule evolvement patterns, which can be applied to evolvements of news and events in reality.

The rest of this paper is organised as follows: Sect. 2 provides an overview of association rules in TRCM. Section 3 discusses the trend analysis of tweets. Section 4 presents our new *TRCM-RTI* algorithm. Section 5 shows a case study used to justify the TRCM methodology. Section 6 concludes the paper.

2 Definition of Rules for TRCM-RTI

Association Rules (*ARs*) find frequent patterns, associations, connections, or underlying structures among sets of items or objects in transaction databases, relational databases and other information repositories [2, 4]. The technique tends to reveal every probable association that satisfies definite boundaries using the lowest support and confidence. The *apriori* method generating *ARs* reveals frequent itemsets that can be used to establish *ARs* that highlight common trends in the database. In this paper we use a combination of the *TRCM* and *apriori* techniques to detect the dynamics of *ARs* present in tweets' hashtags over two consecutive periods of time. Next we define all the rules identified in tweets hashtags as presented in Fig. 1.

An Emerging rule (EM) is discovered when two hashtags in tweets at time t and $t+1$ have similar or greater than the user-defined threshold for the conditional and consequent parts. Hashtags emerge as a result of sudden occurrence (for example breaking news). However, unexpected consequent (*UnxCs*) rule is detected when two rules in r_j^t and r_j^{t+1} have similar conditional parts but different consequent parts

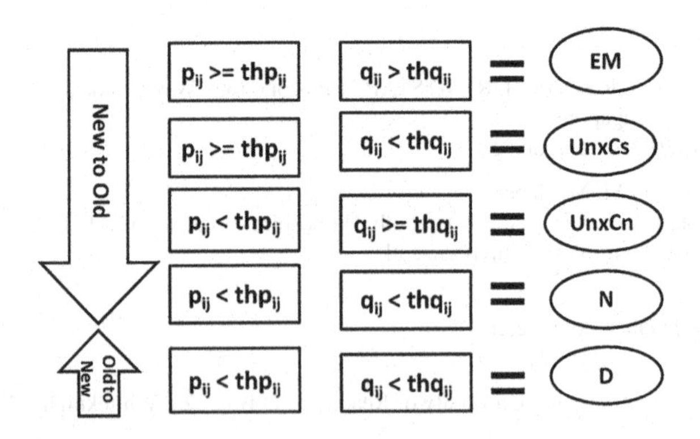

Fig. 1 Transaction-base Rule Change Mining (TRCM)

measurements greater than 0. Unexpected conditional (*UnxCn*) rule is the reverse of *UnxCs* with absolute difference measure of less than 0. Note that in real life occurrence *UnxCs* and *UnxCn* rules are the same. Unexpected change in a rule is always as a result of a twist in an event, an example of this is $\#funeral$ => $\#Thatcher$ as explained in the case study cited in Sect. 5. All rules in $t+1$ are 'new' (N) from those in t until a matching rule is discovered. 'Dead' (D) rule occurrence is the opposite of new rule detection. A rule in t is labelled 'dead' if its maximum similarity measure with all the rules in $t+1$ is less than the user-defined threshold (p_{ij}/q_{ij}).

The four defined rules detected in tweets' hashtags are used for analysing rule trend of Twitter data as explained alongside trend analysis of tweets in Sect. 3.

3 Trend Analysis of Tweets

Trend Analysis (*TA*) of Twitter data in the context of this paper is a way of analysing the rule trend present in tweets' hashtags over a period of time. Rule trend analysis demonstrates the chain/sequence a rule is measured in during its period of evolvement on the Twitter network. A rule may evolve by taking on different status in every time frame window (*TFW*). In some other trend, a rule may evolve back to assume its former status during the course of evolvements. Factors effecting time the frame of rules include unexpected discovery relating to an ongoing event that may elongate or truncate the lifespan of a hashtag. The ultimate goal of *TA* is to be able to trace back the origin of a rule (rule trace). For example a rule $X => Y$ may have actually started up as $A => B$ and over time the rule has evolved unexpectedly.

3.1 Calculating Rule Similarities and Difference

In [1], we calculated rule similarities and differences by adopting [4] and [6] methods, which they used to discover association rules in relational datasets. We defined similarity using the concepts of degree of similarity proposed in both work. The calculations and notations are as follows:

n	Number of hashtags
i	An association rule in dataset 1 presented in binary vector
j	An association rule in dataset 2 presented in binary vector
lh_i/lh_j	Number of hashtags with value 1 in conditional part of rule i/j
rh_i/rh_j	Number of hashtags with value 1 in consequent part of rule i/j
lh_{ij}/rh_{ij}	Number of same hashtags in conditional/consequent part of rules i and j
p_{ij}/q_{ij}	Degree of similarity of features in conditional/consequent part of rules i and j
r_j^t	Rule present at time t
r_j^{t+1}	Rule present at time $t+1$.

3.2 Measuring Similarity

$$p_{ij} = \frac{lh_{ij}}{max(lh_i, lh_j)} \quad (1)$$

$$q_{ij} = \frac{rh_{ij}}{max(rh_i, rh_j)} \quad (2)$$

In Eqs. 1 and 2, p_{ij} and q_{ij} were adopted from [4]. They show the similarities in the conditional and consequent parts of rule i and rule j at different points in time t and $t+1$ respectively. However, unlike the approach taken in [4], we used the transactional view rather than the relational view. Thus, our equations are simpler and faster to compute. More importantly, the transactional view is a better fit for hashtags in Twitter, as existence or absence of the hashtag is what contributes to forming the association rules.

3.3 Notation

$t0$	Tweets retrieved at time t
$t1$	Tweets retrieved at time $t+1$
$r0$	Rules obtained from $t0$
$r1$	Rules obtained from $t1$
$lhsT$	Left hand rule threshold
$lhsMax$	Rules similar in left hand of $R1$ and $R2$
$rhsT$	Right hand rule threshold
$rhsMax$	Rules similar in right hand of $R1$ and $R2$
$ruleSimLH$	Degree of similarity in the left hand as defined in Sect. 3.2
$ruleSimRH$	Degree of similarity in the right hand as defined in Sect. 3.2

4 TRCM-RTI for Tracing Rule Evolvement in Tweets

Algorithm 1 is created to find patterns in the new ruleset (*r1*) extracted (using Association Rule Mining) from the tweets retrieved at time $t+1$ that are similar to those in the old ruleset (*r0*) extracted from the tweets retrieved at time t. The thresholds of the rules are *lhsT* and *rhsT*, which in our experiments are set to 1 for rule similarity to be considered. We find similar patterns of rules, *r0* and *r1*. We then inspect *r1* using TRCM to detect the rule change evolvement at $t+1$. Rule patterns in r1 show the rule trend over the period of evolvement *(r0 to r1)*. We relate the TRCM rule trend of tweets to real events in Sect. 5.

Algorithm 1 TRCM-RTI algorithm

```
Function findPattern(newRule, ruleSet)
Require: lhsT and rhsT
Ensure: lhsMax = 0 and rhsMax = 0
  for all oldRule in ruleSet do
    lhsMax = max(lhsMax, ruleSimLH(newRule,oldRule))
    rhsMax = max(rhsMax, ruleSimRH(newRule,oldRule))
    if (lhsMax >= lhsTandrhsMax >= rhsT) then
      return Emerging
    end if
  end for
  if (lhsMax >= lhsT) then
    return UnexpectedConsequent
  else
    if (rhsMax >= rhsT) then
      return UnexpectedConditional
    end if
  end if
  return New
  EndFunction
Require: r0, r1 rules mined from the tweets retrieved at time t and t + 1
  for all newRule in r1 do
    newRule.TRCM = findPattern(newRule,r0)
  end for
```

5 Case Study of Rule Evolvement in Twitter Data

The news of Margaret Thatcher's death (former first female Prime Minister of Great Britain) on the 8th of April 2013 was widely reported as breaking news all around Europe and in other parts of the world. The occurrence also generated a number of hashtags on Twitter. The TRCM experimental study conducted of $\#Thatcher$ within 25 days of her death revealed different patterns of rule evolvements of hashtags used in tweets. Rules (*r0*) in the first few days of Thatcher's death were mostly EM. However, by the 21st day, 99 % of the rules were dead implying that they had less than 21 day's lifespan on Twitter. $\#Funeral => \#Thatcher$ evolved from EM (TFW) to UnxCn and UnxCs (TFW) within the period under study (2 TFWs). The evolvements could be related to numerous controversies on the £10 million state money spent on her burial which some individuals and groups consider as wasteful spending. The UnxCn evolvement of $\#Unions => \#Thatcher$ could be linked to workers' day celebration on May 1, that bring to mind Thatcher's presumed adverse policy on workers' union during her time in office. However, the UnxCn rule evolvements of $\#AcidParty => \#Thatcher$ can be correlated to Thatcher's negative views on Acid party and hardcore music being embraced by young British in the 80s. Rules such as $\#Channel4 => \#RandomActs$, $\#RandomActs => \#Thatcher$, $\#Channel4 => \#AcidParty, \#RandomActs => \#AcidParty$ became common as a result of Channel 4's (UK public-service television broadcaster) introduction of a peculiar acid party song on their 'Random Acts' programme. Margret Thatcher's speeches were randomly chosen and compose into an 'acid party' song. A number of rules were traced to $\#Thatcher$ in our experiment. Although the rules do not imply $\#Thatcher$, their origin can be easily traced to $\#Thatcher$. An example of such rule is $\#channel4 \; => \; \#RandomActs$ as explained earlier. Rule trace enhances better understanding of hashtags and this emphasises the importance of contextualising hashtag usage in tweets. Our case study portrayed the significance of rule evolvements in tweets and how rule evolvements can be related to unfolding events in reality. *RTI* in tweets using *TRCM* can be used to filter hashtags and shift users' attention to current interesting news and events. Rule evolvement can also be used to update hashtags used in tweets at different evolvement phases. Lastly, rule evolvement in *TRCM-RTI* will enable users and other entities to trace back evolving rules and thus help ascertain the origin of rules.

6 Conclusion

In this paper, we proposed a novel algorithm termed *TRCM-RTI*. Our experiment has been related to real life scenarios which buttress the significance of our algorithm when mining Twitter data. The case study reported in this paper demonstrates how *TRCM-RTI* can be used to analyse a sequence of evolvements of events in reality. The benefits of our method include the enhancement of trend topics on Twitter. This can be achieved by filtering interesting hashtags and captioning them as trending topics. The number of time frame windows of any emerging rule demonstrates the interestingness of its hashtags. *TRCM-RTI* also enables rule trace and assists Twitter users to understand the origin of rules and their evolvement sequence.

References

1. Adedoyin-Olowe, M., Gaber, M., Stahl, F.: TRCM: A Methodology for Temporal Analysis of Evolving Concepts in Twitter. In: Proceedings of the 2013 ICAISC, International Conference on Artificial Intelligence and, Soft Computing. 2013.
2. Agrawal, Rakesh, et al. "Fast Discovery of Association Rules." Advances in knowledge discovery and data mining 12 (1996): 307–328.
3. Grosseck, G., Holotescu, C.: Can We use Twitter for Educational Activities. 4th International Scientific Conference. eLearning and Software for Education, Bucharest, April 17–18, 2008.
4. Liu, D-R., Shih, M-J., Liau, C-J., Lia, C-H.: Mining the Change of Event Trends for Decision Support in Environmental Scanning. Science Direct, Expert Systems with Application, Vol. 36, 972–984. 2009.
5. Ratkiewicz, J., Conover, M., Meiss, M., Goncalves, B., Flammini, A., Menczer, F.: Detecting and Tracking Political Abuse in Social Media. In ICWSM. 2011.
6. Song, H., Kim, J., Kim, S.: Mining the Change of Customer Behaviour in an Internet Shopping Mall. Expert Systems with Applications. Vol. 21, 157–168. 2001.
7. Weng, J., Lim, E. P., He, Q., Leung, C. K.: What do people want in microblogs? Measuring interestingness of hashtags in twitter. Data Mining (ICDM), 2010 IEEE 10th International Conference on. IEEE, 2010.

KNNs and Sequence Alignment for Churn Prediction

Mai Le, Detlef Nauck, Bogdan Gabrys and Trevor Martin

Abstract Large companies interact with their customers to provide a variety of services to them. Customer service is one of the key differentiators for companies. The ability to predict if a customer will leave in order to intervene at the right time can be essential for pre-empting problems and providing high level of customer service. The problem becomes more complex as customer behaviour data is sequential and can be very diverse. We are presenting an efficient sequential forecasting methodology that can cope with the diversity of the customer behaviour data. Our approach uses a combination of KNN (K nearest neighbour) and sequence alignment. Temporal categorical features of the extracted data is exploited to predict churn by using sequence alignment technique. We address the diversity aspect of the data by considering subsets of similar sequences based on KNNs. Via empirical experiments, it can be demonstrated that our model offers better results when compared with original KNNs which implies that it outperforms hidden Markov models (HMMs) because original KNNs and HMMs applied to the same data set are equivalent in terms of performance as reported in another paper.

M. Le (✉) · B. Gabrys
Bournemouth University, School of Design, Engineering and Computing,
Bournemouth, UK
e-mail: mai.phuong@bt.com

B. Gabrys
e-mail: bgabrys@bournemouth.ac.uk

D. Nauck
BT, Research and Technology, Ipswich, UK
e-mail: detlef.nauck@bt.com

T. Martin
Bristol University, Bristol, UK
e-mail: trevor.martin@bristol.ac.uk

M. Bramer and M. Petridis (eds.), *Research and Development in Intelligent Systems XXX*,
DOI: 10.1007/978-3-319-02621-3_21,

1 Introduction

In a very competitive and saturated market, it is important for customer centric companies to keep their existing customers because attracting a new customer is between 6 to 10 times more expensive than retaining an existing customer [1]. Churn prevention is a strategy to identify customers who are at high risk of leaving the company. The company can then target these customers with special services and offers in order to retain them.

Researcher use different approaches and different types of data to predict churn: neural networks and support vector machines [2], hidden Markov models [3], random forests [4], bagging boosting decision trees [5], and sequential KNN [6]; demographic data, contractual information and the changes in call details and usage data [7] and complaint data [8]. Word of mouth has also been studied in churn prediction [9].

Customer behaviour data consists of asynchronous sequences describing interactions between the customers and the company. To investigate the impact of sequential characteristics on churn detection, sequential approaches are of interest. There is a rich available source of mathematical models in data mining [10] but there don't seem to be many efficient sequential approaches. The problem seems to be poorly aligned with any single classical data mining technique.

Our objective is to determine similar sequences, expecting that sequences which behave similarly in earlier steps go to the same final step. To achieve this, an original KNN is combined with sequence alignment to form a sequential KNN. A sequential KNN using Euclidean distance is introduced in [6]. As our data consists of event sequences, in this study we use the matching scores from sequence alignment as distance. The problem can be formulated as follows. A customer behaviour record (S_j) is a composition of discrete events in a time-ordered sequence, $S_j = \left\{s_1^{(j)}, s_2^{(j)}, \ldots, s_{n_j}^{(j)}\right\}$, s_k takes values from a finite set of event types $E = \{e_1, \ldots, e_L\}$. The goal of our models is to predict churn for a given event sequence $S_{N+1} = \left\{s_1^{(N+1)}, s_2^{(N+1)}, \ldots, s_{i-1}^{(N+1)}\right\}$ based on the data from classified sequences $S = \{S_1, S_2, \ldots, S_N\}$ where we know if a customer has churned or not.

The remainder of this paper is organised as follows Sect. 2 presents the proposed model. It is followed by Sect. 3 where we evaluate the performance on a data set. Section 4 provides conclusions and future research directions.

2 Sequential KNNs

KNN is one of the classical approaches in data mining [11]. It can be used as the original non sequential approach [3] or extended into a sequential approach [6]. The core idea in KNNs is to find similar sequences based on distance functions. We adopt sequence alignment to define a distance measurement which is expected to be suitable for event sequences [12]. The obtained approach by coupling KNN with sequence alignment is named KNsSA. In this work, both types of alignment are investigated to verify which one is effective for measuring the similarity between two given sequences in order to predict process outcomes.

Global alignment provides a global optimisation solution, which spans the entire length of all query sequences. Local alignment aims to find the most similar consecutive segments from two query sequences.

- *Global algorithm* One such algorithm was introduced by Needleman and Wunchs [13]. There are three characteristic matrices in this algorithm: substitution matrix, score matrix and traceback matrix.

1. Substitution matrix: in biology a substitution matrix describes the rate at which one amino acid in a sequence transforms to another amino acid over time. In our problem regarding customer behaviour data no mutation occurs. Therefore, we use the simplest form of substitution matrix.

$$s(i,j) = \begin{cases} 0 \; if \quad event\, i \neq event\, j \\ 1 \quad\quad otherwise \end{cases}$$

2. Score matrix: This matrix's elements are similarity degrees of events from the two given sequences considering the event positions.

$$h_{i0} = h_{0i} = -\delta \times i \tag{1}$$

$$h_{ij} = \max\{h_{i-1,j} - \delta, h_{i-1,j-1} + s(x_i, y_j), h_{i,j-1} - \delta\} \tag{2}$$

where $i = \{1, \ldots, len_1\}$, $j = \{1, \ldots, len_2\}$. δ is a specific deletion/insertion penalty value chosen by the users. h_{i0} and h_{0j} are the initial values. x_i and y_j are events at positions i and j from the given sequences. $s(x_i, y_j)$ is the value from the substitution matrix corresponding to events x_i and y_j.

3. Traceback matrix: Elements of this matrix are left, diag or up depending on the corresponding h_{ij} from the score matrix:

$$q(i,j) = \begin{cases} diag \\ \quad if \;\; h(i,j) = h(i-1,j-1) \\ \quad\quad + s(i,j) \\ up \\ \quad if \;\; h(i,j) = h(i-1,j) - \delta \\ left \\ \quad if \;\; h(i,j) = h(i,j-1) - \delta \end{cases} \tag{3}$$

This matrix is used to track back from the bottom right corner to the top left corner, by following the indication within each cell, to find the optimal matching path.

- *Local algorithm* The aim of local algorithms [12–14] is to find a pair of most similar segments, from the given sequences. These algorithms also have substitution matrix and score matrix like global algorithms. However, the initial values for the score matrix in local algorithms are set to be 0:

$$h_{i0} = h_{0j} = h_{00} = 0 \tag{4}$$

The optimal pair of aligned segments is identified by first determining the highest score in the matrix, then tracking back from such score diagonally up toward the left corner until 0 is reached.

3 Evaluation

3.1 Data and Benchmarking Models

We carried out a number of experiments based on records from a customer behaviour data DS from a multi-national telecommunications company. The data consists of 8080 customer records. Each record represents events related to a customer. There are four types of events in the customer sequences: churn, complaint, repair and provision. These sequences are of different lengths and only 161 out of 8080 sequences contain a churn event. Therefore, we artificially created new datasets with different ratios between non-churn and churn sequences in order to investigate its impact on the model performance. To evaluate KnsSA, we benchmarked our models with two other approaches:

- *RM—Random Model*: in order to find the outcome of the process, we randomly generate a number between 0 and 1, if the generated number is greater than 0.5 the outcome is success (1) and vice versa if the generated number is smaller than 0.5 the outcome is failure (0).
- *Original KNN*: we chose K nearest sequences in terms of having common unique tasks. As shown in the work of [3], churn prediction using a sequential approach (HMM) applied to the same data set DS does not outperform non sequential KNNs. Therefore, we benchmark our model only with a non sequential KNN and show that our way of dealing with sequential diverse data is more efficient.

3.2 Results

For the proposed models, we investigate the effect of K as it is important to get the reasonable number of similar sequences. We now present the results of the experiments by applying the two models, global and local KnsSAs to DS. There are three tables which illustrate different aspects of the objectives of the experiments. Table 1 shows the difference obtained by varying K, using local KnsSA and the original churn data. Table 2 demonstrates the results obtained by using local KNsSA when artificial data sets were created by changing the ratio between churn and no churn sequences in order to decrease the skewness of the data. Table 3 shows the performance of global KnsSA applying to the former artificial data sets.

It can be seen from Table 1 that $K = 3$ is the best case. Also, when the original data were modified, the performance of the model in terms of the churn detection objective improved even though the overall performance worsened. Intuitively, when the population of no-churn sequences strongly dominates, it is very likely that our model could not catch the full churn set. It is shown in Table 1 that the precision for churn is 100 % and the corresponding recall is 37.5 %. With the amended data, the overall performance of the model is reduced as well as the precision for churn class. Nonetheless, it is still of interest because the precision and the recall for churn prediction are 92.85 %.

The results in Tables 2 and 3 show that the local KnsSA outperforms the global KnsSAs when applied to the churn data set. This could be caused by the fact that in customer behaviour sequences, only a subset of special segments has strong influence on churn action.

Table 1 Local KnsSA applied on original DS with different K, $K = 3, 5$ and 7

Results/K	3	5	7
Actual tests successful	793	793	793
Actual tests failure	16	16	16
Predicted tests successful	803	805	805
Predicted tests failure	6	4	4
Correct predictions	799	797	797
Failed predictions	10	12	12
Predicted tests success correct	793	793	793
Correct ratio	0.99	0.99	0.99

Table 2 Local KnsSA applied on original DS with different churn and non churn ratios $K = 3$

Results/ratio	0.05	0.10	0.15
Actual tests successful	45	83	120
Actual tests failure	14	17	15
Predicted tests successful	45	84	124
Predicted tests failure	14	16	11
Correct predictions	57	93	127
Failed predictions	2	7	8
Predicted tests success correct	44	80	118
Correct ratio	0.97	0.93	0.94

Table 3 Global KnsSA applied on original DS with different churn and non churn ratios with $K = 3$

Results/ratio	0.05	0.10	0.15
Actual tests successful	43	79	121
Actual tests failure	15	16	16
Predicted tests successful	48	90	134
Predicted tests failure	10	5	3
Corrected predictions	43	82	118
Failed predictions	15	13	19
Predicted tests success correct	38	78	118
Correct ratio	0.74	0.86	0.86

4 Conclusion

In this paper, we propose some extensions to KNNs which were designed in order to capture the temporal characteristics of the data and to profit from KNNs ability to deal with diverse data. These extensions are tested on real customer behaviour data from a multi telecommunications company and the experiments provide some interesting results. Even though churn is a rare event the proposed models correctly capture most of the churn cases and the precision and recall values of churn class are very high. This paper confirms our initial point of view that it is hard to model diverse sequences in a generic way to predict the outcome and that it is important to use the temporal characteristics of the data. Hence, a KNN is a good candidate because KNNs treat certain number of similar sequences in the same way. Combining sequence alignment and KNNs we can achieve better results since this helps to compare two sequences based on the ordered events themselves.

References

1. J. Hadden, A. Tiwari, R. Roy, D. Ruta, International Journal of Intelligent Technology **1**(1), 104 (2006)
2. C. Archaux, H. Laanaya, A. Martin, A. Khenchaf, in *International Conference on Information & Communication Technologies: from Theory to Applications (ICTTA)* (2004), pp. 19–23
3. M. Eastwood, B. Gabrys, in *Proceedings of the KES2009 Conference* (Santiago, Chile, 2009)
4. B. Lariviere, D. Van den Poel, Expert Systems with Applications **29**(2), 472 (2005)
5. A. Lemmens, C. Croux, Journal of Marketing Research **43**(2), 276 (2006)
6. D. Ruta, D. Nauck, B. Azvine, in *Intelligent Data Engineering and Automated Learning IDEAL 2006, Lecture Notes in Computer Science*, vol. 4224, ed. by E. Corchado, H. Yin, V. Botti, C. Fyfe (Springer, Berlin, 2006), pp. 207–215
7. C. Wei, I. Chiu, Expert Systems with Applications **23**(2), 103 (2002)
8. *Churn Prediction using Complaints Data.*
9. S. Nam, P. Manchanda, P. Chintagunta, Ann Arbor **1001**, 48 (2007)
10. R. Duda, P. Hart, D. Stork, *Pattern Classification* (Wiley, New York, 2001)
11. M. Berry, G. Linoff, *Data Mining Techniques: for Marketing, Sales, and Customer Relationship Management* (Wiley, Newyork, 2004)

12. M. Waterman, Philosophical Transactions of the Royal Society of London, Series B: Biological Sciences **344**, 383 (1994)
13. S. Needleman, C. Wunsch, Journal of Molecular Biology **48**, 443 (1970)
14. T. Smith, M. Waterman, Journal of Molecular Biology **147**, 195 (1981)

Applications and Innovations in Intelligent Systems XXI

Best Application Paper

Knowledge Formalisation for Hydrometallurgical Gold Ore Processing

Christian Severin Sauer, Lotta Rintala and Thomas Roth-Berghofer

Abstract This paper describes an approach to externalising and formalising expert knowledge involved in the design and evaluation of hydrometallurgical process chains for gold ore treatment. The objective of this knowledge formalisation effort is to create a case-based reasoning application for recommending a treatment process of gold ores. We describe a twofold approach to formalise the necessary knowledge. First, formalising human expert knowledge about gold mining situations enables the retrieval of similar mining contexts and respective process chains, based on prospection data gathered from a potential gold mining site. The second aspect of our approach formalises empirical knowledge on hydrometallurgical treatments. The latter, not described in this paper, will enable us to evaluate and, where needed, redesign the process chain that was recommended by the first aspect of our approach. The main problems with the formalisation of knowledge in the gold ore refinement domain are the diversity and the amount of parameters used in literature and by experts to describe a mining context. We demonstrate how similarity knowledge was used to formalise literature knowledge. The evaluation of data gathered from experiments with an initial prototype workflow recommender, *Auric Adviser*, provides promising results.

C. S. Sauer · T. Roth-Berghofer (✉)
School of Computing and Technology, University of West London, St Mary's Road, London W5 5RF, UK
e-mail: thomas.roth-berghofer@uwl.ac.uk

C. S. Sauer
e-mail: christian.sauer@uwl.ac.uk

L. Rintala
Department of Materials Science and Engineering, Aalto University, Helsinki, Finland
e-mail: lotta.rintala@aalto.fi

M. Bramer and M. Petridis (eds.), *Research and Development in Intelligent Systems XXX*, 291
DOI: 10.1007/978-3-319-02621-3_22, © Springer International Publishing Switzerland 2013

1 Introduction

Nowadays rich gold ores that can easily be processed with simple metallurgical processes like direct smelting are getting rare. This situation leads to the (re)evaluation of many less rich and difficult to process gold ore deposits, considered too cost intensive for mining before. Refractory, or in other words difficult to process, gold ores are gold ores that in general require complex and cost intensive processes to extract a comparatively small amount of gold from a large volume of ore. Thus the large scale processes involved in the processing of such ores are to be planned carefully to avoid failed investments in ore processing facilities either not adequate for the ore mined at the mining site or not efficient enough to extract sufficient quantities of gold and thus sufficient revenues.

A key problem in today's prospecting for gold mines and in their planning is the estimation of the costs involved, not only in the mining of the ore but mainly in its processing costs given, for example, by the use of certain chemicals. Additional important factors are the ore throughput capacity to reach a necessary volume of ore to be processed per day as well as the amount of gold to be extracted from this ore. An early and exact estimate of these costs and factors allows excluding early on potential mining operations that are not cost efficient and helps to speed up planning of worthwhile mining operations by reusing process knowledge previously employed in successful mining operations.

In this paper we examine the elicitation, formalisation and re-use of expert knowledge about gold mining situations that operate on both, rich and refractory gold ores, hydrometallurgical process design for processing these ores and empirical knowledge focused on the diverse hydrometallurgical processes involved in the process chains.

Hydrometallurgy is a field of science which studies the recovery of metals from ores by using aqueous chemistry. A typical hydrometallurgical process chain is illustrated in Fig. 1. When analysing or designing hydrometallurgical processes, a process

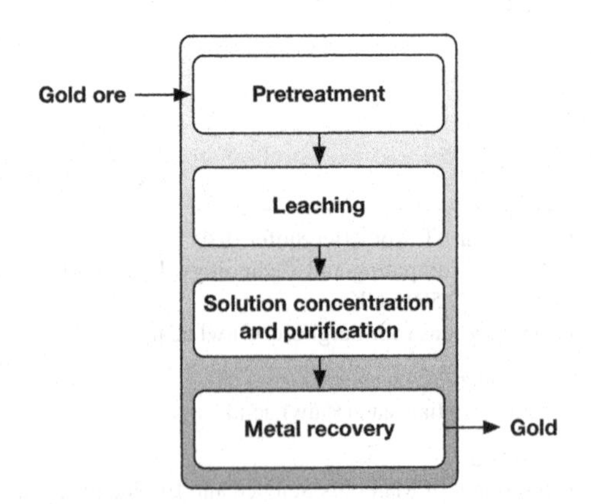

Fig. 1 Basic hydrometallurgical process chain

chain is typically considered to be composed out of smaller stages or single process steps, we call treatments. For example, a commonly used leaching technique for gold extraction is cyanide leaching. However when it is a question of refractory gold ores, for example, simple cyanidation might not be effectively used. In such ores, gold is encapsulated inside a host mineral and thus cannot be reached by the leaching agent. Thus the host mineral must be broken using pre-treatment processes to liberate the gold, before cyanide leaching.

Depending on the level of the refractoriness of the gold ore and the mineralogical characteristics of the ore, a variety of alternative processing routes exist. For example an expert designing a process chain could face the need to choose between oxidising processes, such as roasting, pressure oxidation or bacterial leaching. Typically the design of process chains relies on laboratory tests, which are time consuming and thus cost intensive. Today there already is a large amount of published work available, detailing on experiments about gold extraction from ore. Additionally on-going research is constantly adding to this knowledge. If a process chain designer could easily re-use this existing knowledge more efficiently than manually researching on it, the needed experiments would be selected more precisely or even be made redundant. Furthermore bench and pilot scale experiments, which are testing experiments on a designed process chain before scaling them to production size in a mining operation, would also be achieved more rapidly. Thus the effort for the process chain design could be significantly reduced. The work presented in this paper demonstrates an approach at formalising the existing knowledge on hydrometallurgical process chain design, using CBR and thus make it readily available for re-use by the process chain designer.

The knowledge involved in estimating a mining situation from prospection data and designing a process chain for the specific gold ore expected in this potential mining context is highly encoded. Usually a few experts in the domain are consulted to give their experience-based estimate of a prospected ore deposit and it being worthwhile mining or not. Additionally there exists a great amount of empirical knowledge, mainly from hydrometallurgical experiments on single process steps or treatments. This knowledge about specific treatments is mainly encoded in scientific publications on specific hydrometallurgical treatments.

In this paper we demonstrate our twofold approach to elicit, formalise, and re-use knowledge to facilitate the estimate of a potential mining situation and to help with the design and evaluation of the hydrometallurgical ore refinement process chain for this potential mining site. The first aspect or step in our approach is to reuse knowledge about gold mining situations to retrieve similar mining contexts based upon prospection data gathered from a potential gold mining site. The second aspect or step of our approach is then to evaluate the recommended process chain automatically and recommend a re-design of the process chain where necessary.

In this paper we describe how the knowledge from existing mining contexts was formalised and is used in the *Auric Adviser* workflow recommender software. We further show how *Auric Adviser* recommends on the best process chain to be used in hydrometallurgical treatment of rich and refractory gold ores in a potential mining project. We formalised the knowledge we elicited from human experts and hydrometallurgical publications to re-use it for our case-based reasoning (CBR) approach. Case-based reasoning is an established artificial intelligence technique. It allows for the versatile handling of incomplete and fuzzy knowledge.

The rest of the paper is structured as follows. We interlink our approach with related work on hydrometallurgy, workflow recommendation and knowledge representation in the following section. In Sect. 3 we survey the knowledge sources targeted in Sect. 3.1 and introduce our use of the different knowledge containers of CBR to provide the captured knowledge in Sect. 3.2 and review the resulting knowledge model in Sect. 3.3. Following a brief introduction to the *Auric Adviser* software in Sect. 4 we detail on our experiments in Sect. 5 followed by an evaluation and discussion of feedback on the performance of our knowledge model and the *Auric Adviser* (Sect. 6). A summary and outlook on future aspects of our work concludes the paper.

2 Background and Related Work

Recommending workflows is a research area in CBR (cf. [3, 6]). CBR has been used successfully in a number of workflow recommenders (cf. [5]). The approach to, at least semi-automatise, the design of mining facilities and their accompanying ore process facilities and processing chains is also an established area of research. As, for example, the work of Torres et al. [11, 12] has shown an approach to use decision rules and fuzzy sets to recommend ore processing workflows and calculate an estimate of the associated costs to establish such a process chain. However after an initial small set of case studies this approach seems to have not been followed further. In our view the complexity and variety of the knowledge in the domain of gold mining and refractory gold ore refining is a factor that ultimately could not be covered by a rule based system alone.

Picking up on the initial work of Torres et al. we therefore present our new approach to ore processing workflow recommendation. We were confident in the decision to use CBR as we have already successfully used CBR and its versatile use of similarity based retrieval, in the formalisation of knowledge in the highly complex and encoded domain of audio mixing and subsequently implemented an efficient audio mixing workflow recommender [10]. Additionally the use of the CBR methodology is documented as successful in the field of chemical engineering. It is used, for example, for the separation process synthesis and as a method for combinatorial mixer equipment design from parts and the development of feasible separation process sequences and separation process structures for wood processing [7].

3 Knowledge Formalisation Approach for Use in CBR

As outlined we used case-based reasoning in our approach to reuse the elicited and formalised knowledge of mining experts and empirical knowledge on hydrometallurgical treatments. We chose CBR as a suitable methodology [13] for our task as it is able to handle the inherent vagueness and broad variety of the knowledge present in our domain of interest. For the purpose of modelling and testing the knowledge of our system we used *myCBR* Workbench[1] in its latest version 3.0. We then developed the java-based application, *Auric Adviser*, with the *myCBR* SDK.

Broadly speaking CBR deals with storing/retrieving as well as re-using/adapting experience. It mimics the human approach of re-using problem-solving experience encoded in cases. Cases, in CBR, are problem/solution pairs. Facing a current problem we recall matching past problems and adapt their solution to our current situation. Such a revision might be necessary due to the problem description of the past problem slightly differing from the current problem or some factors necessary for applying the past solution having changed or being not available. After applying the (probably adapted) solution to our current problem we then evaluate if the application of the solution solved our problem. If so, we can retain the pairing of the current problem description and the successfully used solution description as a new episode of experience, or case, in our case base [1].

CBR is able to '*speak the customer's language*', allowing for the use of synonyms and missing terms and likely vague terms describing the amount of, for example, a mineral present within a query to our system. In our case this means that fuzzy amount descriptors such as 'some' or 'traces' can be used to define queries. Additionally CBR is also able to retrieve cases, in *Auric Adviser* descriptions of existing mining operations as well as hydrometallurgical treatments, based on only sparse query data. This is useful while trying to retrieve mining operations only partly specified on sparse prospection data. Furthermore CBR relies on similarities which are comparatively easy to elicit from human experts within our domain of interest. Finally CBR allows for queries that combine retrieval and filtering in the way of using key attributes as selection criteria for a case before calculating the global case similarity, thus reducing the computational effort of an retrieval. An example for such a pre-selection attribute would be the exclusion of all mining operation cases in which the ore processed does includes clay, as the presence of clay is instantly forbidding a number of chemical treatments of the ore.

3.1 Knowledge Acquisition

In this section we review available knowledge sources in our domain, we detail on the kind of knowledge we drew from these sources and why we deemed it to be important. We further describe the techniques we used to elicit the knowledge.

We initially identified four sources from which we gathered the following knowledge:

1. Human experts on gold mining: Necessary attributes and values to describe a gold mining operation

[1] http://mycbr-project.net

2. Scientific publications on gold ore mining operations: Attribute value ranges and applied ore processing chains
3. Communities of human experts on mining: Similarity measures and similarity measure evaluation
4. Scientific publications on specific hydrometallurgical treatments: Attributes and value ranges to describe treatments and knowledge about their applicability

For our initial knowledge elicitation we used interviews with the human experts. We additionally created questionnaires, combined with similarity measure templates, to be completed by the experts. We had to additionally keep in mind that our acquisition of knowledge must follow certain strict guidelines and suit controlled conditions [2]. To optimise our knowledge elicitation process and techniques we employed an iterative elicitation process. In this iterative process we repeatedly asked the experts about their feedback on our knowledge elicitation approach, questionnaires and similarity measure templates. The goal of this effort was to allow for input of the experts on how to best enable them to externalise their tacit knowledge. In short we wanted to know if we 'asked the right questions'. Therefore we asked the experts:

1. Have we asked the right questions?
2. Have we provided the right templates for eliciting the similarity measures (Tables, Taxonomies)?
3. Were our data types and data value ranges correctly set?
4. What input with regard to fundamental knowledge modelling/eliciting did we got from the experts?
5. How would the experts have amended our questions and our way of knowledge gathering?

Based on the feedback, we refined our knowledge elicitation technique and went through a second cycle of knowledge gathering. This iterative knowledge gathering approach is applicable in our work, as the use of *myCBR* Workbench allows us to refine our knowledge model and integrate it in the running *Auric Adviser* software.

Our overall knowledge gathering process was again twofold and focused on two areas: First, knowledge about existing gold ore mining operations and the hydrometallurgical process chains used within these mining operations, and , second, empirical knowledge on single hydrometallurgical treatments and their applicability.

The knowledge on existing gold mining operations [4] was used to create an initial episodical knowledge model of mining operations. This knowledge model is now being used within *Auric Adviser* to realise the similarity based retrieval of mining operations based on a query composed from prospective data of a potential mining site. The retrieved most similar gold mining operation's ore process chain is then recommended for re-use on the potential mining site specified by the prospective data.

The knowledge on specific treatments is to be used to form our second prototypical knowledge model that will be used to realise the similarity based retrieval of single consecutive process steps (treatments), in a very strictly defined and specific context given by a specific ore-constellation and a specific preceding treatment. Thus our second knowledge model will be able to provide treatment recommendations to solve problems such as the need to 'reduce sulphides' for a specific ore/raw product constellation and with regard to a (necessary) preceding treatment already applied to the raw product.

The knowledge necessary for the design of the second knowledge model will be elicited from existing publications on single hydrometallurgical treatments. We assume this source as a valid sources of knowledge as when the quality of published work available on single hydrometallurgical treatments was studied [9], the finding was that, apart from some exceptions, hydrometallurgical publications follow a common principle and contain sufficient information needed to describe a treatment's characteristics.

3.2 Knowledge Formalisation

As already described we aim for a twofold approach of representing the knowledge in our domain. Thus we had to arrange for two different knowledge models serving each of the two aspects of our approach. The first knowledge model which we present in this paper is aimed at holding the knowledge needed to recommend whole ore process chains derived from existing gold mining operations.

Figure 2 shows the approach taken by us to represent and retrieve episodical cases representing existing gold mining operations and their ore refinement process chains. Figure 3 shows the approach which we intent to deploy to represent the case of a preceding and consecutive specific treatment on a specific ore/raw material constellation.

As stated in Sect. 1, we plan to employ these two knowledge models in a consecutive approach in our final version of the *Auric Adviser*. The first knowledge model is

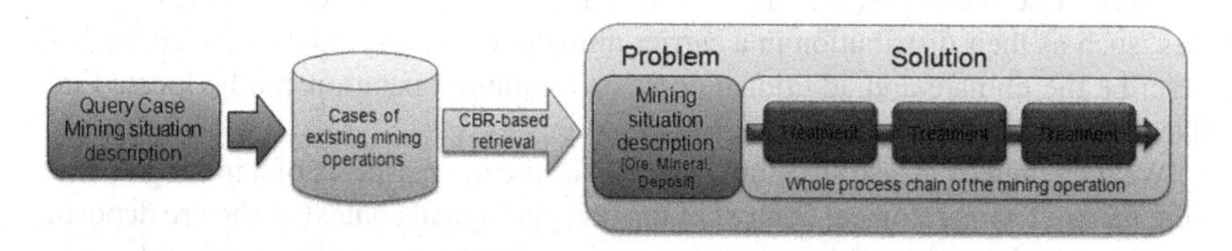

Fig. 2 Case representation of whole ore refinement process chains

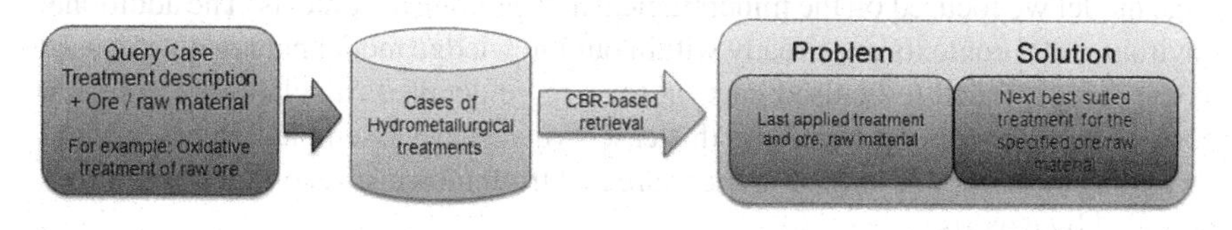

Fig. 3 Case representation of a specific preceding, consecutive pair of treatments

now in use in the *Auric Adviser* in order to retrieve similar existing mining contexts and their ore refinement process chains based on queries composed from prospective data on potential mining sites. Our next development step is now to finalise the second knowledge model to allow for the consecutive automated step-by-step evaluation and, if necessary, re-design of the process chain recommended by the first knowledge model.

As the whole ore process chain of a potential mining site is dependent on a far broader range of variables, starting from the ore- and mineralogical attributes as core description and getting more broader and specific to a given mining situation by adding more attributes describing environmental, economical, climate, political and other factors, we added these attributes to our first knowledge model also. An example for such added attributes, besides the actual composition of the ore deposit within a potential mining operation, is the quantity and quality of water available or even the hindering abundance of water within a potential mining context. Therefore we also added attributes such as *climate* and *rainfall per year* to our knowledge model.

3.3 Initial Knowledge Model

Based on the knowledge gathered from the sources described in Sect. 3.1 we created our initial knowledge model using the approaches described in Sect. 3.2. In the following we describe how we formalised the gathered knowledge into the four knowledge containers of any CBR system: vocabulary, cases, similarity measures, and adaption knowledge [8].

Our vocabulary consists now of 53 attributes, mainly describing the ore and mineralogical aspects of an ore deposit. With regard to the data types used, we used 16 symbolic, 26 floating point, 6 boolean and 5 integer value attributes. The symbolic attributes describe minerals and physical characteristics of minerals and gold particles, such as their distribution in a carrier mineral. Further symbols were elicited to describe the climate and additional contexts a mining operation can be located in, like for example the topography.

We then created a case structure catering for the main aspects of a mining operation, namely: Mineralogical context of the ore, geological context of the ore deposit, environmental context of the mining operation, detailing into: Geographical, topographical, economic and political context of a mining operation. For the initial knowledge model we focused on the mineralogical and geological contexts. The additional environmental contexts are already within our knowledge model but are not yet used as they are intended to be used later on for a more detailed calculation of potential costs of a mining operation. Using this case structure we assembled 25 cases based on mining situations described in literature and from information on such operations provided by experts.

Figure 4 shows a limited excerpt from the case data we generated. Our cases were distinctive mainly with regard to the mineralogical context of the mined ore. Thus

#	Ore type	Sulfide present	IronSulfide present	ArsenicSulfide present	Free milling gold present	Gold ore grade, g/t	Gold recovery %	Troughput, tpd	Au production oz/yr
0	refractory arsenopyritic	1.0	1.0	1.0		8.0	88.0	1,250	100,000
1	refractory arsenopyritic	1.0	1.0	1.0		7.2	91.0	1,000	75,000
2	refractory arsenopyritic	1.0	1.0	1.0		7.0	85.0	465	36,000
3	refractory arsenopyritic	1.0		1.0		12.0	82.0	3,200	360,000
4	refractory arsenopyritic	1.0	1.0	1.0		20.0	90.0	1,100	220,000
5	free milling				1.0	8.0	98.5	8,000	730,000
6	free milling				1.0	2.4	95.0	5,000	130,000
7	free milling				1.0	5.0	96.5	14,000	780,000
8	free milling				1.0	1.1	80.0	41,000	400,000
9	free milling				1.0	2.0	88.0	18,000	360,000
10	silver-rich					3.1	93.0	4,000	130,000
11	silver-rich					70.0	95.0	360	290,000
12	refractory iron sulfide	1.0	1.0	1.0		5.8	91.0	1,650	100,000
13	refractory iron sulfide	1.0	1.0			3.8	89.0	5,700	220,000
14	refractory iron sulfide	1.0	1.0			4.7	90.0	2,700	130,000

Fig. 4 Excerpt from the generated cases

we created 5 cases describing refractory arsenopyritic ores, 5 describing free milling gold ores, 2 on silver rich ores, 6 cases on refractory ores containing iron sulphides, 4 on copper rich ores and one each on antimony sulphide rich ores, telluride ore and carbonaceous ore.

To compute the similarity of a query, composed of prospective data, and a case we modelled a series of similarity measures. We had the choice between comparative tables, taxonomies and integer or floating point functions. For our initial knowledge model we mainly relied on comparative tables.

Our approach includes the idea to model as much of the complex knowledge present in the domain of ore refinement into the similarity measures as possible. This was based on our assumption that the similarity based retrieval approach provided by the use of CBR would allow us to capture and counter most of the vagueness still associated with the selection of the optimal process in the hydrometallurgical treatment of refractory ores domain. For example, we tried to model into the similarity measures such facts as that the ore does not need any more treatment if it contains gold grains greater than 15 mm in diameter. Such facts are easy to integrate into the similarity measure and thus are operational (having an effect) in our knowledge model. We deem this capability of the similarity measures to capture and represent such 'odd' behaviours of the knowledge model very important. These 'odd' facts or bits of knowledge are hard to capture by rules, which may has ultimately kept the mentioned IntelliGold approach from succeeding on a broad scale [11, 12].

For the global similarity measure of our cases we use a weighted sum of the attributes local similarities. We have not yet investigated further on the impact of different weight distributions other than the obvious emphasise of important attributes, such as for example 'Clay Present', as the presence of clay forbids a selection of hydrometallurgical treatments.

As we are mainly aiming for case retrieval the need for adaptation knowledge is not yet pressing. We therefore have not formalised any adaption knowledge. We will however need adaption knowledge for our second knowledge model which will be deployed to enable the process chain validation and possible re-design.

4 Software Prototype *Auric Adviser*

Using our initial knowledge model we implemented a java-based workflow recommender software, the *Auric Adviser*. *Auric Adviser*'s task is it to retrieve a selection of descriptions of existing gold mining operation best matching the ore and mineralogical context described in a query to the *Auric Adviser*. The retrieved case provides the planner of a potential mining operation and subsequent ore refinement process chain with a first draft of what kinds of treatments would be involved in a potential mining on the prospected site. Furthermore the planer gets an insights into how to potentially arrange these treatments in a process chain to most efficiently refine the ore at the potential site. Additionally, *Auric Adviser* provides the planner with a possibility to estimate the effort, the costs and some ore refinement constraints involved with the potential mining site.

The *Auric Adviser*'s straight forward user interface allows a planner to specify the data gathered from a prospection into a problem description part and thus compose the query case to be post to the CBR knowledge model (Fig. 5). Then the process planner can select the number of retrieved cases she wishes to be displayed. The best retrieved cases are then presented in a tabular field and the process chain description and diagram of the best matching case is displayed in a separate UI element called 'Solution view'.

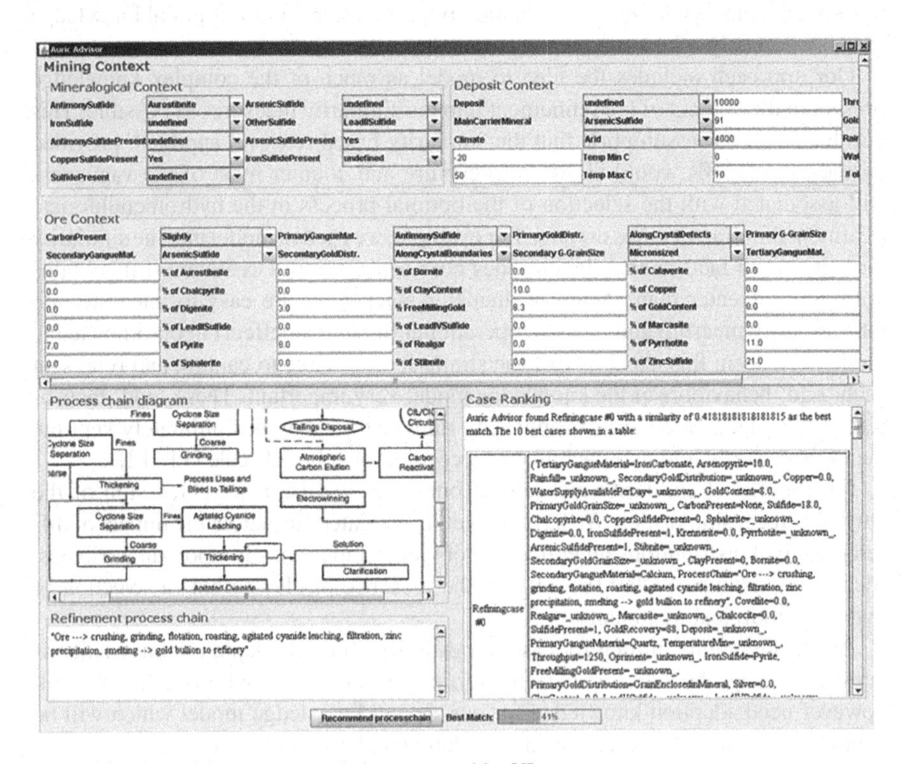

Fig. 5 The *Auric Adviser* process chain recommender UI

We are aware of the complexity of the GUI, offering over 30 elements of input to the planner but we plan to clarify the problem description part of our GUI by breaking it down into the contexts described in our knowledge model, such as *Mineralogical*, *Geological*, *Environmental* etc.

5 Experiments with the *Auric Adviser* and Resulting Refinements to the Initial Knowledge Model

In this section we describe our retrieval experiments with our initial knowledge model. We further detail on the changes we applied to our knowledge model, based on the outcome of the experiments.

For example, we performed retrieval experiments with super weighting single attributes, by outweighing their value against all other local similarities of other attributes, to establish the accuracy of a single discriminant attribute. We did so to establish the effective value ranges for the single attribute being analysed in these experiments. An example for a refined attribute range we could establish that we were able to recalibrate the allowed values for 'gold recovery' as the minimum and maximum values in all cases present are given by 74.5–99.0 %. To enhance the similarity calculation based on the 'gold recovery' attribute we therefore reduced the range of the values from the initial 0.00–100.00 range to the more accurate value range of 74.00–99.00 %. Similar adjustments based on these type of super weighting experiments were made to the attributes: Ore throughput, Gold ore grade, Silver ore grade, Sulphide concentration and Copper ore grade.

During our first set of case retrieval experiments, we ran 13 test queries on our initial 25 cases, we noticed a set of dominant cases being retrieved un-proportionally often as the best matches to any given query. The experts advised us to try and add more data to the cases by providing more attribute values but there was still a set of dominant cases. As we re-questioned the domain experts about this again they advised that we could add even more detail to the cases and refine our selection of the discriminant attributes. By doing so we were able to significantly increase the accuracy and variety of our retrieval results, eliminating the dominant case retrieval.

Based on our experimental data we also noticed that we had to remodel some of the attributes within our knowledge model to represent discriminant Boolean values. These attributes were: *Free milling gold present*, *LeadIISulfide present* and *incSulfide present*. We did so to allow for a quicker *exclusion* of cases, based on these boolean attributes.

We also simplified our initial knowledge model by excluding a number of attributes by setting them as non-discriminant in the calculation of the global similarity. We did so to narrow down our knowledge model to the more important attributes and refine the accuracy of these important attributes representations. See Fig. 6 for the performance data of the 3 versions of our Prototype during the respective retrieval tests.

Query ID	#Attributes specified in query			Quality of retrieved case*		
	Mineralogical context	Deposit context	Ore context	Prototype 1	Prototype 2	Prototype 3
I	2	3	2	5	7	7
II	1	3	2	4	7	7
III	1	3	2	1	1	1
IV	5	2	1	1	1	1
V	0	1	1	7	7	7
VI	0	0	1	7	7	7
VII	2	3	5	4	7	10
VIII	2	3	4	4	7	7
IX	5	3	5	10	10	10
X	5	3	5	10	10	10
XI	2	2	4	4	7	7
XII	2	3	2	5	5	5
XIII	3	3	5	7	7	7
			sum	69	83	86
			average	5.31	6.38	6.62

Fig. 6 Snippet of the performance evaluation of the first three versions of the knowledge model. (* 0 = conflicting, not applicable; 3 = applicable but suboptimal; 5 = applicable, 7 = applicable and well suited; 10 = optimally applicable)

Starting off from a sub optimal accuracy reported by the experts testing our knowledge model we were able to enhance its performance. After the application of the changes described above we are now getting a more positive feedback from the hydrometallurgical experts. Based on the experiments we are now able to see that our initial knowledge model was over engineered with regard to the number of aspects we tried to cover with it. By reducing these details, focusing on important attributes in the mineralogical context and by refining the value ranges of these attributes we were able to significantly increase the performance of our knowledge model. The results for the queries III and IV indicate to us the need to further refine our knowledge model with regard to the representation of sulphides in it. We deduced from our experiments that we can solve the recommendation problem for the queries III and IV by further adjusting the weight of the sulphide attributes and add additional classifying attributes further detailing the presence or absence of specific kinds of sulphides. We further were able to establish that our knowledge model will also benefit from a finer grained similarity measure modelling of the geological context of an ore deposit, which we plan as the next step for its further refinement.

6 Discussion

In this section we will analyse the gathered data from our usage experiments on the *Auric Adviser*. We evaluate the usability and performance/accuracy of our knowledge model and subsequently our *Auric Adviser* software.

Granularity and Scope of the Initial Knowledge Model: Initially the experts we interviewed suggested that all possible attributes that we could assume relevant for our domain and all information which is available in these first cases we gathered should be included in the knowledge model. The reasoning behind that approach was that, when inserting the widest possible range of possibly relevant attributes into our knowledge model we would avoid the hassle of having to add attributes into a working knowledge model later on. The experts based this somewhat cumbersome approach on their concern that, if an attribute was later on added to a working knowledge model, no-one would register this change and thus no use would be made of this change. By not registering such a change in the knowledge model the experts further were concerned that the already existing cases in the working knowledge model would become obsolete or invalid with a growing number of, later, added attributes in the knowledge model not covered or represented within the early cases of the knowledge base. Though the experts initial approach might seem to be very intuitive thinking for a person who is not an expert in knowledge modelling, we had of course to disagree to this intended approach to knowledge modelling on the grounds of a bloated and inefficient resulting knowledge model.

Feedback on the User Interface of the Prototype: The experts suggested that a user should be able to choose if the attributes 'ore throughput rate' (tons per day) and 'gold recovery' are included in the similarity calculation or not. The experts recommended using a fixed set of terms for the display of the description of the process chain within the solution part of our cases. With regard to case ranking the experts advised that we could simplify the case representation in our initial knowledge model. The experts recommended using a table display for the case ranking. Furthermore the readability of our solution display could be further enhanced by relying on colour coding of the local similarities or the use of other visual aids. An additional feature suggested by the experts for our *Auric Adviser* prototype was the ability to save a query and its resulting retrieval result for later reuse and export, for example into an Excel table. We plan to integrate this useful feature as one of the next development steps within our software.

7 Summary and Outlook

We presented the first part of our twofold approach to elicitating and formalising knowledge in the domain of hydrometallurgical processing of gold ore. We demonstrated our processes of formalising the captured knowledge and detailed the resulting first knowledge model and its use in the *Auric Adviser* workflow recommender software. We established its usability and the quality of its recommendations as well as the accuracy and performance of our knowledge model in a set of first experiments.

We now aim at composing the second knowledge model for recommending single treatments, based on 'lessons learned' from the development of the first knowledge model. Parallel to this we will further refine our first knowledge model, used for retrieving process chains based on prospective data. We then will combine both aspects of our approach into a new and complete version of the *Auric Adviser*.

Acknowledgments This work was supported in part by the LOWGRADE project of the ELEMET research program funded by FIMECC Oy. The financial support of TEKES and Outotec Oyj is gratefully acknowledged as well as the financial support of Technology Industries of Finland Centennial Foundation Fund for the Association of Finnish Steel and Metal Producers.

References

1. Aamodt, A., Plaza, E.: Case-based reasoning: Foundational issues, methodological variations, and system approaches. AI Communications **1**(7) (1994)
2. Darke, G.: Assessment of timbre using verbal attributes. In: Conference on Interdisciplinary Musicology. Montreal, Quebec (2005)
3. Madhusudan, T., Zhao, J., Marshall, B.: A case-based reasoning framework for workflow model management. Data & Knowledge Engineering **50**(1), 87–115 (2004)
4. Marsden, J., House, I.: The chemistry of gold extraction. Society for Mining, Metallurgy, and Exploration (2006)
5. Minor, M., Bergmann, R., Görg, S., Walter, K.: Towards case-based adaptation of workflows. Case-Based Reasoning. Research and, Development pp. 421–435 (2010)
6. Minor, M., Tartakovski, A., Bergmann, R.: Representation and structure-based similarity assessment for agile workflows. CBR Research and, Development pp. 224–238 (2007)
7. Pajula, E., et al.: Studies on computer aided process and equipment design in process industry (2006)
8. Richter, M.M.: Introduction. In: M. Lenz, B. Bartsch-Spörl, H.D. Burkhard, S. Wess (eds.) Case-Based Reasoning Technology—From Foundations to Applications, LNAI 1400. Springer-Verlag, Berlin (1998)
9. Rintala, L., Aromaa, J., Forsen, O.: Use of published data in the development of hydrometallurgical flow sheet for gold using decision-support tools. In: Proceedings of the International Mineral Processing Congress, IMPC 2012. CSIR (2012)
10. Sauer, C.S., Roth-Berghofer, T., Auricchio, N., Proctor, S.: Recommending audio mixing workflows. In: Proceedings of the 21st International Conference on Case-Based Reasoning (ICCBR 2013). Springer (2013)
11. Torres, V., Chaves, A., Meech, J.: Intelligold-a fuzzy expert system for gold plant process design. In: Fuzzy Information Processing Society, 1999. NAFIPS. 18th International Conference of the North American, pp. 899–904. IEEE (1999)
12. Torres, V.M., Chaves, A.P., Meech, J.A.: Intelligold-an expert system for gold plant process design. Cybernetics & Systems **31**(5), 591–610 (2000)
13. Watson, I.: Case-based reasoning is a methodology not a technology. Knowledge-Based Systems **12**(5), 303–308 (1999)

Medical Applications

Extracting and Visualising Clinical Statements from Electronic Health Records

M. Arguello, M. J. Fernandez-Prieto and J. Des

Abstract The use of Electronic Health Records (EHRs) standards, like the HL7 Clinical Document Architecture Release Two (CDA R2), presents challenges related to data management, information retrieval, and interactive visualisation. Indeed, it has been widely acknowledged that the usefulness of clinical data within EHRs is limited by the difficulty in accessing it. This paper demonstrates how key components of the Semantic Web (OWL and SPARQL) can tackle the problem of extracting patients' clinical statements from EHRs. The novelty of the work resides in the exploitation of the *terminology binding process*, which specifies how to establish connections between elements of a specific terminology and an information model. The paper argues that in order to find new or better ways to visualise patients' time-oriented information from EHRs, it is essential to perform first adequate EHR-specific searches. Towards this aim, the research presented proposes using SPARQL queries against formal semantic representations of clinical statements within HL7 CDA R2 documents in OWL. To validate the proposal, the study has focused on 1433 clinical statements that are within the physical examination section of 103 anonymised consultation notes in HL7 CDA R2. The paper adopts existing lightweight interactive visualisation techniques to represent patients' clinical information in multiple consultations.

M. Arguello (✉)
University of Manchester, Manchester, UK
e-mail: m.arguello@computer.org; arguellm@cs.man.ac.uk

M. Arguello · M. J. Fernandez-Prieto
University of Salford, Salford, UK

J. Des
DEDIPEGA SL, Monforte de Lemos, Spain

M. Bramer and M. Petridis (eds.), *Research and Development in Intelligent Systems XXX*, 307
DOI: 10.1007/978-3-319-02621-3_23, © Springer International Publishing Switzerland 2013

1 Introduction and Preliminaries

Sweden, Finland, and Denmark are top leaders in the adoption and use of Electronic Health Records (EHRs) systems by primary care physicians and hospitals. As the use of EHRs becomes more widespread, so does the need to search and provide effective information discovery within them [1]. Indeed, EHRs pose a number of challenges that need to be addressed, among them: EHRs generally lack search functionality [2]; and the fact that most EHR standards do not specify how EHR content should be visualised.

On the one hand, the paper explores how to facilitate query building to retrieve EHR data. Given a corpus of consultation notes of Patient-A: (1) how to retrieve certain clinical observations that a clinician (e.g. nurse or general practitioner) is interested in? and (2) is it possible to factor in the variations that happen over time in the clinical observations?

On the other hand, the paper investigates if available third-party visual gadgets related to interactive information visualisation can help clinicians to easily assimilate or understand the clinical data retrieved and aggregated from one or several consultation notes for a given patient.

This paper argues that in order to find new or better ways to visualise a patient's time-oriented clinical data extracted from EHRs, it is essential to perform first adequate EHR-specific searches. In other words, in order to tackle information visualisation from EHRs, it is necessary to solve the implicit problem of data aggregation and organisation. The paper presents pioneer work on exploiting the connections between elements of a specific terminology and an information model (the so called *terminology binding process*) to facilitate query building to retrieve EHR data.

The following subsections offer an overview of the foundations of the research presented.

1.1 EHR Standards and Terminology Binding

There are currently a number of standardisation efforts that are progressing towards providing the interoperability of EHRs such as CEN TC251 EHRcom [3], openEHR [4], and HL7 Clinical Document Architecture (CDA) [5]. These prominent EHRs standards provide different ways to structure and markup clinical content for the purpose of exchange of information.

HL7 Version 3 (HL7 V3 for short) is a lingua franca used by healthcare computers to talk to other computers [6]. The HL7 CDA is a document markup standard based on XML that specifies the structure and semantics of a clinical document, e.g. consultation note, for the purpose of exchange of information [5]. Worldwide HL7 CDA is the most widely adopted application of HL7 V3 [7]. HL7 v3 forms the basis of all clinical communication between NHS CFH systems.[1]

[1] http://www.isb.nhs.uk/use/baselines/interoper. Accessed 15 April 2013.

SNOMED CT has been acknowledged as the most comprehensive, multilingual clinical healthcare terminology in the world [8]. The latest edition of SNOMED CT terminology (January 2013)[2] contains 297,000 active concepts with formal logic-based definitions, which are organized into 19 top-level hierarchies. SNOMED CT is formulated in the description logics EL++ [9], which corresponds to the OWL 2 EL profile [10].

It is now recognized that healthcare terminologies like SNOMED CT [8] and information models like HL7 V3 cannot be separated. There is a *terminology binding process* [11] that specifies how to establish connections between elements of a specific terminology and an information model. The HL7 and IHSTDO report [12] provides guidelines on how to bind SNOMED CT with HL7 V3. And thus, although the *terminology binding process* is properly documented, its exploitation with the aim of facilitating query building has not been studied before.

1.2 Information Discovery in EHRs

Numerous studies have highlighted the problem of extracting information from patients' EHRs [13]. Indeed, Natarajan et al. [2] emphasise: "while search engines have become nearly ubiquitous on the Web, EHRs generally lack search functionality". However, clinicians find search functionality useful for searching within as well as across patient records [14, 15].

Hristidis et al. [16] discusses why the general work on information discovery on XML documents is not adequate to provide quality information discovery on HL7 CDA XML-based documents. The key reasons are the complex and domain-specific semantics and the frequent references to external information sources like dictionaries [16].

There are research studies that use medical ontologies to improve the search quality, such as [17]. XOntoRank system [18] tackles the problem of facilitating ontology-aware keyword search on HL7 CDA XML-based documents, which contain references to clinical ontological concepts. The iSMART prototype system [19] has as key components the XML to RDF transformer and the EL+ ontology reasoner and supports ontology-based semantic query of CDA documents, where the semantic query can be leveraged to provide clinical alerts and finding eligible patients for clinical trials.

More recently, Perez-Rey et al. [20] developed CDAPubMed, an open-source Web browser that uses the contents of HL7 CDA documents to retrieve publications from biomedical literature. CDAPubMed analyses the contents of CDA documents using an open-source Natural Language Processing (NLP) package and uses a keyword identification algorithm. CDAPubMed was tested on 17 CDA documents.

[2] http://www.ihtsdo.org/fileadmin/user_upload/doc/download/doc_ScopeMemo_Current-en-US_INT_20130131.pdf. Accessed 15 April 2013.

1.3 Information Visualisation in EHRs

A significantly important factor for the acceptance of health information systems is to possess intuitive and user-friendly graphical user interfaces [21]. Most EHR standards do not specify how EHR content should be visualised. The CDA specification gives some recommendations on how documents are to be structured and encoded to facilitate the visualisation process. However, a mere list of observations from the physical examination section of a HL7 CDA consultation note may not be good enough for some clinicians.

Obtaining visualisations that take into account a patient's clinical data from one single consultation note (HL7 CDA document) is not hard. Indeed, this can be handled quite efficiently with XSLT style sheets. The difficulties start when shifting from one single CDA document to several CDA documents related to a patient, which is unavoidable from a medical point of view, for example when dealing with chronic diseases.

Gershon et al. [22] define information visualisation as: "the process of transforming data, information, and knowledge into visual form making use of humans' natural visual capabilities". Information visualisation has a long tradition of dealing with medical data, most particularly in the field of medical imaging. Recently, Roque et al. [23] drew attention to six information visualisation systems designed specifically for gaining an overview of *electronic medical records*: Lifelines [24], Lifelines2 [25], KNAVE-II [26], CLEF Visual Navigator [27], Timeline [28], and AsbruView [29]. However none of these have adopted the efforts around prominent EHR standards, such as CEN TC251 EHRcom [3], openEHR [4], and HL7 CDA [5].

More recently, Tao et al. [30] presented a preliminary study on connecting: (a) the LifeFlow system, which is designed to support an interactive exploration of event sequences using visualisation techniques; and (b) the Clinical Narrative Temporal Reasoning Ontology (CNTRO)-based system, which is designed for semantically representing, annotating, and inferring temporal relations and constraints for clinical events in EHRs represented in both structured and unstructured ways. The connection of these two systems will allow using LifeFlow for interactively visualising the data represented in and inferred by the CNTRO system. The *new* interface allows CNTRO to function with RDF triple stores [31] and SPARQL queries [32]; however, it is unclear if this work also adopts the efforts around prominent EHR standards, i.e. CEN TC251 EHRcom, openEHR, or HL7 CDA.

2 How to Bind SNOMED CT with HL7 V3

HL7 CDA documents derive their machine processable meaning from the HL7 Reference Information Model (RIM) [33] and use the HL7 V3 [34] data types. The RIM and the V3 data types provide a powerful mechanism for enabling CDA's incorporation of concepts from standard coding systems such as SNOMED CT [8] and

Table 1 Examples of connections established between SNOMED CT and HL7 V3 RIM

<table>
<tr><th>HL7 V3 RIM attribute</th><th>SNOMED CT attribute</th></tr>
<tr><td>Observation.targetSiteCode</td><td>363698007|Finding site|</td></tr>
<tr><td>Observation.methodCode</td><td>418775008| Finding method|</td></tr>
</table>

Table 2 Example of common observation patterns incorporating the connections from Table 1

<table>
<tr><th>Observation code/value</th><th>Common observation pattern</th></tr>
<tr><td>Observable entity with physical quantity as result</td><td><observation classCode="OBS" moodCode="EVN">
<code code="41633001 | Intraocular pressure|" codeSystem="2.16.840.1.113883.6.96" />
<value xsi:type="PQ" value="15" unit="mm[Hg]"/>
<methodCode code="252832004 | Intraocular pressure test|" codeSystem="2.16.840.1.113883.6.96" />
<targetSiteCode code="8966001 | Left eye structure|" codeSystem="2.16.840.1.113883.6.96" />
</observation></td></tr>
</table>

LOINC [35]. The HL7 and IHSTDO report [12] provides guidelines on how to bind SNOMED CT with HL7 V3, and presents common patterns.

The HL7 and IHSTDO report [12] defines common patterns as: "clinical statements that are used frequently, often in many different applications, for a wide variety of communication use cases". Table 1 provides examples of the equivalences defined between attributes in HL7 RIM and relationships in SNOMED CT, which are part of the *terminology binding process* performed. Table 2 shows a common pattern for HL7 RIM observation code/value.

Taking into account the work described in [36], it is possible to map CDA sections and entries as well as HL7 RIM Observations from HL7 CDA documents to instances of an OWL 2 ontology. This research reuses two ontologies: (1) an OWL ontology for HL7 CDA that appears in [37]; and (2) a SNOMED CT ontology in OWL 2, which was created by means of the Simple SNOMED Module Extraction [38]. Additionally an ontology for LOINC [35] has been created. These three ontologies are imported into each HL7 CDA document (e.g. consultation note), i.e. OWL 2 file that contains the clinical information about a particular patient.

3 Approach Overview

Visualisation of time-oriented clinical data from a patient's record implies accessing and aggregating clinical data obtained in different consultations of the patient. As data aggregation and organisation are crucial, this research adopts ontologies due to their suitability to integrate heterogeneous information sources and manage infor-

```xml
<HL7_CDA_R2:Observation_Entry rdf:ID="15184_200502031000">
  <HL7_CDA_R2:isAssociatedWith rdf:resource="http://localhost/ontology/LOINC#Eye,_Physical_findings" />
  <HL7_CDA_R2:observation rdf:resource="http://localhost/ontology/SNOMED_CT#distance vision 20/20" />
  <HL7_CDA_R2:displayName rdf:datatype="&xsd;string">Visual acuity testing [RE]: distance vision 20/20</HL7_CDA_R2:displayName>
  <HL7_CDA_R2:statusCode rdf:datatype="&xsd;string">completed</HL7_CDA_R2:statusCode>
  <HL7_CDA_R2:effectiveTime rdf:datatype="&xsd;string">200502031000</HL7_CDA_R2:effectiveTime>
  <HL7_CDA_R2:targetSite rdf:resource="http://localhost/ontology/SNOMED_CT#Right_eye_structure" />
  <HL7_CDA_R2:method rdf:resource="http://localhost/ontology/SNOMED_CT#Visual_acuity_testing" />
</HL7_CDA_R2:Observation_Entry>
```

Fig. 1 OWL 2 instance: formal semantic representation of a CDA observation entry

mation resources and services. This research proposes to use: (1) OWL to obtain a formal semantic representation of clinical statements (common patterns – see previous section) within CDA specification, and to take advantage of the OWL's ability to express ontological information about instances appearing in multiple documents; and (2) SPARQL to query the OWL model to filter out individuals (ontological instances) with specific characteristics.

Figure 1 shows an OWL 2 instance, which is a formal semantic representation of a CDA *observation* entry for the physical examination section of a HL7 CDA R2 consultation note.

From the implementation point of view, the research relies on Web services that at the moment of writing exhibit different levels of maturity. Thus, the work presented follows a service-oriented approach that promotes re-usability of functionality. The current approach considers three Web services: (1) *OWL converter service* that provides functionality to perform a semantic enhancement of the HL7 CDA documents and obtain a formal semantic representation of clinical statements within CDA specification in OWL. (2) *SPARQL query processing service* that provides functionality for query processing, where the query results comply with a W3C XML Schema. (3) *XML converter service* that provides functionality for transforming SPARQL query results in XML format into proprietary XML-based data files that are used by specific visual gadgets. This service performs the XML-to-XML transformation; so third-party visual gadgets can directly process the data.

4 Information Extraction: SPARQL SELECT Queries for EHRs

The key components of the Semantic Web include the Resource Description Framework (RDF) as the basic data model, the Web Ontology Language (OWL) for expressive ontologies, and the SPARQL query language. SPARQL is the most commonly used Semantic Web query language, which is defined in terms of the W3C's RDF data model and will work for any data source that can be mapped into RDF. But there is no query language specifically for OWL ontologies. However, SPARQL can be used to query an OWL model to filter out individuals with specific characteristics.

The OWL's ability to express ontological information about instances appearing in multiple documents facilitates the use of SPARQL to query simultaneously clinical statements (common patterns) within CDA specification from multiple consultation notes in a principled way. To validate the proposal, the study has focused on 1433 clinical statements that are within the physical examination section of 103 anonymised consultation notes in HL7 CDA R2.

The common pattern presented in Table 2, once formalised in OWL 2, can be easily queried with SPARQL SELECT queries. This section presents three SPARQL SELECT queries (Q1, Q2, and Q3) that exemplify the exploitation of different common patterns. Indeed, the SPARQL SELECT queries Q1, Q2, and Q3 allow extracting a patient's clinical data (ontology instances) from multiple documents (formal semantic representations of HL7 CDA consultation notes).

Q1: a SPARQL SELECT query to aggregate procedures and observations for multiple consultations of a patient – the SPARQL query depicted in Fig. 2 can be effectively used to retrieve the observations for a particular set of procedures performed in Right Eye (RE) and Left Eye (LE) that may appear within a physical examination section of a single HL7 CDA consultation note.

Protégé[3] is a widespread ontology-design and knowledge acquisition tool that can be used to build ontologies in OWL. Protégé has a SPARQL query panel (see Fig. 2) where it is possible to put forward a SPARQL query. After pressing the Execute Query button, the results for the formulated SPARQL query appear.

A SPARQL query may specify the dataset to be used for matching by using the FROM clause. Therefore, it makes possible to query the OWL ontology instances related to HL7 CDA document entries from several HL7 CDA consultation notes of a patient. This ability of querying simultaneously clinical statements (e.g. observations) within CDA specification from multiple consultation notes in a principled way allows retrieving a patient's clinical data needed to populate graphical clinical timelines.

Q2: a SPARQL SELECT query to aggregate observations for a single procedure for multiple consultations of a patient – the SPARQL query depicted in Fig. 3 can retrieve the observation(s) for a single procedure (ophthalmoscopy) performed in either RE or LE that may appear within a physical examination section of multiple HL7 CDA consultation notes.

Fig. 2 Protégé snapshot: SPARQL query panel

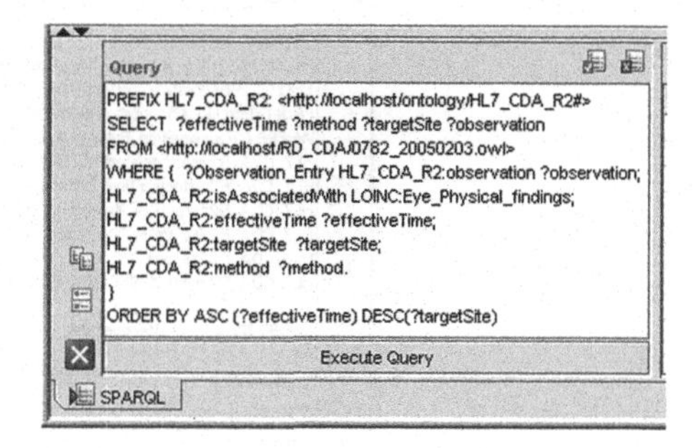

[3] http://protege.stanford.edu. Accessed 15 April 2013.

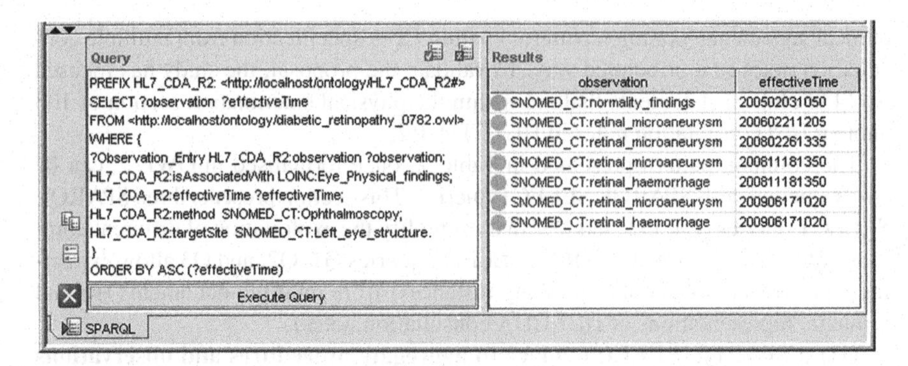

Fig. 3 SPARQL query to retrieve the observation(s) for a single procedure

As it can be seen from Fig. 3, with one SPARQL query it is possible to obtain the observation(s) for a single procedure (ophthalmoscopy) performed during multiple consultations (HL7 CDA documents) that span over time (from 2005 to 2009), where CDA targetSiteCode is Left Eye structure (in short LE) and CDA methodCode is ophthalmoscopy.

Q3: a SPARQL SELECT query to retrieve numeric values and units associated with observations for a single procedure from multiple consultations of a patient – the SPARQL query depicted in Fig. 4 can retrieve the numeric values (physical quantities) and units associated with observations like "raised intraocular pressure" or "normal intraocular pressure" that may appear in multiple HL7 CDA consultation notes within a physical examination section for the procedure intraocular pressure test.

Observations like "body temperature" or "intraocular pressure" have an associated range of numeric values and units (e.g. mmHg). The interpretation of these values by healthcare professionals may lead to clinical findings that can contribute to make a diagnosis. The numeric values obtained after executing the query from Fig. 4 can

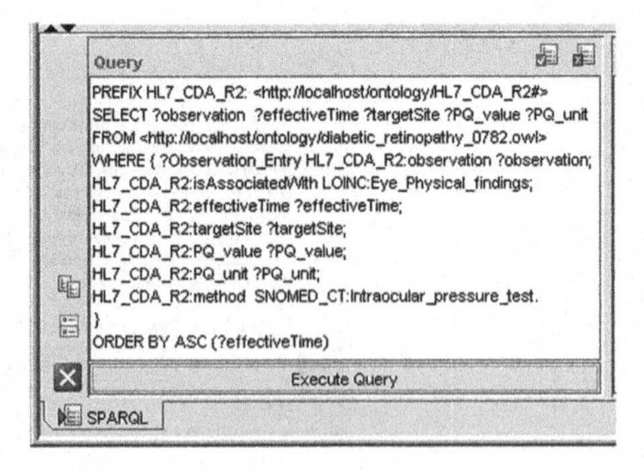

Fig. 4 SPARQL query to retrieve numeric values and units associated with observations

have an associated interpretation, such as: "normal intraocular pressure" or "raised intraocular pressure".

5 Information Visualisation: Using Third-Party Visual Gadgets

This section illustrates how the time-oriented clinical data retrieved from the SPARQL SELECT queries (see previous section) is subsequently formatted according to customised interactive graphical representations. The current approach relies on Web services that act as XML converters, which transform SPARQL SELECT query results in XML format into proprietary XML-based data files that are visual gadget-specific. These *XML converter services* are data-driven and are used to read and dynamically transform data. The data transformation allows populating the XML-based data files for Timeline [39] and AmCharts [40].

5.1 Timeline

Timeline [39] can be configured to show two bands that can be panned infinitely by dragging with the mouse pointer. The lower band shows the years and when there is data a dash line appears.

A SPARQL SELECT query like Q1 that appears in Fig. 2 in conjunction with a XML converter, which uses a Timeline-specific template, facilitates populating the proprietary XML-based data file with key information extracted from CDA observation entries belonging to the physical examination section of multiple HL7 CDA consultation notes that go from 2005 to 2009.

By dragging with the mouse pointer along the lower band (quicker as it covers years) or the upper band (slower as it covers days and for consultation day even covers hours), Timeline can show procedures performed during five consultations of a patient, covering a time period that goes from 2005 to 2009. Initially, the procedures performed for a particular consultation day are displayed as shown in Fig. 5. By mouse clicking over a particular procedure a bubble appears (see Fig. 5) that displays

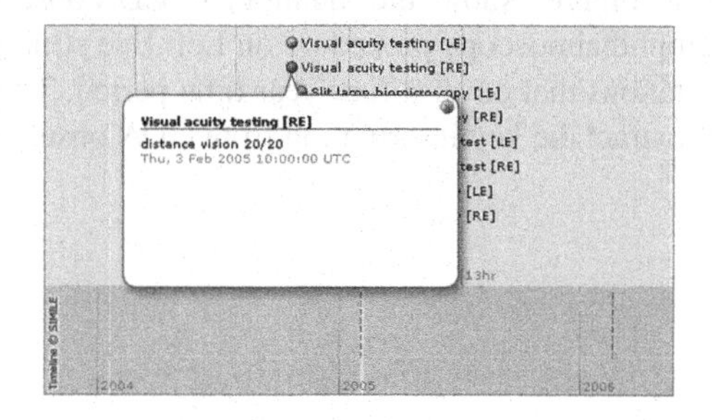

Fig. 5 Using Timeline to show procedures performed during one consultation

the associated observation(s) for the procedure selected. Figure 5 shows the CDA observation associated with the procedure (CDA methodCode) visual acuity testing, where CDA targetSiteCode is Right Eye structure (in short RE) and the effectiveTime is 10:00am on Thursday 3rd February 2005.

The interactive information visualisation technique introduced focuses on a patient's clinical data from physical examination sections of HL7 CDA consultation notes. Timeline can be particularly useful to create automatic "problem-centric" views from a myriad of CDA documents, where procedures can be easily replaced for diseases or any other condition-specific views that may facilitate medical tasks.

5.2 AmCharts

AmCharts [40] has a set of fully customizable animated 2D or 2D Flash charts that are useful to graphically represent "qualitative" CDA observations as well as numeric values (physical quantities) associated to CDA observation entries, and therefore, further support can be provided to clinicians who need to cross-compare clinical data obtained in different consultations of a patient

The visual gadget from AmCharts that is used in Fig. 6 allows end-user interaction as depicted, where a green box appears when the end-user points the mouse over a green point in the graph. The light grey dots and line in Fig. 6 correspond to numeric values for the observation "intraocular pressure" (IOP) for Left Eye structure (LE), which are labelled for short as: LE IOP. The SPARQL SELECT query that appears in Fig. 4 in conjunction with a XML converter, which uses an AmCharts-specific template, facilitates populating the proprietary XML-based data file with key information extracted from CDA observation entries belonging to the physical examination section of multiple HL7 CDA consultation notes that go from 2005 to 2009.

To provide further support to physicians, not only numeric values (physical quantities) associated to CDA observations, but also signs that typically correspond to "qualitative" CDA observations need to be graphically represented. In other words, it is necessary to find a way to represent how "qualitative" CDA observations change over time, and even allow physicians to cross-compare them for a time interval that clinicians may find interesting or relevant.

Figure 7 shows the "qualitative" CDA observations obtained when performing the ophthalmoscopy procedure on Left Eye structure (for short LE) during five consultations that cover a four year time period. The SPARQL SELECT query needed to extract the key information from CDA observation entries appears in Fig. 3.

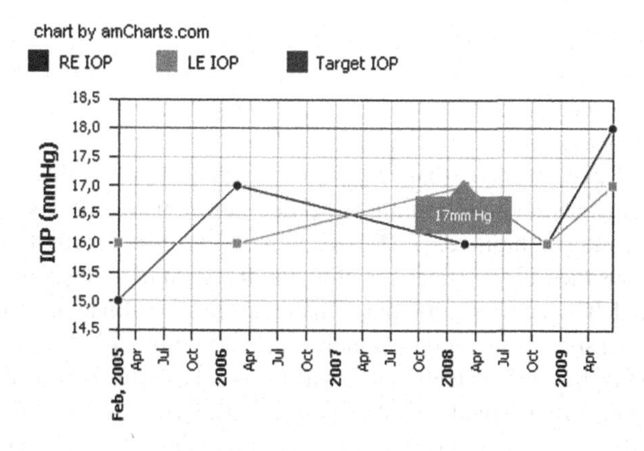

Fig. 6 Using AmCharts to represent intraocular pressure values for multiple consultations

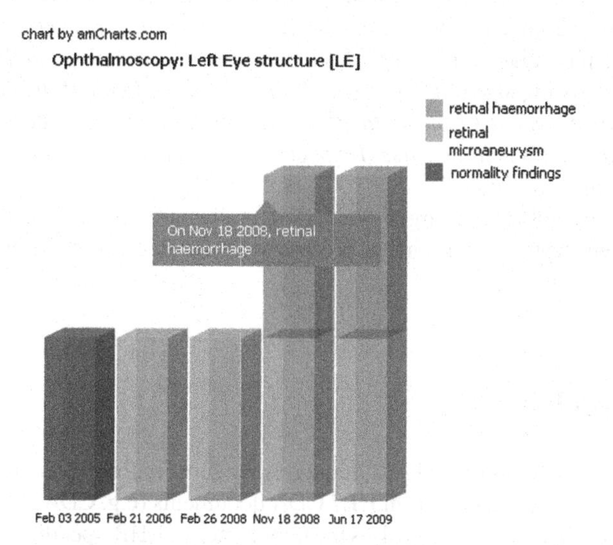

Fig. 7 Using AmCharts to represent CDA observation entries for multiple consultations

6 Approach Evaluation

To validate the proposal, the study has focused on 1433 clinical statements that are within the physical examination section of 103 anonymised consultation notes, which are formatted according to HL7 CDA R2. The acceptability of the approach presented is assessed by means of evaluation sessions with five clinical professionals who are not familiar with key components of the Semantic Web, such as OWL and SPARQL. During those sessions, the participants were encouraged to "think aloud". This practice, Think-Aloud Protocol [41], plays an important role in usability data collection. From the verbal responses it is possible to get valuable insights around key points, such as: How easy is it to use the visual gadgets (e.g. Timeline [39] to access multiple consultations of a patient)? How clear is the data presented? How helpful are the visualisations in assessing a patient's condition? How comfortable is the clinician when using the visual gadgets?.

Taking into account the verbal responses received, it is possible to draw some tentative conclusions: (1) No training is needed before using the visual gadgets as they are seen as fairly intuitive. (2) The Timeline [39] representation of procedures and observations (see Fig. 5) gives the clinician freedom to go back and forward in time, and switch between consultations. This seems to be useful to him/her. However, cross-comparing signs from two consultations can be tedious. (3) The AmCharts [40] representation of IOP values (see Fig. 6) was assessed as intuitive due to its similarity to the graphical representations of vital signs, such as body temperature. (4) The AmCharts [40] representation of signs for one procedure (e.g. ophthalmoscopy in Fig. 7) was assessed as novel, with the main advantage of allowing in one glimpse to foresee the presence or absence of changes based on the (changed or unchanged) colours. One physician commented: "*as a chronic disease evolves adding signs and aggravating existing signs, it is easy to observe from the figure if the patient's condition remains the same (same rectangle colours in different consultations) or if the patient's condition is getting worse (I expect an increment in the number of rectangles or a change in their colour)*".

It should be noted that larger evaluation studies are currently being set up to promote larger testing with clinical professionals in order to obtain more detailed feedback.

7 Conclusion

The paper demonstrates how SPARQL queries can retrieve formal semantic representations of clinical statements from HL7 CDA documents (e.g. CDA R2 consultation notes) in OWL. The study emphasises the importance of EHR-specific searches when dealing with chronic diseases, and also to obtain a patient's time-oriented clinical data organised in a suitable way to enable interactive information visualisation (e.g. graphical clinical timelines) that goes beyond the simple rendering of narrative clinical reports. The main novelty of the work lies in creating SPARQL SELECT queries that exploit the *terminology binding process* and the common patters defined by HL7 and IHSTDO, which to the best of our knowledge, have not been exploited before. Clinicians and clinical researchers can benefit from getting clinical statements organised and aggregated across NHS patient records, which can lead to better support for decision-making.

References

1. Hristidis, V., Varadarajan, R.R., Biondich, P., Weiner, M.: Information Discovery on Electronic Health Records Using Authority Flow Techniques. BMC Medical Informatics and Decision Making, doi:10.1186/1472-6947-10-64 (2010).
2. Natarajan, K., Stein, D., Jain, S., Elhadad, N.: An Analysis of clinical queries in an electronic health record search utility. International Journal of Medical Informatics, Vol. 79, No. 7, pp. 515–522 (2010).
3. CEN TC251 EHRcom, http://www.CENtc251.org. Accessed January 2013.
4. OpenEHR, http://www.OpenEHR.org. Accessed January 2013.
5. Dolin, R.H., Alshuler, L., Boyer, S., Beebe, C., Behlen, F.M., Biron, P.V., Shabo, A.: HL7 Clinical Document Architecture, Release 2. Journal of the American Medical Informatics Association, doi:10.1197/jamia.M1888 (2006).
6. Benson, T.: Principles of Health Interoperability HL7 and SNOMED, HI, doi:10.1007/978-1-84882-803-2_12 (2010).
7. HL7 v3 Message Development Framework, http://www.hl7.org/library/mdf99/mdf99.pdf. Accessed January 2013.
8. IHTSDO, SNOMED CT Editorial Guide (January 2013 International Release), IHTSDO (2013).
9. Baader, F., Brandt, D., Lutz, C.: Pushing the EL envelope. In: Proceedings of the 19th International Joint Conference on, Artificial Intelligence, pp. 364–369 (2005).
10. OWL 2 Web Ontology Language, Primer (Second Edition), W3C Recommendation 11 December 2012. Available at http://www.w3.org/TR/owl2-primer/. Accessed January 2013.
11. Markwell, D., Sato, L., Cheetham, E.: Representing clinical information using SNOMED Clinical Terms with different structural information models. In: Proceedings of the 3rd international conference on Knowledge Representation in Medicine (2008).
12. Cheetham, E., Markwell, D., Dolin, R.: Using SNOMED CT in HL7 Version 3; Implementation Guide, Release 1.5 (2009).
13. Sucurovic, S.: An approach to access control in electronic health record. Journal of Medical Systems, Vol 34, No. 4, pp. 659–66 (2010).
14. Hanauer, D.A.: EMERSE: the electronic medical record search engine. In: Proceedings of the AMIA Annual, Symposium, pp. 941 (2006).
15. Schulz, S., Daumke, P., Fischer, P., Müller, M.: Evaluation of a document search engine in a clinical department system. In: Proceedings of the AMIA Annual, Symposium, pp. 647–651 (2008).
16. Hristidis, V., Farfán, F., Burke, R.P., Rossi, A.F., White, J.A.: Information Discovery on Electronic Medical Records. In: National Science Foundation Symposium on Next Generation of Data Mining and Cyber-Enabled Discovery for Innovation (2007).
17. Segura, N.A., Sanchez, S., Garcia-Barriocanal, E., Prieto, M.: An Empirical analysis of ontology-based query expansion for learning resource searches using MERLOT and the Gene ontology. Knowledge-Based Systems, Vol. 24, No. 1, pp. 119–133 (2011).
18. Farfán, F., Hristidis, V., Ranganathan, A., Weiner, M.: XOntoRank: Ontology-Aware Search of Electronic Medical Records. In: Proceedings of the International Conference on Data, Engineering, pp. 820–831 (2009).
19. Liu, S., Ni, Y., Mei, J., Li, H., Xie, G., Hu, G., Liu, H., Hou, X., Pan, Y.: iSMART : Ontology-based Semantic Query of CDA documents. In: Proceedings of AMIA Annual, Symposium, pp. 375–379 (2009).
20. Perez-Rey, D., Jimenez-Castellanos, A., Garcia-Remesal, M., Crespo, J., Maojo, V.: CDA-PubMed: a browser extension to retrieve EHR-based biomedical literature. BMC medical informatics and decision making, Vol. 12, No. 1 (2012).
21. Sittig, D., Kuperman, G., Fiskio, J.: Evaluating physician satisfaction regarding user interactions with an electronic medical record system. In: Proceedings of AMIA Annual, Symposium, pp. 400–404 (1999).

22. Gershon, N., Eick, S.G., Card, S.: Information visualization. ACM Interactions, Vol. 5, No. 2, pp. 9–15 (1998).
23. Roque, F.S., Slaughter, L., Tkatsenko, A.: A Comparison of Several Key Information Visualization Systems for Secondary Use of Electronic Health Record Content. In: Proceedings of the NAACL HLT 2010 Second Louhi Workshop on Text and Data Mining of Health, Documents, pp. 76–83 (2010).
24. Plaisant, C., Milash, B., Rose, A., Widoff, S., Shneiderman, B.: LifeLines: Visualizing personal histories. In: Proceedings of ACM Conference on Human Factors in, Computing Systems, pp. 221–227 (1996).
25. Wang, T.D., Plaisant, C., Quinn, A., Stanchak, R., Shneiderman, B., Murphy, S.: Aligning Temporal Data by Sentinel Events: Discovering Patterns in Electronic Health Records. In: Proceedings of SIGCHI Conference on Human Factors in, Computing Systems, pp. 457–466 (2008).
26. Martins, S.B., Shahar, Y., Goren-Bar, D., Galperin, M., Kaizer, H., Basso, L.V., McNaughton, D., Goldstein, M.K.: Evaluation of an architecture for intelligent query and exploitation of time-oriented clinical data. Artificial Intelligence in Medicine, Vol. 43, pp. 17–34 (2008).
27. Hallett, C.: Multi-modal presentation of medical histories. In: Proceedings of the 13th international conference on Intelligent user, interfaces, pp. 80–89 (2008).
28. Cousins, S.B., Kahn, M.G.: The visual display of temporal information. Artificial Intelligence in Medicine, Vol. 6, No. 3, pp. 341–357 (1991).
29. Miksch, S., Kosara, R., Shahar, Y., Johnson, P.D.: AsbruView: visualization of time-oriented, skeletal plans. In: Proceedings of the 4th International Conference on Artificial Intelligence Planning Systems, pp. 11–18 (1998).
30. Tao, C., Wongsuphasawat, K., Clark, K., Plaisant, C., Shneiderman, B., Chute, C.G.: Towards event sequence representation, reasoning and visualization for EHR data. Proceedings of the 2nd ACM SIGHIT International Health Informatics, Symposium, pp. 801–806 (2012).
31. RDF, http://www.3.org/RDF. Accessed January 2013.
32. SPARQL, http://www.w3.org/TR/rdf-sparql-query/. Accessed January 2013.
33. HL7 RIM, http://www.hl7.org/implement/standards/rim.cfm. Accessed January 2013.
34. HL7 V3 data types, http://www.hl7.org/implement/standards/product_brief.cfm?product_id=264. Accessed January 2013.
35. LOINC, http://www.loinc.org. Accessed January 2013.
36. Arguello, M., Des, J., Fernandez-Prieto, M.J., Perez, R., Paniagua, H.: Executing Medical Guidelines on the Web: Towards Next Generation Healthcare. Journal Knowledge-Based Systems, Vol. 22, pp. 545–551 (2009).
37. HL7 CDA OWL, http://www.w3.org/wiki/HCLS/ClinicalObservationsInteroperability/HL7CDA2OWL.html. Accessed January 2013.
38. Simple SNOMED Module Extraction, http://owl.cs.manchester.ac.uk/snomed/. Accessed January 2013.
39. Timeline, http://simile.mit.edu/timeline/. Accessed January 2010.
40. Am Charts, http://www.amcharts.com/. Accessed January 2013.
41. Ericsson, K.A., Simon, H.A.: Protocol analysis: verbal reports as data. MIT Press, Cambridge (1984).
42. Medjahed, B., Bouguettaya, A.: A Multilevel Composability Model for Semantic Web services. IEEE Transactions on Knowledge and Data Engineering, Vol. 17, No. 7, pp. 954–968 (2005).

Evaluation of Machine Learning Techniques in Predicting Acute Coronary Syndrome Outcome

Juliana Jaafar, Eric Atwell, Owen Johnson, Susan Clamp and Wan Azman Wan Ahmad

Abstract Data mining using machine learning techniques may aid in the development of prediction models for Acute Coronary Syndrome (ACS) patients. ACS prediction models such as TIMI and GRACE have been developed using traditional statistical techniques such as linear regression. In this paper, different machine learning techniques were evaluated to present the potential use of machine learning techniques in classification tasks as basis for future medical prediction model development. A dataset of 960 of ACS patients from the Malaysian National Cardiovascular Disease Database registry was employed and trained on three popular machine learning classifiers i.e. Naïve Bayes, Decision Tree and Neural Network to predict ACS outcome. The outcome being evaluated was whether the patient is dead or alive. An open source tool—Waikato Environment for Knowledge Analysis (WEKA) were used in executing these classification tasks. A 10-folds cross validation technique was used to evaluate the models. The performance of classifiers was presented by their accuracy rate, confusion matrix and area under the receiver operating characteristic curve (AUC). Naïve Bayes and Neural Network show generally convincing results with an average of 0.8 AUC values and 90 % accuracy rate.

J. Jaafar (✉) · O. Johnson · S. Clamp
Yorkshire Centre for Health Informatics (YCHI), University of Leeds, Leeds, UK
e-mail: umjja@leeds.ac.uk

O. Johnson
e-mail: O.A.Johnson@leeds.ac.uk

S. Clamp
e-mail: S.Clamp@leeds.ac.uk

E. Atwell
School of Computing, University of Leeds, Leeds, UK
e-mail: E.S.Atwell@leeds.ac.uk

W. A. Wan Ahmad
Department of Medicine, University Malaya Medical Centre, Kuala Lumpur, Malaysia
e-mail: wanazman@ummc.edu.my

M. Bramer and M. Petridis (eds.), *Research and Development in Intelligent Systems XXX*,
DOI: 10.1007/978-3-319-02621-3_24,

1 Introduction

Acute Coronary Syndrome (ACS) is defined as a heart problem condition that is due to the blockage of blood supply to the heart commonly due to *atherosclerosis*. Due to the high risk of short term death, more powerful strategies are recommended by major cardiology practice guidelines for high risk ACS patients [1, 2]. Therefore, it is essential to identify those with high risk in improving overall ACS outcome. Common ACS prediction models such as TIMI risk scores and GRACE score have been widely used in clinical practice to assist in identifying patients with high risk [3].

In medical domains, prediction models are generally derived from data sample collected in a population (identified/randomized) and have been constructed using traditional statistical methods such as liner regression and Cox Hazard regression [4, 5]. These linear based statistical methods are based on the assumption of linear relationships. Machine learning offers a spectrum of learning algorithms comprises of non-linear methods, linear methods and ensemble methods (combination of multiple learning methods) which may better describe the relationships between predictors in developing prediction models. Machine learning techniques have been used in predicting the outcome of coronary artery bypass surgery [6], cancer diagnosis and prognosis [7] and risk identification of 8 chronic diseases [8], among others. There is also evidence showing that machine learning algorithms may provide additional improvement in prediction outcomes over statistical methods as exemplified by [9, 10] studies. For best results, machine learning techniques were also used to compliment statistical method in improving overall classification and prediction tasks [11].

In addition, the increasing amount of medical records and the unique nature of medical records have urged the need for more advanced computational methods in data analysis and exploration [12, 13]. Medical data which were kept in medical databases are valuable assets to enhance the development of clinical prediction modelling. QRISK2 was derived from electronic health record has proved to be more accurate in comparison with Framingham [14]. Chia et al. [15] has developed prediction model using historical data by introducing a novel modern approach—1.5SVM. Using existing medical records also promote evidence-based medicine i.e. a new medical practice which uses clinical results as an evidence for prognosis, diagnosis and making clinical decisions for treatment [16, 17]. Thus, data mining using machine learning is one of the best computational solutions for classification mechanism and prediction model development.

Exploring prediction model development for ACS using modern machine learning techniques would potentially enhance the benefits of data mining in the medical field. In this evaluation study, the use of machine learning techniques is evaluated in the development of a prediction model. The aim is to evaluate different machine learning techniques as the basis for further analysis in the prediction model development particularly using ACS dataset.

2 Machine Learning Classifiers

Naïve Bayes, Decision Tree and Neural Network are the basic and most commonly used by machine learning community [18]. Due to their common use, these classifiers were chosen for this study.

2.1 *Naïve Bayes (NB)*

NB is a probabilistic classifier based on Bayes theorem. The theorem is expressed as below equation:-

$$P(H|E) = P(E|H)P(H)/P(E)$$

Applying the theorem on training dataset, NB classifier calculates the outcome P (H|E) by analyzing the relationship of attributes (E) and the class (H). The probability is calculated by observing a set of attributes (E = $\{e_1, e_2 \ldots e_n\}$) occurring in specific class (H) with the assumption that each attribute is independent of other attributes [19].

2.2 *Decision Tree (DT)*

DT generally categorizes classes by recursively building up a tree with nodes that consist of branches and leaves which represent the attributes and classes respectively. The node is built up by measuring the information gain of attributes in the training example and picking up the attribute with the highest information gain in each iteration of building up the tree.

In J48 algorithm, the tree is built up using pruning method to reduce the size of the tree by minimizing over-fitting leading to a better accuracy in prediction [20].

2.3 *Neural Network (NN)*

NN is an artificial intelligence learning whose main concept is derived from how information is being processed in a human brain. The structure is made up from numerous interconnected (network) neurons divided into 3 main layers namely input layer, hidden layer and output layer. The learning is based on the input neurons and how the neurons are complexly connected to each other to produce the output.

Multilayer Perceptron algorithm is of type Feed-forward network that is claimed to be the simplest type where the information is transferred in one direction from input layer to hidden layer and finally to the output layer [21].

3 Method

3.1 Dataset

The National Cardiovascular Disease Database (NCVD) is a national (Malaysian) registry for Acute Coronary Syndrome (ACS). It was initiated to strategically manage the cardiovascular disease (CVD) threat as to improve the overall cardiac services in Malaysia [22]. The dataset contains unidentifiable ACS patients which include ST-elevation myocardial infraction (STEMI), non-STEMI and unstable angina (UA), aged from 18 years and above, admitted to 18 participating sites in Malaysia between 2006 and 2010. It is a life registry that is being continuously updated with new ACS cases in Malaysia.

A total record that was extracted from the registry in September 2012 was 13,592 records with 214 attributes. The attributes include non-clinical elements, clinical elements and database specific elements. As a preliminary study, only 959 random records were utilized; with 64 attributes that were manually selected.

Among the selected attributes, 62 are input attributes, 1 identification attribute (*patientid*) and 1 outcome attribute (*ptoutcome*). The attribute type has either nominal or numeric values. The outcome attribute has a nominal value of *'Died'* or *'Discharge'*. Table 1 summarizes the selection of attributes.

The dataset is severely imbalanced with only 8 % of the instances belonging to 'Died' class.

3.2 Experimental Setup

Classification task was conducted using the Waikato Environment for Knowledge Analysis (WEKA 3.6.9)—an open source tool for data mining [23] (available at http://www.cs.waikato.ac.nz/ml/weka/downloading.html); utilizing their available learning schemes i.e. NB, DT using j48 algorithm and NN using Multilayer Perceptron algorithm; and following the standard data mining methodology.

Figure 1 presents the workflow of the experimental studies. Total of 959 records were randomly selected from the registry as the input data for the classification task. Preprocessing phase processed the selection of attributes and handled the missing values. Then, all data were organized in *arff* file format before executing the classifiers. The models were developed using 3 classifiers i.e. NB, DT and NN which were train and test using the 10-folds cross validation technique. The performance of the models was measured by its accuracy rate, specificity, sensitivity and development time. The illustration of the results was further presented using confusion matrix and area under the receiver operating characteristic curve (AUC).

Table 1 Description of attributes

Attributes	Attribute description	Attribute type
Non-Clinical attributes		
patientid	Patient unique ID	Numeric—Auto generated id
sdpid	Cardiac center ID	Nominal—coded ID
yradmit	Year admitted	Numeric—yyyy
Clinical attributes—Demographic		
ptsex	Gender	Nominal—Male/Female
ptagenotification	Age at notification	Numeric—in year
ptrace	Race	Nominal—e.g. Malay/ Indian/ Chinese/Iban
Clinical attributes—Status Before ACS Event		
smokingstatus	Smoking status	Nominal—Current/Former/Never
statusaspirinuse	Status of aspirin use	Nominal—None/Used less than 7 days/ Used more or equal than 7 days
cdys, cdm, chpt, cpremcvd, cmi, ccap, canginamt2wk, canginapast2wk, cheartfail, clung, crenal, ccerebrovascular, cpvascular,	ACS comorbidities—e.g. Dyslipidemia, Hypertension, chronic lung disease, and heart failure	Nominal—Yes/No
Clinical attributes—Clinical Presentation & Examination		
anginaepisodeno	Number of Distinct episodes of angina in past 24 h	Numeric
heartrate	Heart rate	Numeric—in beats/min
bpsys, bpdias	Blood pressure (systolic and diastolic) at present	Numeric—in mmHg
height, weight, bmi	Anthropometric—height, weight and BMI	Numeric—in cm, kg and kg/m^2
killipclass	Killip Classification code	Nominal—I/II/III/IV
Clinical attributes—ECG		
ecgabnormtypestelev1, ecgabnormtypestelev2, ecgabnormtypestdep, ecgabnormtypetwave, ecgabnormtypebbb	Different ECG Abnormalities types	Nominal—Yes/No
ecgabnormlocationil, ecgabnormlocational, ecgabnormlocationll, ecgabnormlocationtp, ecgabnormlocationrv	Different ECG Abnormalities locations	Nominal—Yes/No

(continued)

Table 1 (continued)

Clinical attributes—Baseline investigation obtained within 48 hours from admission		
ckmb, ulckmb, ck, ulck	Peak CK-MB and CK	Numeric—in Unit/L
troponini, troponinive, troponint, troponintve, ultroponini, ultroponint	Peak Tropinin—TnT and TnI	Nominal—Positive/Negative Numeric—in mcg/L
Tc, hdlc, ldlc, tg	Fasting lipid profile—Total cholesterol, HDLC, LDLC and triglycerides	Numeric—in mmol/L
Fbg	Fasting blood glucose	Numeric—in mmol/L
Lvef	Left ventricular ejection fraction	Numeric—% of fraction
Clinical attributes—Fibrinolytic therapy and Invasive therapeutic procedure		
fbstatus_new	Fibrinolytic therapy status	Nominal—Given/Not Given-Contraindicated/Not Given-Missed Thrombolysis/ Not given-proceeded directly to primary angioplasty
cardiaccath, pci, cabg	Patient undergo invasive therapeutic procedure such as Cardiac catheterization, PCI and CABG	Nominal—Yes/No
disvesselno	Number of diseasesd vessel	Nominal—0/1/2/3
culpritartery	Culprit artery	Nominal—LAD/LCx/RCA/LM/Bypass Graft
Outcome attribute		
ptoutcome	Patient outcome	Nominal—Discharge/Died

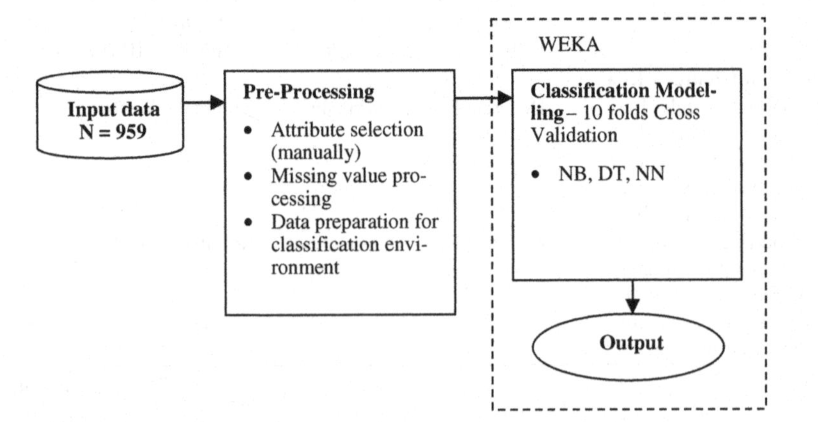

Fig. 1 The Flowchart of the experiment

3.2.1 Data Pre-Processing

All attributes used for this classification task were manually selected. The attributes selected include demographic information, status before the event, clinical presentation and baseline examination, status of fibrinolytic therapy and therapeutic procedures. All duplicate features and most of database elements attributes were removed from the input file. Removing irrelevant attributes is expected to minimize over-fitting problem and thereby produce more reliable model [24].

The first filtering was to remove all irrelevant attributes. Database specific elements such as *sdpid* (represent the code for the cardiac center), *outcomeid* (generated ID upon specifying the outcome of a patient) and *patientnotifid* (generated id for each notification of ACS event) were considered irrelevant for the classification task. Additionally, the information of death which was recorded by National Registration Department of Malaysia was also removed. The information was considered not relevant as it is related to after death events.

The relevant attributes are mainly the clinical elements which contribute to ACS outcome. Clinical elements cover the information on demographic, past medical history, clinical investigation and examination, procedures, pharmacotherapy and diagnosis. In the study, we have excluded all pharmacotherapy procedures and details of ACS procedures to simplify the classification task. The selected attributes mainly focus on the information captured within 48 hours of admission plus some basic information on fibrinolytic therapy and invasive therapeutic procedures. The selected attributes are generally basic information captured for ACS patients at least in Malaysian setting. We also included the basic information of admission registry which is considered as non-clinical element.

Medical data is known to be incomplete, incorrect and missing, and complex by nature [25, 26]. NCVD dataset contains mean of 30 % missing data in 40 attributes in the dataset. However, in this preliminary experiment, no strategy had been applied to accommodate the missing values of all input attributes.

On the other hand, missing values in the outcome attribute (*ptoutcome*) were further considered for imputation strategy. The value used for imputation was based on the value in *jpnmatchingstatus. jpnmatchingstatus* is an attribute that holds the current life status of all Malaysian citizens. It has a value of *'Died'* or *'Not Died'*. Since *'Died'* value in *jpnmatchingstatus* can be due to any reasons (Not specific due to ACS event), only *'Not Died'* cases were considered for the imputation. Therefore, all missing values in *ptoutcome* attribute that contains *'Not Died'* value in *jpnmatchingstatus* were imputed with *'Discharge'* value. Fortunately, all missing values in *ptoutcome* were having *'Not Died'* value in its *jpnmatchingstatus* attribute, and they were all being imputed with *'Discharge'* value.

The data file consisting inputs for classification have been converted to the standard WEKA format called the standard attribute-relation file format (ARFF). The file was used as input file to run the selected classifiers in WEKA.

3.2.2 Classification Modeling

The models were developed using 3 common machine learning techniques i.e. NB, DT and NN [27, 28], and evaluated using the k-folds cross validation technique. In k-folds cross validation technique, data is divided into k-subsets of equal size and is being processed and evaluated k- times. For this experiment, we have chosen 10 as the number of folds. Theoretically it has been proven that within 10-folds optimum estimate of error can be achieved [28]. In each iteration, one subset will be used as a testing dataset (once) and the remaining subsets as training datasets. The final result is calculated by averaging the results from the k- iterations. Using the new version of WEKA, the subsets are stratified i.e. all subsets have the same class distribution as the initial data leading to more reliable results [29].

3.3 Performance Measure

The performance of classification tasks was evaluated using commonly referenced performance measures i.e. accuracy rate, sensitivity, specificity, confusion matrix and AUC. Accuracy rate is a simple measure which calculates the percentage of correctly classified from the total number of instances. Thus, accuracy rate cannot be the solely mechanism to measure the classifiers especially when dealing with imbalanced datasets. It is a need to also evaluate its sensitivity and specificity to accurately asses the two classes [29]. Below equations define sensitivity and specificity.

$Sensitivity = TP/TP + TN$
$Specificity = TN/TP + TN$

where TP—True Positive, TN—True Negative, FP—False Positive and FN—False Negative.

The predicted instances resulting from classification task were tabulated in confusion matrix to get a clearer overview of the results. Moreover, the performance also measures how well the classes are discriminate by assessing the ROC Curve (AUC). AUC which calculates the area under the curve of ROC graph is being accepted by the imbalance dataset community as a standard measure in evaluating imbalance datasets [30].

In addition to accuracy, the execution time was examined for each classifier. This is important as the dataset is relatively big in both the actual registry and possibility of getting more data for further analysis.

Table 2 Comparison of accuracy rate, AUC, sensitivity, specificity and execution time for NB, DT and NN

ML classifiers	Accuracy rate (%)	AUC	Sensitivity	Specificity	Execution time (s)
NB	88.42	0.824	0.395	0.930	0.01
DT	91.45	0.517	0.012	0.998	0.03
NN	90.82	0.821	0.407	0.961	55.76

Table 3 Comparison of confusion matrix for NB, DT and NN

Outcome	Predicted					
	NB		DT		NN	
	Discharge	Died	Discharge	Died	Discharge	Died
Discharge	816	62	876	2	844	34
Died	49	32	80	1	48	33

4 Results

The results of this preliminary exploratory study are illustrated in Table 2: Comparison of Accuracy Rate, AUC, Sensitivity, Specificity and Execution Time for NB, DT, and NN and Table 3: Comparison of Confusion Matrix for NB, DT and NN.

Comparing the accuracy rate, DT shows higher accuracy rate of 91.45 % compared to NB (88.42 %) and NN (90.82 %). However, looking further on AUC and sensitivity, DT is the worst classifier for the dataset. The AUC value of 0.517 indicates random guessing while the low sensitivity value of 0.012 shows poor ability in predicting 'Died' class. This can been seen in confusion matrix table which clearly demonstrates that only 1 out of 81 instances of 'Died' class was correctly predicted.

On the other hand NB and NN show promising accuracy rates as well as AUC values on average of 0.8. In principle, AUC value which closes to 1 signifies better classifier. However, both Sensitivity and precision rate for NB and NN is still considered relatively low with average of 0.4. Although NB and NN achieve about the same results, NN takes longer time in developing the model.

5 Discussion

This preliminary experiment have shown promising potential for using machine learning in predicting ACS outcome. NB and NN are both appropriate classifier to predict ACS outcome. Even though the sensitivity value of both classifiers seems small, they are still considered acceptable as the most observed highly imbalance dataset has almost 0 rate in predicting the minority class [31].

On the contrary, DT is not a reliable classifier to discover 'Died' cases. The reason may be due to pruning process in j48 algorithm. [32] argues that the pruning strategy embedded in c4.5 algorithm (which was originally developed by Ross Quinlan [20] and being implemented in Java that included in WEKA) can possibly shrink the nodes in minority class and hence reduce the coverage. It also supports the argument that accuracy rate cannot be the only measuring factor in choosing a classifier. It is more crucial when the problem deals with imbalance datasets. Accuracy rate only calculates the total number of the correctly specified without looking at the distribution of classes. In imbalance datasets, there is high possibility of presenting a biased result towards the dominant class. Thus, it is essential to extend the analysis onto other performance measure such as AUC, sensitivity and specificity which accounts for the discrimination of classes in their measurement.

Imbalance dataset issues have attracted researchers to work on improving the problem such as the work done by [33] and [34]. The aim for next iteration is to selectively incorporate those solutions and suit them with the dataset in order to improve the overall performance of the classifier.

Moreover, there is also a need to explore other available machine learning techniques. It is important to evaluate other machine learning techniques as there is no specific technique that is relatively better than others. Each machine learning technique has different ways of searching, extracting and representing the pattern. Thus, the nature of the dataset and the specific problem to be solved highly influence the performance of a classifier [10]. For example, in [35], a study on ACS dataset from Switzerland registry based on 7 input attributes have found Averaged One Dependence Estimators as the most accurate classifier.

To our current knowledge, this is the first initiative of using ACS dataset for Malaysian cohorts in developing prediction model using machine learning. Furthermore, not many available ACS prediction models were developed using machine learning. The dataset is rich with information promoting further exploration on how other risk factors contribute to ACS outcome. But, for machine learning to be accepted in medical and ACS domain in particular, the issues of using machine learning in medical domains must be minimized [36, 37]. It is crucial for the model development methodology to comply with the needs of medical domain [12, 38]. For example, in validating prediction models, calibration is one of the important measures to be evaluated which were not applied in our current study. And, it is crucial that any generalizations made from the model development are medically explainable and accepted by the medical community.

6 Conclusions and Further Work

A reliable prediction model for ACS is essential in providing on-time treatment and care for patients with ACS. The advancement of technology and the increase of medical datasets have triggered the need for machine learning techniques in medical field. The study has generally presented the potential of utilizing machine learning technology in developing prediction models specifically for ACS dataset. From the results, NB and NN classifiers perform comparably better than DT. The study will be the basis to continue improving the model performance by exploring other machine learning techniques and investigating more suitable techniques and strategies to be applied on the said dataset e.g. handling the imbalance dataset. The study will also be used as the basis to ensure that a proper acceptable clinical prediction model paradigm is followed.

Acknowledgments The authors would like to thank the Governance Board member of Malaysian National Cardiovascular Disease Database Registry for providing us the Acute Coronary Syndrome (ACS) dataset to be used for the research. Also, University Kuala Lumpur (UniKL), Malaysia for funding the PHD of the main author.

References

1. Bassand, J.-P., Hamm, C. W., Ardissino, D., Boersma, E., Budaj, A., Fernandez-Aviles, F., Fox, K. A., Hasdai, D., Ohman, E. M. & Wallentin, L. 2007. Guidelines for the diagnosis and treatment of non-ST-segment elevation acute coronary syndromes The Task Force for the Diagnosis and Treatment of Non-ST-Segment Elevation Acute Coronary Syndromes of the European Society of Cardiology. *European Heart Journal*, 28, 1598–1660.
2. SIGN 2007 (Updated 2013). Acute coronary syndromes : A national clinical guideline. Edinburgh, UK: Scottish Intercollegiate Guidelines Network.
3. Chin, C. T., Chua, T. & LIM, S. 2010. Risk assessment models in acute coronary syndromes and their applicability in Singapore. *Ann Acad Med Singapore*, 39, 216–20.
4. Antman, E. M., Cohen, M., Bernink, P. J., Mccabe, C. H., Horacek, T., Papuchis, G., Mautner, B., Corbalan, R., Radley, D. & Braunwald, E. 2000. The TIMI risk score for unstable angina/non-ST elevation MI. *JAMA: the, journal of the American Medical Association*, 284, 835–842.
5. Cooney, M. T., Dudina, A. L. & Graham, I. M. 2009. Value and limitations of existing scores for the assessment of cardiovascular risk: a review for clinicians. *Journal of the American College of Cardiology*, 54, 1209–1227.
6. Delen, D., Oztekin, A. & Tomak, L. 2012. An analytic approach to better understanding and management of coronary surgeries. *Decision Support Systems*, 52, 698–705.
7. Cruz, J. A. & Wishart, D. S. 2006. Applications of machine learning in cancer prediction and prognosis. *Cancer Informatics*, 2, 59.
8. Khalilia, M., Chakraborty, S. & Popescu, M. 2011. Predicting disease risks from highly imbalanced data using random forest. *Bmc Medical Informatics and Decision Making*, 11, 51.
9. Westreich, D., Lessler, J. & Funk, M. J. 2010. Propensity score estimation: machine learning and classification methods as alternatives to logistic regression. *Journal of clinical epidemiology*, 63, 826.
10. Song, X., Mitnitski, A., Cox, J. & Rockwood, K. 2004. Comparison of machine learning techniques with classical statistical models in predicting health outcomes. *Medinfo*, 11, 736–40.
11. Oztekin, A., Delen, D. & Kong, Z. J. 2009. Predicting the graft survival for heart-lung transplantation patients: an integrated data mining methodology. *International Journal of Medical Informatics*, 78, e84.
12. Shillabeer, A. & Roddick, J. F. Establishing a lineage for medical knowledge discovery. 2007. Australian Computer Society, Inc., 29–37.

13. Li, J., Fu, A. W.-C., He, H., Chen, J., Jin, H., Mcaullay, D., Williams, G., Sparks, R. & Kelman, C. Mining risk patterns in medical data. Proceedings of the eleventh ACM SIGKDD international conference on Knowledge discovery in data mining, 2005. ACM, 770–775.
14. Hippisley-Cox, J., C. Coupland, Y. Vinogradova, J. Robson and P. Brindle. 2008a. Performance of the QRISK cardiovascular risk prediction algorithm in an independent UK sample of patients from general practice: a validation study. Heart, 94(1), pp. 34–39.
15. Chia, C. C., Rubinfeld, I., Scirica, B. M., McMillan, S., Gurm, H. S. & Syed, Z. 2012. Looking Beyond Historical Patient Outcomes to Improve Clinical Models. *Science Translational Medicine*, 4, 131ra49-131ra49.
16. Stolba, N. and A. M. Tjoa. 2006. The Relevance of Data Warehousing and Data Mining in the Field of Evidence-based Medicine to Support Healthcare Decision Making. In: C. ARDIL, ed. Proceedings of World Academy of Science, Engineering and Technology, Vol 11. pp. 12–17.
17. Horvitz, E. 2010. From Data to Predictions and Decisions: Enabling Evidence-Based Healthcare. Computing Community Consortium, 6.
18. Kotsiantis, S., Zaharakis, I. & Pintelas, P. 2007. Supervised machine learning: A review of classification techniques. *Frontiers in Artificial Intelligence and Applications*, 160, 3.
19. Mitchell., T. M. 1997. *Machine Learning New York*; London, McGraw-Hill.
20. Quinlan., J. R. 1993. C4.5 : *Programs for Machine Learning*, San Mateo, California, Morgan Kaufmann.
21. Lau., C. 1992. Neural networks : theoretical foundations and analysis, New York, IEEE Press.
22. Chin, S., Jeyaindran, S., Azhari, R., Wan Azman, W., Omar, I., Robaayah, Z. & SIM, K. 2008. Acute coronary syndrome (ACS) registry-leading the charge for National Cardiovascular Disease (NCVD) Database. *Med J Malaysia*, 63, 29–36.
23. Hall, M., Frank, E., Holmes, G., Pfahringer, B., Reutemann, P. & Witten, I. H. 2009. The WEKA data mining software: an update. *ACM SIGKDD Explorations Newsletter*, 11, 10–18.
24. Saeys, Y., Inza, I. & Larranaga, P. 2007. A review of feature selection techniques in bioinformatics. *Bioinformatics*, 23, 2507–2517.
25. Cios, K. J. & Moore, G. W. 2002. Uniqueness of medical data mining. *Artificial Intelligence in Medicine*, 26, 1–24.
26. Bellazzi, R. & Zupan, B. 2008. Predictive data mining in clinical medicine: Current issues and guidelines. *International Journal of Medical Informatics*, 77, 81–97.
27. OlsonLSON, D. L. & DELEN, D. 2008. *Advanced data mining techniques*, Springer Verlag.
28. Ian H. Witten, E. F. 2005. *Data mining : practical machine learning tools and techniques*, Amsterdam; London : Elsevier, c2005.
29. Han, J. & Kamber, M. 2001. *Data mining : concepts and techniques*, San Francisco Morgan Kaufmann Publishers.
30. Fawcett, T. 2004. ROC graphs: Notes and practical considerations for researchers. *Machine Learning*, 31, 1–38.
31. Guo, X., Yin, Y., Dong, C., Yang, G. & Zhou, G. On the class imbalance problem. Natural Computation, 2008. ICNC'08. Fourth International Conference on, 2008. IEEE, 192–201.
32. Chawla, N. V. C4. 5 and imbalanced data sets: investigating the effect of sampling method, probabilistic estimate, and decision tree structure. Proceedings of the ICML, 2003.
33. Folorunso, S. & Adeyemo, A. 2013. Alleviating Classification Problem of Imbalanced Dataset. *African Journal of Computing & ICT*, 6.
34. Weng, C. G. & Poon, J. A new evaluation measure for imbalanced datasets. Proceedings of the 7th Australasian Data Mining Conference-Volume 87, 2008. Australian Computer Society, Inc., 27–32.
35. Kurz, D. J., Bernstein, A., Hunt, K., Radovanovic, D., Erne, P., Siudak, Z. & Bertel, O. 2009. Simple point-of-care risk stratification in acute coronary syndromes: the AMIS model. *Heart*, 95, 662–668.
36. Kononenko, I., Bratko, I. & Kukar, M. 1997. Application of machine learning to medical diagnosis. *Machine Learning and Data Mining: Methods and Applications*, 389, 408.

37. Wyatt, J. C. & Douglas G Altman, H. 1995. Commentary: Prognostic models: clinically useful or quickly forgotten? *BMJ*, 311, 1539.
38. Kononenko, I. 2001. Machine learning for medical diagnosis: history, state of the art and perspective. *Artificial Intelligence in Medicine*, 23, 89–109.

Applications in Education and Information Science

An AI-Based Process for Generating Games from Flat Stories

Rosella Gennari, Sara Tonelli and Pierpaolo Vittorini

Abstract TERENCE is an FP7 ICT European project that is developing an adaptive learning system for supporting poor comprehenders and their educators. Its learning material are books of stories and games. The so-called smart games serve to stimulate story comprehension. This paper focuses on the analysis of flat stories with a specific annotation language and the generation of smart games from the analysed texts, mixing natural language processing and temporal constraint-reasoning technologies. The paper ends commenting on the approach to the automated analysis and extraction of information from stories for specific users and domains, briefly evaluating the benefits of the semi-automated generation process in terms of production costs.

1 Introduction

From the age of 7–8 until the age of 11, children develop as independent readers. Nowadays, more and more children in that age range turn out to be poor (text) comprehenders: they demonstrate difficulties in deep text comprehension, despite well developed low-level cognitive skills like vocabulary knowledge, e.g., see [25]. TERENCE is a Collaborative project funded by the European Commission and developing the first intelligent *adaptive learning system* (ALS) [24] for poor comprehenders from primary schools, that is, the TERENCE *learners*, and their educators. The system proposes stories, organised into difficulty categories and collected into books, and

R. Gennari (✉)
Free University of Bozen-Bolzano, Piazza Domenicani 3, 39100 Bolzano, BZ, Italy
e-mail: gennari@inf.unibz.it

S. Tonelli
Free University of Bozen-Bolzano, FBK-irst-Via Sommarive 18, 38100 Bolzano, BZ, Italy
e-mail: stonelli@fbk.eu

P. Vittorini
University of L'Aquila, Piazzale S. Tommasi 1, 67100 Coppito, AQ, Italy
e-mail: pierpaolo.vittorini@univaq.it

M. Bramer and M. Petridis (eds.), *Research and Development in Intelligent Systems XXX*, DOI: 10.1007/978-3-319-02621-3_25,

smart games for reasoning about events of a story and their relations. The TERENCE smart games are serious games [13] and, like the entire ALS, are designed within a stimulation plan for the TERENCE learners. See Sect. 2 of this paper for an overview of the games. One of the goals of TERENCE is to generate textual components for games semi-automatically, starting from flat stories for primary-school children, in Italian and in English, so as to improve development performances, given that we deal with 256 different stories, each having c. 12 games. The starting point for such a generation process is the definition of a language that allows for the annotation of stories' key features for the smart games for the TERENCE learners. The annotations are then used (1) for automatically generating textual components of smart games (2) and for rating these into difficulty levels. In particular, the paper shows how we combine *natural language processing* (NLP) techniques and tools, that recognise events and (causal-) temporal relations among them in stories, with *temporal constraint* (TC) algorithms and tools, that reason about events and temporal relations for generating textual components of games. Firstly, the paper outlines the requirements for the language, resulting from contextual inquiries with c. 70 experts of poor compreheneders and educators. See Sect. 3. Given such a language, the automated analysis process can take place on stories by mixing NLP and constraint-based algorithms and tools, see Sect. 4. The resulting annotated stories are then fed to the generation modules of TERENCE that, again integrating NLP and reasoning techniques, automatically generates games as explained in Sect. 5. This paper concludes with an analysis of the approach to the generation of games and proposes new working directions.

2 Background on the TERENCE Smart Games

According to [1], a game should specify (at least) the *instructions*, the *states* of the game, with the initial and terminal states, and the legal *actions* of the players. For specifying the data for the TERENCE smart games, we analysed the requirements for smart games resulting from contextual inquiries with c. 70 experts of poor comprehenders and educators as well as from field studies with c. 500 poor comprehenders. The results allowed us to classify and design smart games as follows. Following experts of stimulation plans, the TERENCE smart games were classified into 3 main difficulty macro-levels as follows: (1) at the entry macro-level, *character* games, that is, either who the agent of a story event is (WHO), or what a character does in the story (WHAT); (2) at the intermediate macro-level, *time* games, for reasoning about temporal relations between events of the story, purely sequential (BEFORE-AFTER) or not (all others); (3) at the last macro-level, *causality* games, namely, concerning causal-temporal relations between events of the story, that is, the cause of a given event (CAUSE), the effect (EFFECT), or the cause-effect relations between two events (CAUSE-EFFECT). All games were specialised into *levels* arranged in a so-called linear layout, and designed as puzzle casual games [1]. In particular, the data for all levels were structured via the TERENCE game framework. See [3, 7] for a description of the framework and the design process.

Fig. 1 Screenshots of WHO and BEFORE-AFTER game instances

Relevant fields of the framework for this paper are the following ones: (1) the *question* of the game, (2) a *central event* from the story, (3) the *choices* for learner's actions, so that the availability of choices depend on the state the game is in, and (4) which choices form a correct or wrong *solution*. The fields are rendered with illustrations and textual components that vary according to the level of the game and the specific game instance, as explained below.

In case of WHO games, the question is related to a single central event that depends on the game instance, if the central event is "Aidan runs fast" then the question is "Who runs fast?". The choices are three characters and only one is the correct solution, i.e., "Aidan", the agent of the central event. See the left screenshot in Fig. 1.

In all the other levels, the question is related to the instructions for the game only, and is the same for all game instances. Choices are textual and visual descriptions of events from the story, and are placed at the bottom of the interface. In WHAT games, the central event is also the correct solution, and is placed at the bottom with the other choices that are wrong solutions. In time and causality games, the central event is at the centre of the interface. The choices are at the bottom and are story events that the learner has to correctly correlate with the central event. See the right screenshot in Fig. 1, a BEFORE-AFTER time game instance.

3 The TERENCE Annotation Language

The events of a story, their temporal and causal-temporal relations, and the characters that participate in events are at the heart of the TERENCE smart games. Therefore TERENCE needs an annotation language that can specify them. TimeML [20, 22] is a good starting point for that: it covers events and qualitative temporal information, relevant for stories, and is the de-facto standard markup language for temporal information in the NLP community, which allows for the re-use of existing NLP and TC tools for qualitative temporal reasoning. See Sect. 3.1 for an introduction to TimeML. The analysis of the limitations lies at the heart of the TERENCE annotation language, built on top of TimeML, as outlined in Sect. 3.2.

3.1 The TimeML Annotation Format

TimeML [20, 22] is an international markup language for annotating events and temporal information in a text. TimeML includes three major data structures, namely, a qualitative time entity (e.g., action verbs) called `EVENT`, a quantitative time entity (e.g., dates) called `TIMEX`, and relations among them called generically links. In particular, in TimeML events can be expressed by tensed and untensed verbs, but also by nominalizations (e.g. *invasion, discussion, speech*), predicative clauses (e.g. *to be the President of something*), adjectives (e.g. *dormant*) or prepositional phrases (e.g. *on board*). TimeML is by now a standard in the NLP community and is used in the annotation of linguistic resources such as the TimeBank corpus [21], the Ita-TimeBank [5] and of news events in the 2010 TempEval competition [28].

As for `TIMEX` temporal expressions, they include specific dates (e.g. *June 11, 1989*), and durations (*three months*). TimeML includes also the concept of time normalisation, where a normalised form is assigned to each expression based on consistency and interchange format in line with the ISO 8601 standard for representing dates, times, and durations. Time entities (`EVENT` and `TIMEX`) can be connected through a qualitative temporal relation called `TLINK`, making explicit it the two entities are simultaneous, identical or in sequential order. Specifically, several types of relations have been introduced in TimeML modelled on *Allen interval algebra* of qualitative temporal relations [2]. Other types of links are also present in TimeML standard, for instance subordination links between specific verbs of reporting and opinion and their event arguments, or aspectual links between aspectual events and their argument events. However, since they are not included in the TERENCE framework, we do not discuss them in details.

3.2 The TERENCE Annotation Format (TAF)

TimeML is the de-facto standard language on which the NLP communities have grown their experience and tools in the last decade. However, TimeML is sometimes underspecified and sometimes too detailed for the TERENCE purposes. For instance, the semantics of `TLINKs` in terms of Allen relations is not uniquely specified but, in the working practice, is pretty context dependent. Key information like the participants in an event have already been proposed [23] but neither their attributes nor their relations with events have been clearly pinpointed. Since knowing which the characters in a story are, their role and how they interact is crucial for the TERENCE smart games, such information is essential for analysing the TERENCE stories. Another main problem for TERENCE with the TimeML language is the lack of causal links when explicitly signalled by linguistic cues such as "because" and "since". In several other cases, the level of detail foreseen in TimeML is not needed to process children's stories such as in TERENCE. For instance, while in TimeML modal and hypothetical actions are annotated as events, the primary interest of TERENCE are the restricted events that actually take place in the story, namely, so-called factual events.

To take advantage of the pros of TimeML and cope with its limitations, we built the *TERENCE Annotation Format* (TAF) on top of TimeML. See [16]. Hereby we outline the main design choices of the language, relevant for generating diverse smart games, their wrong and correct solutions, as well as for classifying smart games into fine-grained difficulty categories. The requirements for the language and the analyses leading to such requirements are in [25]. For instance, well-placed signals like connectives tend to easy their understanding of relations between events. Thus a requirement of TAF is to trace whether, for instance, a temporal or causal-temporal relation between events is rendered by a connective, and if the events occur in the same sentence, or otherwise, keeping track of where.

For explaining TAF, we use the *Extended Entity-Relationship* (EER) [26] diagram in Fig. 2. Therein, classes are coloured in pink. Relations among classes are yellow. Attributes are in balloons. Each class has a unique key identifying attribute. Cardinality constraints between an entity and an attribute are equal to (1, 1) unless differently set. The diagram in Fig. 2 specifies the classes of `Event` and `Pos`, denoting an event and the lemma of the word evoking it, respectively. The attributes of `Pos` specify relevant information, in particular, if it is a verb and, e.g., its regularity. In the diagram in Fig. 2, we also have `TIMEX` and `TLINKs`. Since children's stories, like in TERENCE, are not usually anchored to a specific date, time normalisation is often not possible, therefore temporal expressions are not normalised as opposed to TimeML standard. Therefore we focus on the analysis of a `TLINK`, which is relevant for time games (see Sect. 2). A `TLINK` denotes a temporal binary relation between events, and has the following attributes:

1. `reverse`, with Booleans "true" or "false", capturing if the order of occurrence of the two events in the story text is the temporal order ("false") or not ("true");
2. `signalID`, with string values the id of the annotated signal (e.g., "30" for the signal "and then") if the two `Events` are related through an explicit temporal signal, or "none" otherwise;
3. `rel_type`, with string value the TimeML relations "before", "after", "overlaps", "is_included", "includes", "identity", "all";
4. `local`, with Boolean values "true" if the two `Events` are in the same sentence or in consecutive sentences, or "false" otherwise.

Except for `rel_type`, such attributes are not present in TimeML. The `hasTTarget` and `hasTSource` relations serve to specify, respectively, the source event and the target event of the relation, which is done differently than in TimeML. Similar remarks apply for the `CLINK` class, its attributes, the `hasTTarget` and `hasTSource` relations. In particular, in TAF and not in TimeML we have causal links between events when explicitly signaled by linguistic cues such as "because", "since". The `CLINKs` are used for causality games (see Sect. 2).

The diagram in Fig. 2 also introduces classes and relations for specifying the characters or other entities participating in events of a story, as well as their role within the story (see Sect. 2). In particular, `Entity` refers to the unique identifier of

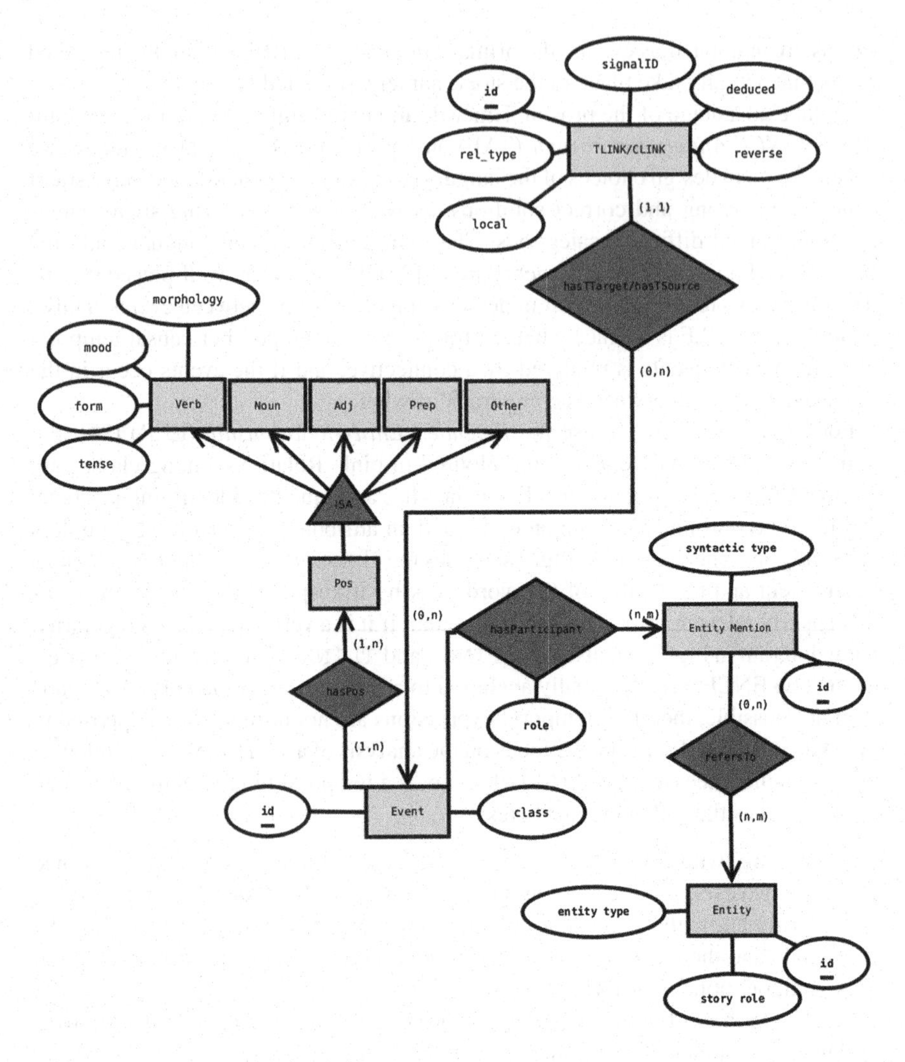

Fig. 2 EER of events and their relations

a participant, like a character in a story (e.g., the proper name "Ben" of a male character), where as the `Entity Mention` stands for any expression that correlates via `refersTo` to a unique `Entity` (e.g., "the boy"); such a distinction between entities and their story mentions has been introduced to account for coreference in TAF, which is not covered in TimeML. The relation `hasParticipant` correlates each `Event` to all the `Entity Mentions` involved in such event. The relation has the composite attribute named `role`, which stands in for the role that a correlated entity mention has in the event. One of its atomic attributes is `semantic role`, and among its values we have "agent" for the agents of an event. While `Entity`

`Mentions` are described through the syntactic type they have in the sentence in which they occur (e.g. "subject", "object"). `Entities` are characterised at the semantic level with the following attributes with string values:

1. `entity type`, with values "person", "location", "animal", "pseudo-person", "vegetable", "artifact", "other";
2. `story role`, with values "protagonist", "secondary", "minor".

Such information is relevant for the generation of more plausible wrong solutions in games and for organising them into fine-grained difficulty categories.

4 The Automated Analysis Process

The NLP module and the TC module perform the annotation process of stories of TERENCE with the TAF language described above. Firstly, the NLP module detects relevant information and annotates them with TAF. See Sect. 4.1. Secondly, the TC reasoning module transform `EVENTs` and `TLINKs` into a qualitative temporal constraint problem, mapping `TLINKs` into Allen relations; in case the problem is detected inconsistent, the mapping is relaxed. If the problem is returned consistent, then the TC reasoning module deduces further relations, annotating their distance in the text. See Sect. 4.2. The resulting stories are stored as XML files in the annotated story repository.

4.1 The NLP Analysis of Flat Stories

The goal of this processing step is to take in input a flat story and output an XML-based version of the story annotated with the TAF language. We describe here the workflow for Italian stories, and a similar approach has been tested also for English. While the tools used for processing the stories vary from language to language, the approach is language-independent, and also TAF has been formulated so as to minimise language-specific adaptations. The story is first processed with existing NLP tools for Italian. Specifically, it is analysed with the TextPro suite [19] which performs part-of-speech tagging, lemmatization, morphological analysis, named entity recognition and annotation of temporal expressions. The story is then processed with the Malt parser for Italian [15], which outputs labeled dependency relations between tokens in the story (for instance, it identifies the subject and the object of a sentence and links them to the verb). This intermediate analysis is then passed to the TAF annotation module that adds information on events, temporal links, causal links and participants to the pre-processed data in a cascaded fashion. The performed steps are the following ones.

Annotation of factual events. Factual `Events` are selected among tensed verbs that are not copulas and modals, and that are in the indicative mood. We discard verbs in the future tense, and expressing wish or conditionality.

Annotation of event participants. For each selected `Event`, we retrieve the syntactic dependents from the parsing analysis, labelled as `Entity Mentions`. For each mention, we link its head to the most frequent sense in MultiWordNet [18] and then get the corresponding type label from the Suggested Upper Merged Ontology [17], which we further mapped to the TAF `entity types`. For instance, "the cat" may be linked to the *cat#n#1* synset in MultiWordNet, which corresponds to the *Feline* class in SUMO. *Feline* is then mapped to the *Animal* `entity type`.

Annotation of participants' coreference. If two or more participants' mentions show a (partial) string match, they are annotated as co-referring via an `Entity` to which both are linked through a `refers_to` relation. This concerns for instance mentions such as "Ernesta Sparalesta" and "Ernesta". More sophisticated algorithms for coreference annotation are currently being evaluated within TERENCE, for instance supervised approaches to treat pronominal anaphora and zero-pronouns. But current results are still too low to be integrated in the final NLP pipeline.

Annotation of temporal links. Events that are in the same sentence or in contiguous sentences are local. For each pair of local events, a `TLINK` is created and its `rel_type` specified based on the tense and mood information associated with the events. Since events in children's stories tend to follow a chronological order, the default value for `rel_type` is "before", indicating that the event mentioned first (e_1) precedes the other (e_2). However if e_1 is at present tense and e_2 is expressed in a past tense, the relation is likely to be "after". Cases of "includes" and "is_included" are annotated when one of the two events is in a continuous form (e.g., "was playing") and the other is a punctual event (e.g., "noticed"). In this manner, a temporal chain is created for the story.

Annotation of causal links. Temporal links between events are also marked with causal links (`CLINKs`) if the events occur in two clauses connected by a causal marker such as "because", "for this reason", "so that".

The annotation module outputs an XML file with stand-off annotation which is then fed to the reasoning module.

4.2 The TC Enrichment of Annotated Stories

The TC reasoning module takes in input a story annotated by the NLP module from the annotated story repository, and processes it as explained hereby. First of all, the architecture of the reasoning module is made up of: (1) a REST service, that contains two main operations: consistency checking and deduction; (2) a Java library `tml2gqr`, which is used by the service for implementing the operations, but can be also embedded in other applications for performances' reasons; (3) the GQR software [10], invoked by the library for performing the above operations. The operations are outlined and executed as follows.

Consistency checking. The static method `consistency` of class `GQR` takes in input a Java `String` and returns a Java `boolean`. The input is the TAF document annotated by the NLP module. The TAF document is converted into a TC problem file, with disjunctive Allen relations [2]. The most critical task is to choose the 'right' semantics for `TLINKs`, that is, how to convert `TLINKs` into relations of a tractable subalgebra of the Allen interval algebra [2, 14] for the TERENCE stories. The chosen mappings were two and agreed upon with the NLP partner as those returning inter-consistency among manual annotators, within pilot studies for English and Italian children stories. The hard mapping is as follows:

before	$\mapsto$ *before*
after	$\mapsto$ *before*$^{-1}$
overlaps	$\mapsto$ *equal or during or during*$^{-1}$ *or overlaps or overlaps*$^{-1}$ *or starts or starts*$^{-1}$ *or finishes*$^{-1}$ *or finishes*$^{-1}$
includes	$\mapsto$ *during*$^{-1}$
is_included	$\mapsto$ *during*
identity	$\mapsto$ *equal*
all	$\mapsto$ the disjunction of all Allen atomic relations

The relaxed mapping differs from the hard one in that "before" is mapped into "before or meets" and "after" into its inverse "before^{-1} or meets^{-1}". For both mappings the range is a subalgebra of the *continuous algebra* (CA), for which consistency checking and deduction take at worst cubic time in the number of events [11]. In fact, both consistency checking and deduction are done via the optimised path consistency algorithm of `GQR`, which runs in time cubic in the number of events. The output is then parsed and whether the document is consistent is returned as a Boolean. In case not, human interventions is invoked. Else, deduction is invoked.

Deduction. The static method `deduction` has in input the TAF document checked for consistency, with either the hard or relaxed mapping. The TAF document is converted into a TC file in the format of `GQR`, with the mapping for which the document is consistent. Then `GQR` is executed. The output of `GQR` is then parsed, and the deduced relations are added with `TLINKs` via the inverse mapping. In case a relation r deduced by the reasoner corresponds to no one in the range of adopted mapping for `TLINKs`, then r is approximated by the smallest (for inclusion) relation in the range of the mapping and that contains r, and a comment is added with the deduced relation for further processing—human or automated. The input TAF document is updated, and the updated TAF document is returned as XML file in the annotated story repository.

The story in the annotated story repository is referred to as the *enriched* story, used as input for the generation process described below.

5 The Automated Generation Process

The reasoning module and the NLP also collaborate for generating textual components for smart games, to which we refer as *textual games*. More precisely, the TC reasoning module takes in input the enriched stories, described in the previous section, and ranks events according to their relevance for smart games. Textual game instances, for each level (see Sect. 2), are generated for the top ranked events. The process is described in [7, 12]. In particular, for each textual game instance, according to its level, the TC reasoning module sets its data structure, the available choices, wrong and correct solutions, as outlined in Sect. 5.1. Then, for each textual game instance, for each event in it, the TC module invokes the NLP module that generates the necessary textual information for the event, as outlined in Sect. 5.2. The TC module then places the generated text in the pertinent textual game instances. See Fig. 3 for the overall workflow.

5.1 The Generation of Game Instances from Enriched Stories

In the following, we explain how choices, correct and wrong solutions are generated by giving examples of generated game instances for each macro-level. The generation process takes as input an enriched story and proceeds as follows.

Character games. In order to create a WHO game for an event, we exploit the existence, or not, of a chain that, starting with the event marked with `Event`, continues to the participant related via `hasParticipant` and marked with `Entity Mention`. This is then resolved by correlating it via `refersTo` to its `Entity`. Figure 4 depicts the case of two entities, Ben and Kate, mentioned in the story as "The boy", and "The girl", respectively, and their participation in two events, i.e., "swim" and "run".

For instance, to generate the WHO game for the central event "swim", the module exploits the links that correlate (1) "swim" to the respective mentions, then to the actual entities (in the example, Ben), for the correct choice, (2) other entities that are not mentioned as participants in the "swim" event as wrong (e.g., Kate).

A similar process takes place with WHAT character games. Given an event in the story, this is taken as the correct choice, whereas the wrong choices are from either a different story, or from the same story but using a different entity as event participant.

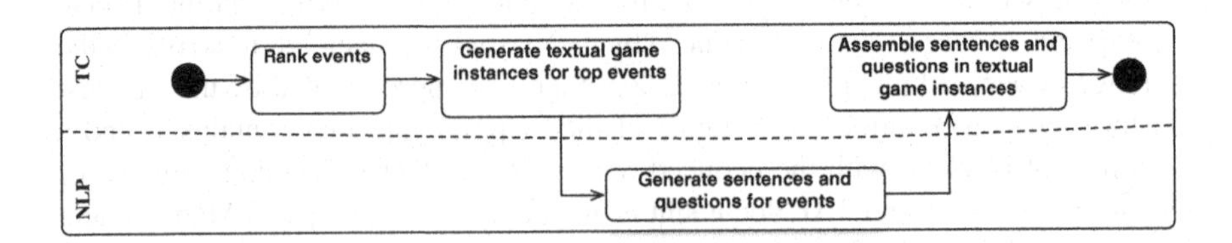

Fig. 3 The generation workflow

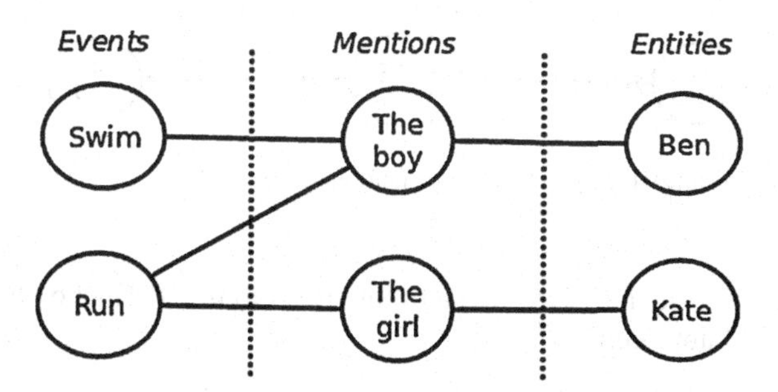

Fig. 4 TERENCE TimeML tags for a WHO smart game instance

In the case depicted in Fig. 4, the correct choice would be "Ben swims", and a wrong choice would be "Kate swims".

Time games. In order to create a time game with a given central event, we exploit the `Event` and `TLINK` tags, as well as the `local` and `rel_type` attributes. Figure 5 shows a portion of the temporal structure of a story, where the event "swim" occurs before "run", "jump" occurs at the same time as "run", and "rest" occurs after "run". Furthermore, two TLINKs are not local and were deduced by the TC module in the story annotation process, i.e., that "swim" occurs before "rest".

For instance, in order to create a BEFORE-AFTER-game instance difficult for a poor comprehender, we can: (1) exploit the `TLINK` that has a `rel_type` equal either to "before" or "after", (2) consider relations with `local` set to false (i.e., jump-rest, swim-rest).

Causality games. In order to create causality games with a given central event, we exploit `CLINKs` and the `reverse` attribute, mainly. Figure 6 depicts causal relations in a story where a race was lost because the participant did not sleep enough, and that caused the participant to become sad.

For instance, to create a CAUSE-game, we can exploit the `CLINKs` stating that the cause for "The boy is sad" is that "He lost the race", and properly swap the cause

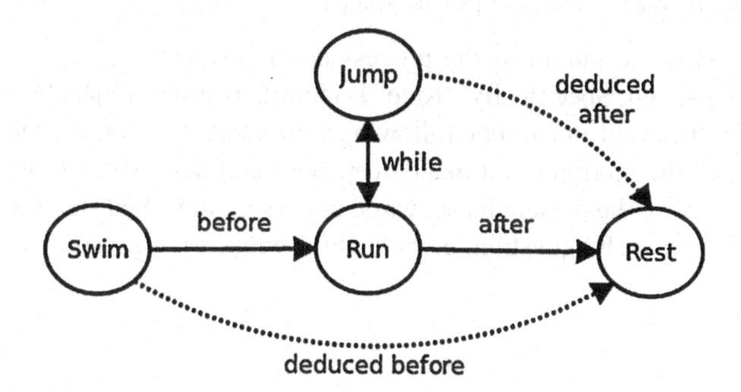

Fig. 5 TERENCE TimeML tags for time smart game instances

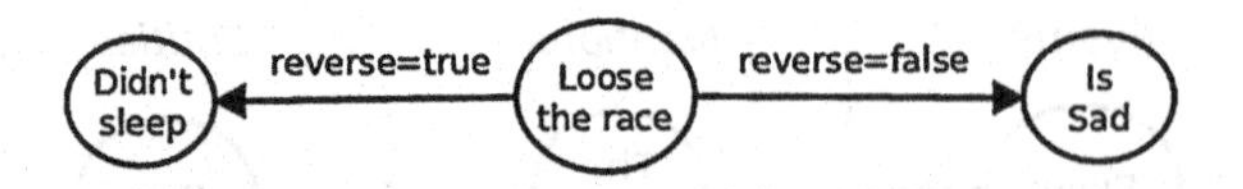

Fig. 6 TERENCE TimeML tags for causality smart game instances

and the effect, in case the `reverse` attribute is set to true ("The boy lost the race" because "He did not sleep").

5.2 The Generation of Textual Components for Game Instances

The NLP generation task, given in input the identifier of a story event, generates *(i)* the question for WHO games for that event, and *(ii)* the sentence for the event used as choice of the games of all the other levels. The generation task is performed with a modified version of the sentence simplification system presented in [4]. Specifically, a module has been developed that, taking a TAF-annotated text in input, outputs a sequence of simplified statements for each identified event, and WHO-questions on the event subjects. The module removes from the dependency tree the event arguments that are not mandatory, i.e., that do not correspond to participants. For instance, in case of transitive verbs, it only retains subject and object. Other arguments labeled as RMOD (modifiers) are removed. Then, the NLP module puts the verb in the present indicative tense using a generator of morphological forms included in TextPro [19]. Finally, it generates questions on the subject of the simplified clause: it retrieves the corresponding entity type saved in TAF, and uses it to generate a WHO question. We describe the four steps performed by the NLP module in the light of the following example:

(1) [Kate$_{Subj/Person}$] [was eating$_{event}$] [a cake$_{Obj}$] [with her friends$_{RMod}$]
(2) [Kate$_{Subj/Person}$] [was eating$_{event}$] [a cake$_{Obj}$]
(3) [Kate$_{Subj/Person}$] [eats$_{event}$] [a cake$_{Obj}$]
(4) [Who$_{Subj/Person}$] [eats$_{event}$] [a cake$_{Obj}$] ?

Sentence (1) is the output of the process described in Sect. 4.1, in which TAF annotations are added. Specifically, "Kate" is identified as the subject being a person, "was eating" is the event and the two following arguments are an object and a modifier. In Sentence (2), the modifier, not mandatory, has been deleted. In Sentence (3) the event is expressed in the present tense, while in (4) the simplified statement has been transformed into a WHO question, given that the subject is a person.

6 Conclusions

The paper shows how textual components of smart games for learners with specific text comprehension problems are generated via an automated process, mixing NLP and qualitative constraint-based technologies and tools. In the remainder, we briefly assess such a generation approach. Firstly, the automated generation requires an annotation language defined on top of the learners' requirements, e.g., the language allows for specifying whether the events are not in the same sentence. That is per se a novelty element. Secondly, releases of the system and, in particular, its games are also evaluated iteratively, in four main evaluations; the second release for the TERENCE learners is available at [27]. One of such evaluations was concerned with the generated textual components of smart games, and it was done manually for all the TERENCE games by experts of education or poor comprehenders. According to this evaluation, the work of developing manual games was longer than for generating and revising them—on average 23 versus 13 min. That indicates that the generation process is promising for cutting development costs. See [8] for such a revision process. Another evaluation was instead conducted as a field study with 168 learners for detecting usability issues and gaining further indications concerning the pedagogical effectiveness of smart games, see [6, 9]. For instance, this evaluation indicated the need of optimising the generation of smart games with heuristics for more plausible wrong solutions, an on-going work done with NLP and constraint-based reasoning techniques. Another on-going work is the optimisation of the annotation process of stories, by interleaving NLP and temporal constraint reasoning enhanced with explanation capabilities, according to the specific application domain and intended users.

Acknowledgments The authors' work was supported by the TERENCE project. TERENCE is funded by the European Commission through the Seventh Framework Programme for RTD, Strategic Objective ICT-2009.4.2, ICT, Technology-enhanced learning. The contents of the paper reflects only the authors' view and the European Commission is not liable for it. Gennari work was also supported by the DARE project, financed by the Province of Bozen-Bolzano.

References

1. E. Adams. Fundamentals of Game Design. New Riders, 2010.
2. J. F. Allen. Maintaining Knowledge about Temporal Intervals. ACM Comm, 26:832–843, 1983.
3. M. Alrifai and R. Gennari. Deliverable 2.3: Game Design. Technical Report D2.3, TERENCE project, 2012.
4. G. Barlacchi and S. Tonelli. ERNESTA: A Sentence Simplification Tool for Children's Stories in Italian. In A. Gelbukh, editor, Computational Linguistics and Intelligent Text Processing, volume 7817 of Lecture Notes in Computer Science, pages 476–487. Springer, Berlin Heidelberg, 2013.
5. T. Caselli, V. B. Lenzi, R. Sprugnoli, E. Pianta, and I. Prodanof. Annotating Events, Temporal Expressions and Relations in Italian: the It-TimeML Experience for the Ita-TimeBank. In Proceedings of LAW V, Portland, Oregon, USA, 2011.

6. M. Cecilia, T. D. Mascio, and A. Melonio. The 1st Release of the TERENCE Learner GUI: the User-based Usability Evaluation. In Proc. of the 2nd evidence-based TEL workshop (ebTEL 2013), Advances in Intelligent and Soft Computing. Springer, 2013.
7. V. Cofini, F. de la Prieta, T. D. Mascio, R. Gennari, and P. Vittorini. Design Smart Games with Context, Generate them with a Click, and Revise them with a GUI. Advances in Distributed Computing and Artificial Intelligence Journal, 3, Dec. 2012 2012.
8. V. Cofini, R. Gennari, and P. Vittorini. The Manual Revision of the TERENCE Italian Smart Games. In Proc. of the 2nd evidence-based TEL workshop (ebTEL 2013). Springer, 2013.
9. F. de la Prieta, T. D. Mascio, R. Gennari, I. Marenzi, and P. Vittorini. User-centred and Evidence-based Design of Smart Games. International Journal of Technology Enhanced Learning (IJTEL), Submitted for PDSG special issue in 2013.
10. Z. Gantner, M. Westphal, and S. Wölfl. GQR - A Fast Reasoner for Binary Qualitative Constraint Calculi. In AAAI'08 Workshop on Spatial and Temporal Reasoning, 2008.
11. R. Gennari. Temporal Constraint Programming: a Survey. CWI Quarterly, Report, 1998.
12. R. Gennari. Generation Service and Repository of Textual Smart Games. Technical report, TERENCE project, 2012.
13. M. S. Jong, J. H. Lee, and J. Shang. Reshaping Learning, chapter Educational Use of Computer Games: Where We Are, And What's Next. Springer, 2013.
14. A. Krokhin, P. Jeavons, and P. Jonsson. Reasoning about Temporal Relations: The Tractable Subalgebras of Allen's Interval Algebra. Journal of ACM, 50(5):591–640, 2005.
15. A. Lavelli, J. Hall, J. Nilsson, and J. Nivre. MaltParser at the EVALITA 2009 Dependency Parsing Task. In Proceedings of EVALITA Evaluation Campaign, 2009.
16. S. Moens. Deliverable 3.1: State of the Art and Design of Novel Annotation Languages and Technologies. Technical Report D3.1, TERENCE project, 2012.
17. A. Pease, I. Niles, J. Li. The Suggested Upper Merged Ontology: A Large Ontology for the Semantic Web and its Applications. Working Notes of the AAAI-2002 Workshop on Ontologies and the, Semantic Web, 2002.
18. E. Pianta, L. Bentivogli, and C. Girardi. MultiWordNet: developing an aligned multilingual database. In First International Conference on Global WordNet, pages 292–302, Mysore, India, 2002.
19. E. Pianta, C. Girardi, and R. Zanoli. The TextPro tool suite. In Proc. of the 6th Language Resources and Evaluation Conference (LREC), Marrakech, Morocco, 2008.
20. J. Pustejovsky, J. Castano, R. Ingria, R. Saurí, R. Gaizauskas, A. Setzer, G. Katz, and D. Radev. TimeML: Robust specification of event and temporal expressions in text. In IWCS-5 Fifth International Workshop on Computational Semantics, 2003.
21. J. Pustejovsky, P. Hanks, R. Sauri, A. See, R. Gaizauskas, A. Setzer, D. Radev, B. Sundheim, D. Day, L. Ferro, et al. The TimeBank corpus. In Corpus Linguistics, volume 2003, page 40, 2003.
22. J. Pustejovsky, K. Lee, H. Bunt, L. Romary. ISO-TimeML: An International Standard for Semantic Annotation. In Proceedings of the Fifth International Workshop on Interoperable Semantic, Annotation (ISA-5), 2010.
23. J. Pustejovsky, J. Littman, and R. Saurí. Arguments in TimeML: events and entities. Annotating, Extracting and Reasoning about Time and Events, pages 107–126, 2007.
24. J. Santos, L. Anido, M. Llamas, L. Álvarez, and F. Mikic. Computational Science, chapter Applying Computational Science Techniques to Support Adaptive Learning, pages 1079–1087. Lecture Notes in Computer Science 2658. 2003.
25. K. Slegers and R. Gennari. Deliverable 1.1: State of the Art of Methods for the User Analysis and Description of Context of Use. Technical Report D1.1, TERENCE project, 2011.
26. T. J. Teorey, D. Yang, and J. P. Fry. A Logical Design Methodology for Relational Databases using the Extended Entity-relationship Model. In ACM Computing Surveys. ACM, 1986.
27. TERENCE Consortium. The 2nd release of the TERENCE system for poor comprehenders.
28. M. Verhagen, R. Sauri, T. Caselli, and J. Pustejovsky. SemEval-2010 Task 13: TempEval-2. In Proceedings of the 5th International Workshop on Semantic Evaluation, pages 57–62, Uppsala, Sweden, July 2010. Association for Computational Linguistics.

Partridge: An Effective System for the Automatic Cassification of the Types of Academic Papers

James Ravenscroft, Maria Liakata and Amanda Clare

Abstract Partridge is a system that enables intelligent search for academic papers by allowing users to query terms within sentences designating a particular core scientific concept (e.g. Hypothesis, Result, etc). The system also automatically classifies papers according to article types (e.g. Review, Case Study). Here, we focus on the latter aspect of the system. For each paper, Partridge automatically extracts the full paper content from PDF files, converts it to XML, determines sentence boundaries, automatically labels the sentences with core scientific concepts, and then uses a random forest model to classify the paper type. We show that the type of a paper can be reliably predicted by a model which analyses the distribution of core scientific concepts within the sentences of the paper. We discuss the appropriateness of many of the existing paper types used by major journals, and their corresponding distributions. Partridge is online and available for use, includes a browser-friendly bookmarklet for new paper submission, and demonstrates a range of possibilities for more intelligent search in the scientific literature. The Partridge instance and further information about the project can be found at http://papro.org.uk.

1 Introduction

Since the advent of the 'Digital Age', the amount of information available to researchers has been increasing drastically, relevant material is becoming progressively more difficult to find manually and the need for an automated information

J. Ravenscroft (✉) · A. Clare
Department of Computer Science, Aberystwyth University, Aberystwyth SY23 3DB, UK
e-mail: ravenscroft@papro.org.uk

A. Clare
e-mail: afc@aber.ac.uk

M. Liakata
Department of Computer Science, University of Warwick, Aberystwyth, UK
e-mail: M.Liakata@warwick.ac.uk

M. Bramer and M. Petridis (eds.), *Research and Development in Intelligent Systems XXX*,
DOI: 10.1007/978-3-319-02621-3_26,

retrieval tool more apparent. There are already a large number of information retrieval and recommendation systems for scientific publications. Many of these systems, such as AGRICOLA,[1] the Cochrane Library[2] and Textpresso[3] index only publications from predefined journals or topics (for the above examples, Agriculture, Biology and Bioinformatics respectively). Unfortunately, these domain specific indexing systems usually only contain a small subset of papers, excluding potentially crucial literature because it does not quite fit into the subject domain.

The value of these systems to their users is often restricted by the small proportion of available literature that they index, forcing researchers to use multiple, domain specific, search engines for their queries. In contrast, there are also a number of interdisciplinary indexing systems and online journals such as arXiv[4] and PloSOne[5] that try to incorporate wide ranges of papers from as many disciplines as possible. The traits of these systems often complement those of their domain-specific counterparts; they provide a comprehensive collection of literature but insufficient filtering and indexing capabilities, usually based on title and abstract.

However, the document title is just one of the crucial parts of a scientific paper's structure. Liakata et al. describe a system for automatically processing and classifying sentences in a research paper according to the core scientific concept (CoreSC) that they describe [1]. There are 11 CoreSCs, including *Hypothesis*, *Goal*, *Background*, *Method*, *Result* and *Conclusion*. CoreSC labels can be allocated to all sentences in a scientific paper in order to identify which scientific concept each sentence encapsulates. SAPIENTA[6] is a publicly available machine learning application which can automatically annotate all sentences in a scientific paper with their CoreSC labels. It was trained using a corpus of physical chemistry and biochemistry research papers whose sentences were manually annotated using the CoreSC [2] scheme. An intelligent information retrieval system can use this data to provide better filtering and search capabilities for researchers. Partridge implements such context-aware keyword search, by allowing researchers to search for papers where a term appears in sentences with a specific CoreSC label (e.g only in *Method* sentences). This can be used to greatly improve both the precision with which researchers are able to perform searches for scientific literature and the accuracy of those searches in terms of relevance to the reader.

The type of a paper (*Review*, *Case Study*, *Research*, *Perspective*, etc) is another useful feature through which a user can narrow down the results of a search. The type of a paper can then be used to augment queries. For example, a user may search for a *Review* paper containing the keywords "PCR microfluidics", or a *Research* paper with a *Hypothesis* containing the keywords "cerevisiae" and "glucose". Such paper types are not yet standardised by journals. We expect the structure of a paper

[1] http://agricola.nal.usda.gov/

[2] http://www.thecochranelibrary.com/

[3] http://www.textpresso.org/

[4] http://arxiv.org

[5] http://plosone.org

[6] http://www.sapientaproject.com

to reflect its paper type. For example, review papers would be expected to contain a large amount of background material. In this article, we describe the application of machine learning (using random forests) to create predictive models of a paper's type, using the distribution of CoreSC labels found in the full text of the paper.

This model of paper type is currently in use in our Partridge system, which has been created as an intelligent full-text search platform for scientific papers. Partridge (which currently holds 1288 papers and is constantly expanding) makes use of automatically derived CoreSC sentence labels and automatically derived paper types, to allow deeper information queries. We discuss the reliability of this model of paper types and the insights that have been gained for the authorship of papers.

2 Methods

2.1 Collection of Scientific Articles

Partridge allows users to upload any paper which is free of copyright restrictions. For the purpose of this study we needed a large set of papers that we could label with CoreSC to investigate how this information assists classification into paper type. Open Access (OA) journals provide free read access to their papers, but many do not permit the user of articles for data mining purposes. The Public Library of Science (PLoS) journals contain large volumes of OA literature under a permissive license that allows data mining. They also use the PubMed Central markup schema, which is compatible with SAPIENTA, for papers published through their journal. The PLoS journals advanced search offers approximately 50 types of paper through which to restrict the search. Many of these paper type categories contain too few papers to be useful, others are too ad hoc (e.g. *Message from PLoS*). Indeed it is not clear how these paper types have been identified. We chose to look at a range of types, some of which we expect to overlap or have an unclear distinction. These are namely: Essay, Correspondence, Case Study, Perspective, Viewpoint, Opinion, Review, Research.

A script called *plosget.py*[7] was written, in order to download the papers via the PloS RESTful search API.[8] The number of papers downloaded per paper type category was as follows: 200 Essay, 99 Correspondence, 107 Case Study, 200 Perspective, 74 Viewpoint, 93 Opinion, 312 Review, 200 Research. These formed a corpus of 1285 check the numbers add up papers.

Figure 1 shows the CoreSC content of a review paper and a research paper randomly selected from the corpus. The review papers tend to be made up almost entirely from Background CoreSC sentences. However, research papers are much more evenly spread, made up of several different types of CoreSC. This

[7] https://github.com/ravenscroftj/partridge

[8] http://api.plos.org/solr/faq/

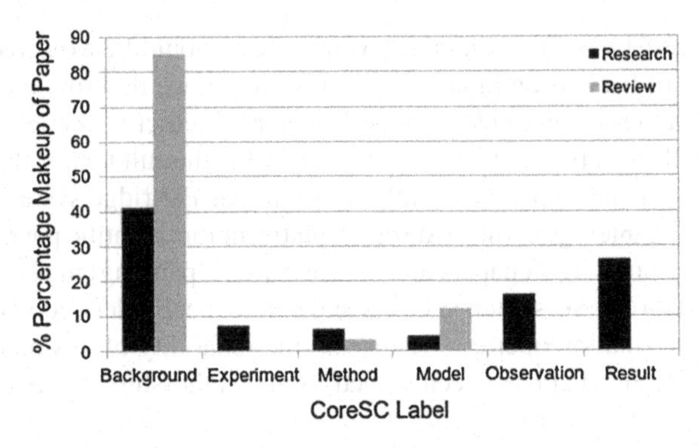

Fig. 1 The CoreSC makeup of a research paper and a review paper randomly selected from the corpus

investigation suggested that there is almost certainly a discriminative relationship between CoreSC categories and a paper's type.

2.2 Paper Processing and Classification

In order to obtain automatic labeling of sentences in a paper with CoreSC concepts we first needed to convert the papers to a format that the the CoreSC classifier, SAPIENTA can analyse. Currently, SAPIENTA supports the SciXML and PubMed Central DTDs. Papers in PDF format are first converted to XML using PDFX, a free service hosted by the University of Manchester [3]. Once the paper has been split into sentences with Liakata's SSplit tool [4], SAPIENTA is used to annotate each sentence with a relevant CoreSC label. These labels are assigned using a conditional random fields model based upon a number of features such as the sentence's location within the paper and pairs and triplets of words found consecutively within the sentence [1]. The SAPIENTA system has been trained on a corpus of 265 chemistry and biochemistry papers and we have done no domain adaptation for the papers in PLoS.

Random forest learning [5] was chosen for the construction of the paper type classifier. This is because random forests are a fast, accurate and widely accepted learning technology, and the underlying decision trees can be inspected to understand some of the reasoning behind the predictions made by the final paper type classifiers. The feature set consisted of the percentage composition of each of the 11 respective CoreSC labels assigned to the sentences in each paper i.e. (the percentage of *Background* sentences, the percentage of *Hypothesis* sentences, etc.). The random forest learning was conducted using the Orange data mining library for Python [6]. The parameters for the random forest learner were as the defaults, except with min_subsets = 5 and same_majority_pruning = true. We used 10-fold cross validation to estimate the precision, recall and F-measure.

3 Results and Discussion

The results of the random forest learning as recall, precision and F-measure for each paper type, averaged over a 10-fold cross validation are shown in Table 1. Paper types *Research*, *Review* and *Correspondence* are the most accurate classes, and *Viewpoint* is the most difficult paper type to predict.

A confusion matrix for the paper types is given in Table 2. Research papers are usually predicted to belong to the Research class, and are sometimes confused with Case Study. However, Case Study papers are often predicted to be Research or Review. The ratio between Case Study, Research and Review papers is 1:2:3 so the outcome is not entirely surprising in terms of the data size effect. We expected Research and Case Study to share similar paper structures but perhaps the fact that Case Study is equally confused with Research and Review suggests two distinct types of Case Study. We expected the classes Essay, Opinion, Perspective and Viewpoint to be confused, as the four labels all indicate a paper containing an author's personal thoughts on an issue, rather than experiment-driven science. Opinion is confused with Perspective and Viewpoint. Perspective is confused with almost all classes, except Research. Viewpoint is a small class and mostly confused with Opinion and Perspective. From this it would seem that the paper types {Opinion, Perspective, Viewpoint} are very similar and should be grouped together into one category. Indeed when we trained the model on a super class consisting of the previous three categories, performance was improved overall with the following F-measures: OpinionSuper: 59, Research: 67.6, Review: 63.7, Essay: 44.9, Correspondence: 46 and Case Study 24.8.

We inspected the detail of a single decision tree, grown on the entire dataset, to gain further insight into which CoreSC class decisions were responsible for the paper

Table 1 Per-class recall, precision and F-measure micro-averaged over 10-fold cross validation, reporting the results on the held-out validation segment

Classes	Recall (%)	Precision (%)	F-measure (%)
Case study	24.3	24.5	24.4
Correspondence	**54.5**	**50.5**	**52.4**
Essay	45.0	46.2	45.6
Opinion	24.7	23.5	24.1
Perspective	25.5	30.4	27.7
Research	**70.0**	**61.9**	**65.7**
Review	**63.1**	**62.5**	**62.8**
Viewpoint	14.9	15.7	15.3

Top paper types in bold

Table 2 Confusion matrix summed over 10-fold cross validation, reporting the results on the held-out validation segment

	Case study	Correspondence	Essay	Opinion	Perspective	Research	Review	Viewpoint
Case study	26	2	11	2	5	26	25	10
Correspondence	4	54	3	9	15	8	4	2
Essay	8	5	90	2	39	5	46	5
Opinion	4	7	2	23	27	10	3	17
Perspective	11	16	49	30	51	6	26	11
Research	19	6	2	8	5	140	12	8
Review	27	8	35	5	13	21	197	6
Viewpoint	7	9	3	19	13	10	2	11

Rows represent true classes, columns represent predicted classes type predictions. This was a very large tree, with a depth of 37 nodes in places. The first decision was based upon the number of Background sentences in the article. Low Background percentages, of less than 0.694 indicates a Correspondence paper. 28 out of the 31 Correspondence papers were correctly classified by this decision.

The next decision, for higher amounts of Background, was based on the percentage of Experiment sentences in the paper. For a very low percentage of Experiment sentences (<0.061), the papers then branched into a long side chain of detailed classification decisions, to separate mostly the Opinion, Viewpoint and Perspective papers from other Correspondence papers, and a few examples of the remaining categories. For a higher percentage of Experiment sentences, Research papers were classified as those that had Observations >5.6 %, or Conclusions <=1.4 %, whereas Case Studies had fewer Observations and more Conclusions.

The largest node classifying Essay did so via a route after the low Experiment decision that asked for a Background >48 %, but then low values for Goal, Hypothesis, Result, Observation, Model, Conclusion and Object. These decisions seem reasonable and agreed with our expectations of the content of an Essay.

4 Summary and Conclusions

To summarise, we have demonstrated that paper type can largely be predicted from the distribution of the automatically obtained CoreSC sentence labels in the full text of a paper. We have described some of the particular CoreSC features that determine a paper type (e.g. Correspondence characterised by very little Background, Opinions characterised by low percentage of Experiment) and discussed which of the paper types are not easily separable in this way. We recommended for example merging the Opinion, Viewpoint and Perspective articles into a single category, which increased overall classification performance. It is also potentially useful to distinguish between two types of Case study. Analysis of an example tree shows the decision making process to be complex, but agrees with our general understanding of paper types. Partridge allows refinement of paper search using CoreSC and paper type, both of which can be intelligently determined using machine learning methods.

The potential for automatic extraction of useful features from scientific papers to assist researchers in their knowledge queries is now an exciting area for research. Open Access journals that permit full text mining lead the way in allowing this research to expand and flourish. In future work we aim to develop and implement a range of further useful properties that will uncover more of the information that is hidden within the text of articles, and to use Partridge as a working engine to demonstrate their usefulness in practice.

Acknowledgments We thank the Leverhulme Trust for the support to Dr Liakata's Early Career Fellowship and also EMBL-EBI, Cambridge UK for the facilities offered to Dr Liakata.

References

1. M. Liakata, S. Saha, S. Dobnik, C. Batchelor, D. Rebholz-Schuhmann, Bioinformatics pp. 991–1000 (2012)
2. M. Liakata, S. Teufel, A. Siddharthan, C. Batchelor, in *Proceedings of LREC'10* (2010)
3. A. Constantin, S. Pettier, A. Voronkov, in *Proceedings of the 13th ACM Symposium on Document Engineering (Doc Eng)* (2013)
4. M. Liakata, L.N. Soldatova, et al., in *Proceedings of the Workshop on Current Trends in Biomedical Natural Language Processing* (Association for, Computational Linguistics, 2009), pp. 193–200
5. L. Breiman, Machine, Learning pp. 5–32 (2001)
6. J. Demšar, B. Zupan, G. Leban, T. Curk, in *Knowledge Discovery in Databases PKDD 2004, Lecture Notes in Computer Science*, vol. 3202, ed. by J.F. Boulicaut, F. Esposito, F. Giannotti, D. Pedreschi. Faculty of Computer and Information Science, University of Ljubljana (Springer, 2004), *Lecture Notes in Computer Science*, vol. 3202, pp. 537–539. DOI 10.1007/b100704. URL http://www.springerlink.com/index/G58613YV08BX48QJ.pdf

Aggregation Semantics for Link Validity

Léa Guizol, Madalina Croitoru and Michel Leclère

Abstract In this paper we address the problem of link repair in bibliographic knowledge bases. In the context of the SudocAD project, a decision support system (DSS) is being developed, aiming to assist librarians when adding new bibliographic records. The DSS makes the assumption that existing data in the system contains no linkage errors. We lift this assumption and detail a method that allows for link validation. Our method is based on two partitioning semantics which are formally introduced and evaluated on a sample of real data.

1 Introduction

Since 2001, ABES (French Bibliographic Agency for Higher Education) has been managing SUDOC[2] (University System of Documentation), a French collective catalog containing over ten million bibliographic records. In addition to *bibliographic records* that describe the documents of the collections of the French university and higher education and research libraries, it contains nearly 2.4 million *authority records* that describe individual entities (or named entities) useful for the description of documents (persons, families, corporate bodies, events etc.). Bibliographic records contain *links* to authority records that identify individuals with respect to the document described.

A typical entry of a book, by a librarian, in SUDOC takes place as follows. The librarian enters the title of the book, ISBN, number of pages and so forth (referred later on as the attributes of the bibliographic record corresponding to the book in question). Then (s)he needs to indicate the authors of the book. This is done by

[1] http://en.abes.fr/Sudoc/The-Sudoc-catalog

L. Guizol (✉) · M. Croitoru · M. Leclère
LIRMM, University of Montpellier II, Montpellier, France
e-mail: guizol@lirmm.fr

M. Bramer and M. Petridis (eds.), *Research and Development in Intelligent Systems XXX*, 359
DOI: 10.1007/978-3-319-02621-3_27, © Springer International Publishing Switzerland 2013

searching in the SUDOC base the authority record corresponding to that name. If several possibilities are returned by the system (e.g. homonyms), the librarian decides based on the bibliographic information associated to each candidate which one is most suitable for to choose as author of the book at hand. If none of the authors already in the base is suitable, then the librarian will create a new authority record in the system and link the book to this new record. The lack of distinguishing characteristics in the authority records and the lack of knowledge about the identity of the book's author imply that the librarian's decision is based on consultation of previous bibliographic records linked to each considered candidate. So any linkage error will entail new linkage errors.

In the SudocAd project [1] a decision support system was proposed to assist librarians choosing authority records. However, the SudocAd project relies on the assumption that the existing data in the SUDOC is clean, and namely:

- there are no distinct authority records describing one real world person,
- each contributor's name in a bibliographic record is linked to the "correct" authority record, and
- for each record, there is no big mistake in its attributes values.

In this work we lift the first two assumptions and aim to assess the quality of the data in the SUDOC. Preliminary work towards this goal was proposed in [2], where a general decision support system methodology was proposed to assess and repair such data. The method was based on partitioning contextual authorities (bibliographic records from the point of view of the authority record) according to various criteria. The global method, as represented in Fig. 1, consists of:

1. Allowing the domain expert to enter an appellation (name and surname). The system returns the authority records in Sudoc data with an appellation syntactically close to the one entered by the expert. For each authority record a set of corresponding bibliographic records (the books written by the author in question) are also returned.
2. Constructing all contextual authorities for all returned authority records. A contextual authority is the union of an authority record with one of the bibliographic records pointing to it. Intuitively it corresponds to an author in the context of one written book.

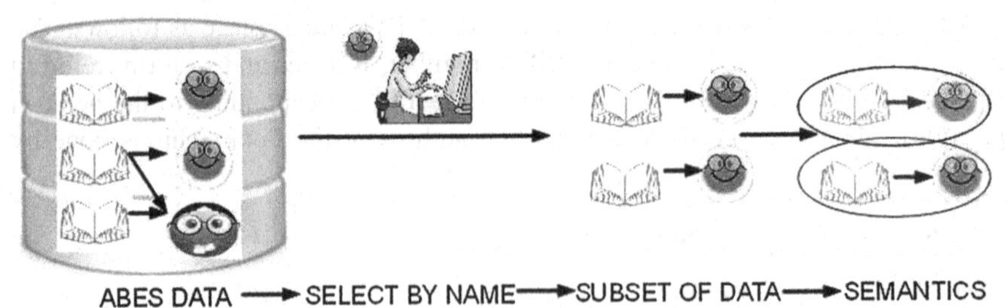

Fig. 1 First steps of the global approach

3. Partitioning the set of contextual authorities according to a partitioning semantics and a set of criteria. The set of criteria all return symbolic values. The aim of the partitioning semantics is to obtain a partition of the set of contextual authorities which "makes sense" from the point of view of the set of criteria. The obtained partition can also be compared with the initial partition (where contextual authorities belonging to the same authority record are in the same class of partition). Such comparison could provide paths for eventual repairs of the SUDOC data set.

The contribution of the paper is addressing the third item above. Namely we address the research question of "*how to propose a partitioning semantics for a set of criteria returning symbolic values*". We propose two semantics (a local and a global one) and evaluate them in terms of quality of returned partitions as well as algorithmic efficiency.

The paper is structured as follows. After presenting the SUDOC data as well as the different criteria currently implemented (Sect. 2), we present in Sect. 3 a motivating example showing the limitations of existing work. We then introduce two partitioning semantics (Sect. 4): while the first semantics can discard a criteria value overall on the dataset to be partitioned, the second semantics refines the first semantics by introducing the notion of local incoherence. We evaluate our approach on a sample of SUDOC data, then discuss its execution time in Sect. 5.

2 SUDOC Data and Criteria

An authority record is used to represent a person in SUDOC. In addition to an identifier, it contains at least a set of names used to designate the person and, possibly, dates of birth/death, sex, nationality, titles and any comments in plain text. All the other information regarding his (her) contribution to some works (what (s)he wrote, what domains (s)he has contributed to etc.) are only available from the bibliographic records of the documents (s)he has (co-)authored. A bibliographic record is used to represent a particular document in SUDOC. Most information (such as title, publication date, language, domain) is reliable. The contributor information is added by searching the system for (person) authority records corresponding to each of the names indicated as contributing to the document.

We compute for each document a contextual description of each of its contributors. Such a description, denoted a *contextual authority*, will contain a set of selected information extracted from the bibliographic record and the reliable information about the contributor from the linked authority record. We select from the bibliographic record the title, the publication date, the domain, the language, the co-contributors and the role. From the authority record, we consider that only names are reliable.

The contextual authorities will be then compared amongst each other in order to group together the contextual authorities which are similar. The comparison will be done according to various criteria. This process has to mirror the decision process of a domain expert when deciding if a contextual authority is close to another. This is the reason why in our project, one of the most important requirements is to consider symbolic criteria (i.e. the domain values of criteria considered are symbolic). The set of symbolic values for each criteria is also equipped with a total order.

As previously mentioned, given a set of contextual authority records we are interested to use the criteria (provided by the domain experts, the librarians) to "cluster" the authority records. The date of publication criteria, for instance, will provide a partitioning of the contextual authority records (publications close by date, far by date, very far by date etc.). We then want to combine the obtained partitions according to the different criteria and provide one or more overall partition(s) corresponding to the whole set of criteria.

In the reminder of the section we briefly describe the criteria considered and currently implemented in our system.

Let $\mathbb{O}$ be the objects set to partition. A criterion $C \in \mathbb{C}$ is a function that gives a comparison value for any pair of objects in $\mathbb{O}^2$. This comparison value is discrete and in a totally ordered set $V = \{never\} \cup V^C_{far} \cup \{neutral\} \cup V^C_{close} \cup \{always\}$. V^C_{far} and V^C_{close} are two totally ordered values sets.

Closeness values are denoted $+$, $++$, ... and farness values are denoted $-$, $--$, ... such as : $always \geq ... \geq ++ \geq + \geq neutral \geq - \geq -- \geq ... \geq never$ (where $\geq$ stands for "is stronger" than).

The criteria we have currently implemented are: *domain*, *date*, *title*, *appellation*, *contributor* and *language*. The "domain" attribute is represented in SUDOC as a set of domain codes (see example in Table 1). The distance on domain codes and their aggregation function was provided by domain experts (omitted here for lack of space). The publication dates are compared using a distance based on the intervals between dates (intervals of 60, respectively 100 years). The titles are compared using a Levenstein adapted distance. The *contributor* criterion gives closeness values if there is common contributor(s) (without the contributor designed by the appellation). The *appellation* criterion is based on expert comparison function for names and surnames. The *language* criterion gives a farness comparison value if publications languages are distinct and none of them is English.

Table 1 Example of real contextual authorities

Id	Title	Date	Domains	[...]	Appellations
1	Le banquet	1868			"Platon"
2	Le banquet	2007			"Platon"
3	Letter to a Christian nation		[320,200]		"Harris, Sam"
4	Surat terbuka untuk bangsa kristen	2008	[200]		"Harris, Sam"
5	The philosophical basis of theism	1883	[100,200,150,100]		"Harris, Samuel"
6	Building pathology	2001	[720,690,690,690]		"Harris, Samuel Y."
7	Aluminium alloys 2002	2002	[540]		"Harris, Sam J."
8	Dispositifs GAA en technologie SON	2005	[620,620,530,620]		"Harrison, Samuel"

3 Motivating Example

As previously mentioned, partitioning semantics based on numerical values are not interesting for our problem since one of the requirements of the Decision Support System for the librarians is to use symbolic valued criteria (justified by the need of modelling human expert reasoning).

Even if the clustering methods such as those of [3] seemingly deal with symbolic values, the symbolic values are treated in a numerical manner. Let us consider a real world example from the SUDOC data and see how the Dedupalog semantics of [3] behaves.

In Dedupalog, the criteria (denoted C) return a comparison value between two objects as follows: $C : \mathbb{O} \times \mathbb{O} \rightarrow \{close, far, always, never\}$. To decide whether two objects represent the same entity, the first step is to check if there is at least a criterion returning $always$ or $never$.[2] If it is not the case, we simply count: vote = (criteriareturningclose) − (criteriareturningfar). If $vote \geq 0$, we consider the comparison value $close$, else far.

We are interested in a best partition on the object set $\mathbb{O}$. Semantically, two objects represent the same entity if and only if they are in a same partition class. A partition is valid if and only if there are no two objects with an $always$ comparison value (respectively $never$) in distinct classes (respectively in a same class) of the partition. A partition P is a best partition if P is valid, and there is the fewest possible number of pairs of objects with a $close$ comparison value (respectively far) in distinct classes (respectively in a same class) of the partition.

Let us consider the SUDOC data subset shown in Table 1. We consider the records set of "Harris, Sam" appellation (denoted $\mathbb{O}_s = \{3, 4, 5, 6, 7, 8\}$). The expert-validated partition on $\mathbb{O}_s$ is $Ph_s = \{\{5\}, \{3, 4\}, \{7\}, \{6\}, \{8\}\}$. The "domain" attribute is composed of a set of domains codes. Two objects are considered by $domain$ criterion as $close$ if they have at least a common domain, and far if not. Two objects are considered by $date$ criterion as far if there is more than 59 years between their publication dates. Therefore, the $date$ criterion returns that object 5 is far with all other objects. However, the $domain$ criterion says than objects 3, 4 and 5 are pairwise $close$ together because of common 200 domain code (= religion): 3, 4 and 5 should be in a same class. The $domain$ criterion says than 6, 7 and 8 are pairwise far together and far from respectively 3, 4 and 5, so the only best partition is $\{\{5, 3, 4\}, \{7\}, \{6\}, \{8\}\}$. Unfortunately, this best partition is not the expert-validated partition. We claim that the reason for this unsatisfactory result is the way symbolic values are treated by such approaches: considering them as numerical.

In this paper, we propose two partitioning semantics that improve the state of art by allowing for:

---

[2] Dedupalog forbids the possibility of a criterion returning $always$ and another returning $never$ for the same pair of objects.

- several levels of far and $close$ values;
- non interference of $close$ and far comparison values (for example, a $close$ comparison value cannot erase a far comparison value).

The proposed semantics are detailed in Sect. 4.

4 Proposed Semantics

4.1 Partitioning

The set of contextual authorities is partitioned according to different criteria (at least a common name, closeness of domains, dates of publication, languages of publication etc.). The result is a partition of compared objects based on closeness criteria. Intuitively, objects in a same class are close from the point of view of the respective criteria, and far from objects in another class.

In Table 1, the contextual authority number 4 has been written 125 years after the contextual authority number 5. So, their authors should be different persons. With respect to the $date$ criterion, these contextual authorities should then be in different classes of partitions. However, they have a common publication domain, the domain number 200 (= religion), so we could be tempted to put them in a same class with respect to the $domain$ criterion. To decide whether these contextual authorities represent or not a same person, we should aggregate criteria, i.e. decide which value of which criteria is meaningless in this case. Once this is done, as explained in [2], we compare the aggregated partitions with the initial partition.

4.2 Preliminary Notions

In the following we solely consider valid partitions. A valid partition is a partition such that there are no two objects with an $always$ comparison value (respectively $never$) in distinct classes (respectively in a same class) of the partition.

Of course, the ideal case consists of never having objects in the same class with a farness value, respectively objects in different classes with a closeness value. If such a partition does not exist, there are *incoherences* with respect to our criteria set. The incoherence (property of an objects subset such that they must be in a same class according to some criteria and must be in separated classes according to other criteria) notion is central to the definitions of the two semantics detailed in the Sects. 4.3 and 4.4.

From the point of view of a single criterion at a time, we prefer to group objects linked by a higher closeness comparison value than a smaller one. Similarly, we prefer to separate objects with a greater farness value than objects with a lesser farness value. In the case of incoherences one has to make choices in order to satisfy closeness or farness values which, by definition, could lead to distinct partitions.

The value of a partition with respect to a criterion is given by a couple of values (v_p, v_n) such that v_p (p for "positive") is the smallest closeness or *always* value such that all objects pairs with a criterion value bigger or equivalent value than v_p are in the same class (denoted "satisfied" objects pairs in [4]). The bipolar condition on v_n (n for "negative") also applies: v_n is the biggest farness or *never* value such that all objects pairs with a criterion value smaller or equivalent value than v_n are in distinct classes.

The partition values are then used to order partitions. One partition is better than another if and only if its v_p value is smaller and its v_n value is bigger. We denote $v(P, C)$ the P partition value with respect to criterion C and $v(P, \mathbb{C})$ the P partition value with respect to criteria set $\mathbb{C}$. For several criteria, it is possible that there are several best partitions on an object set according to the partition order. Due to domain expert requirements we cannot employ a criterion preference order.

In Sects. 4.3 and 4.4 we present two partitioning semantics: a global semantics and a local semantics. The global semantics, in an incoherent case, will give the best partitions that respects a criterion value in the same manner overall. The local semantics tries to localise the incoherence sources and treat it separately.

4.3 Global Semantics

Let us consider the example in Table 1 and the two records sets of the "Harris, Sam" appellation (denoted $\mathbb{O}_s = \{3, 4, 5, 6, 7, 8\}$) and "Platon" appellation ($\mathbb{O}_p = \{1, 2\}$). The expert-validated partitions are respectively (for the two record sets) $Ph_p = \{\{1, 2\}\}$ and $Ph_s = \{\{5\}, \{3, 4\}, \{7\}, \{6\}, \{8\}\}$. We compute the best partition of the two separate sets.

Let us apply global semantics on $\mathbb{O}_s = \{3, 4, 5, 6, 7, 8\}$. This objects set is not coherent with respect to our criteria. The Ph_s value is such that:

- $v(Ph_s, domains) = (+++ +, -)$,
- $v(Ph_s, date) = (always, -)$.

Ph_s has a best partition value on $\mathbb{O}_s$. However, partitions $P'_s = \{\{5, 3, 4\}, \{7\}, \{6\}, \{8\}\}$ with $(+, -)$ value for *domain* criterion and $(always, --)$ value for *date* criterion is also a best partition. The plurality of best partitions values comes from incoherence between the *date* and *domain* criteria.

Let us now apply global semantics on $\mathbb{O}_p = \{1, 2\}$.The expert-validated partition is $Ph_p = \{\{1, 2\}\}$. There is an incoherence in the *date* criterion. Ph_p value is such that:

- $v(Ph_p, title) = (+, never)$,
- $v(Ph_p, date) = (always, never)$.

Ph_p is the only best partition possible on $\mathbb{O}_p$ because $\{\{1\}, \{2\}\}$ is not valid with respect to *title* criterion.

Let us now illustrate how global semantics will affect the whole set of objects by computing the best global partition on the union of records of the two appellations. We now apply global semantics on all our selected contextual authorities: $\mathbb{O} = \mathbb{O}_p \cup \mathbb{O}_s$. The expert-validated partition is $Ph_{ps} = \{\{1, 2\}, \{5\}, \{3, 4\}, \{7\}, \{6\}, \{8\}\}$. We also encounter incoherences and this partition has the worst of Ph_p and Ph_s values for each criterion, in particular:

- $v(Ph_{ps}, title) = (+, never)$
- $v(Ph_{ps}, domains) = (+ + ++, -)$
- $v(Ph_{ps}, date) = (always, never)$.

Ph_{ps} has not the best partition value because we could improve partition value for *domain* criterion. This does not affect the *date* criterion value because it is already as bad as possible. For example, partition $P'_{ps} = \{\{1, 2\}, \{5, 3, 4\}, \{7\}, \{6\}, \{8\}\}$ has a best value.

A way to fix this problem is to propose the local semantics detailed in the Sect. 4.4. We complete this subsection by presenting the algorithms used to find all best partitions values according to global semantics. Please check [5] for more details of the algorithms presented in this paper.

4.3.1 Global Semantics Algorithm for One Criterion

The input data (SUDOC data) structure is represented internally by our system as a multiple complete graph $\mathbb{G}_\mathbb{C}$: the $\mathbb{G}_\mathbb{C}$ vertex set is $\mathbb{O}$ and $\mathbb{G}_\mathbb{C}$ edges are labelled by the comparison value between linked objects according to a specified criterion. We denote G_C the criterion graph of a single criterion C.

Let C be a criterion on an object set $\mathbb{O}$. In order to find the best partition values on $\mathbb{O}$ with respect to C, we have to find and to evaluate reference partitions on $\mathbb{O}$ with respect to C for all closeness values $v_i \in V^C_{close} \cup \{always\}$. To define a reference partition we need to use the notion of a refined partition as explained below.

Definition 1 (Refined partition). Let P_i, P_j, be two partitions on an object set $\mathbb{O}$. P_i is more refined than P_j if and only if $\forall$ c_i class $\in P_i$ $\exists c_j$ class $\in P_j | c_i \subseteq c_j$.

P_j partition is said to be less refined than P_i partition.

Definition 2 (Reference partition for a criterion). Let C be a criterion on an object set $\mathbb{O}$ and v_i a closeness value so that $v_i \in V^C_{close} \cup \{always\}$. The reference partition P_{ref} for C with respect to v_i, is the most refined partition P such as $v(P, C) = (v_p, v_n)$ and $v_p \leq v_i$.

We denote $ref(v_i)$ the reference partition for a criterion C with respect to closeness value v_i.

Evaluating the reference partitions is enough to calculate and evaluate all the best partition values. Since the reference partition with respect to a closeness value v_i for criterion C is the most refined partition with $v_p \leq v_i$, it is the partition with the less possible farness edges such as both vertexes are inside the same class (edges inside a single class). This makes it a partition with the best possible v_n value with respect to the v_p such that it is $\leq v_i$.

To calculate a reference partition with respect to a closeness value v_i for criterion C comes to calculating connected components on G_C considering only v'_i labelled edges such as $v'_i \geq v_i$. We can then simply use Kruskal's algorithm (complexity $\mathcal{O}(m \log n)$ for n vertexes and m edges). Please note that the connected component idea has been already explored in [6] and [4]. However the authors do not consider incoherence problems or even more levels of farness and closeness values. In the worst case (when there is no valid partitions with a v_n value such as $v_n = max(V^C_{far} \cup \{never\})$) we have $k+1 = |V^C_{close} \cup \{always\}|$ references partitions to find and evaluate. The complexity of the global semantics for one criterion algorithm is $\mathcal{O}((k+1) * m \log n)$, and it is depicted below (Algorithm 1).

Algorithm 1 BestsValuesForASingleCriteria

Require: C: criterion on an objects set $\mathbb{O}$; G_C: criterion graph of C on $\mathbb{O}$;
Ensure: set of best partitions values with respect to C on $\mathbb{O}$

1: best partitions values set $bestV = \{\}$;
2: **for all** value $v_i \in V^C_{close} \cup \{always\}$ in $<$ order **do**
3: Partition $P = ref(v_p)$;
4: Partition value $v = v(P)$;
5: **if** P is valid and $\nexists v' \in bestV | v' \succeq v$ **then**
6: add v to $bestV$;
7: **end if**
8: **if** $v(P) = (v'_p, v'_n)$ such as $v_n = max(V^C_{far} \cup \{never\})$ **then**
9: **return** $bestV$;
10: **end if**
11: **end for**
12: **return** $bestV$;

4.3.2 Global Semantics Algorithm for Several Criteria

Let us now consider the global semantics when there are more than one criteria to consider. We will first need three notions: closeness values set, ascendant closeness values set and reference partition. The reference partition, as explained above, is the actual test to be performed by the algorithm. The cardinality of the closeness value set represents the number of tests that the algorithm will need to perform in the worse case. Finally, the ascendant closeness values notion will allow us to skip some tests, and optimise the algorithm.

Definition 3 (Closeness values set). A closeness values set $\mathbb{VC}$ for a criteria set $\mathbb{C}$ is a set of v_i such as $v_i \in V^{C_i}_{close} \cup \{always\}$ and $C_i \in \mathbb{C}$, with one and only one closeness value v_i for each criterion $C_i \in \mathbb{C}$.

Definition 4 (Ascendant closeness values set). Let $\mathbb{VC}1$ and $\mathbb{VC}2$ be two closeness values sets for the same criteria set $\mathbb{C}$. $\mathbb{VC}1$ is an ascendant of $\mathbb{VC}2$ if and only if $\mathbb{VC}1$ has a best or equivalent (smaller) closeness value than $\mathbb{VC}2$ for each criterion in $\mathbb{C}$.

$\mathbb{VC}2$ is a descendant of $\mathbb{VC}1$.

Definition 5 (Reference partition with respect to a criteria set). Let $\mathbb{C}$ be a criteria set on an object set $\mathbb{O}$, and $\mathbb{VC}$ a closeness values set for $\mathbb{C}$. The reference partition P_{ref} for $\mathbb{C}$ with respect to $\mathbb{VC}$ (denoted $ref(\mathbb{VC})$) is the most refined (please see Definition 1) partition such as $v(P_{ref}) = \{v(P_{ref}, C_i) \forall C_i \in \mathbb{C}\}$ with $\forall C_i$ criterion: $v(P_{ref}, C_i) = (v_p, v_n) | v_p \leq v_i \in \mathbb{VC}$.

The global semantics algorithm for several criteria is an extension of the one for one criteria (Algorithm 1). The best partition values are also reference partitions values, so we calculate, evaluate and compare them.

First, we find all closeness values set (Definition 3) for $\mathbb{C}$. We compute the reference partition (Definition 5) for each closeness values set $\mathbb{VC}$ by searching for connected components with Kruskal's algorithm (complexity $\mathcal{O}(m \log n)$) on $\mathbb{G}_\mathbb{C}$ considering only v_p labelled edges such as $v_p \geq v_i | v_i \in (V_{close}^{C_i} \cup \{always\}) \cap \mathbb{VC}$ and $C_i \in \mathbb{C}$. We then evaluate reference partition values and only keep best ones.

If a reference partition $ref(\mathbb{VC})$ has $v(P, C_i) = (v_p, v_n)$ for each criteria $C_i \in \mathbb{C}$ such as $v_n = max(V_{far}^{C_i} \cup \{never\})$, then reference partitions $ref(\mathbb{VC}')$ with $\mathbb{VC}'$ descendants (Definition 4) of $\mathbb{VC}$ have a worse or same value than $ref(\mathbb{VC})$, so we do not need to evaluate them.

For a criteria set $\mathbb{C}$ of c criteria, we have to calculate and evaluate $|V_{close}^{C_1} \cup \{always\}| * ... * |V_{close}^{C_c} \cup \{always\}|$ reference partitions in the worst case, namely $(k+1)^c$ reference partitions with $k = max(|V_{close}^{C_i}| \forall C_i \in \mathbb{C})$. So, this algorithm has $\mathcal{O}((k+1)^c * m \log n)$ as complexity (see Algorithm 2).

Algorithm 2 BestPartitionsValuesForCriteriaSet

Require: $\mathbb{C}$, a criteria set on an objects set $\mathbb{O}$; $\mathbb{G}_\mathbb{C}$ criteria graph of $\mathbb{C}$
Ensure: set of best partitions values with respect to $\mathbb{C}$ on $\mathbb{O}$.
1: best partitions values set $bestV = \{\}$;
2: set of closeness values set to test $toTest = \{\mathbb{VP} | \mathbb{VP}$, closeness values set for $\mathbb{C}\}$;
3: **while** $toTest \neq \{\}$ **do**
4: pick up $\mathbb{VP}$ from $toTest$ such as $\mathbb{VP}$ has no ascendant in $toTest$;
5: Partition $P = ref(\mathbb{VP})$;
6: Partition value $v = v(P, \mathbb{C})$;
7: **if** P is valid and $\nexists v' \in bestV | v' \succeq v$ **then**
8: add v to $bestV$;
9: **end if**
10: **if** $\forall C_i \in \mathbb{C}, v(P, C_i) = (v_p, v_n) | v_n = max(V_{far}^{C_i} \cup \{never\})$ **then**
11: remove all descendants of $\mathbb{VP}$ from $toTest$;
12: **end if**
13: **end while**
14: **return** $bestV$;

4.4 Local Semantics

Local semantics do not consider incoherence for the whole treated object set (denoted $\mathbb{O}$) but only for pairs of objects that cause incoherence. The pairs of objects that cause incoherence represent the objects that are to be put in the same class by some criteria and kept separate according to others (e.g. objects 4 and 5 described in Table 1 : *date* criterion returns than they must be in distinct classes but *domain* criterion returns than they must be in a same class).

We consider in $\mathbb{O}$ the objects in incoherent parts. A minimal incoherent subset $\mathbb{I}$ of $\mathbb{O}$ is a subset of $\mathbb{O}$ such that:

- it contains a pair of objects that causes incoherences,
- there are no closeness comparison values between an object of $\mathbb{O} \setminus \mathbb{I}$ and an object in $\mathbb{I}$, and
- there is no subset of $\mathbb{I}$ which is a minimal incoherent subset of $\mathbb{O}$.

In the previous examples, we saw than 4 and 5 contains an incoherence. However, $\{4, 5\}$ is not a incoherent subset of $\mathbb{O}_s = \{3, 4, 5, 6, 7, 8\}$ because 3 is linked by a closeness value (*domain* criterion) to 4 and 5. $\{3, 4, 5\}$ is an incoherent subset of $\mathbb{O}_s$.

The coherent part contains all $\mathbb{O}$ objects but considers that the comparison value for every pair of objects that are occurring in the same minimal incoherent subset is *neutral*. We denote the incoherent part of a $\mathbb{G}_\mathbb{C}$ criteria graph according to the set $\mathbb{IP}$ of incoherent parts of $\mathbb{G}_\mathbb{C}$: $coherentPart(\mathbb{G}_\mathbb{C}, \mathbb{IP})$.

A partition on $\mathbb{O}$ is better than another partition if it has a best value for the coherent part and for each incoherent part. The values of each (in)coherent part are determined by global semantics.

Let us consider an example and apply local semantics on all the selected contextual authorities in Table 1: $\mathbb{O} = \mathbb{O}_p \cup \mathbb{O}_s$. The expert-validated partition is $Ph_{ps} = \{\{1, 2\}, \{5\}, \{3, 4\}, \{7\}, \{6\}, \{8\}\}$. There are two minimal incoherent subsets: $\{5, 3, 4\}$ and $\{1, 2\}$. For $\{5, 3, 4\}$ we have a best partition as for $\{1, 2\}$. Since in this semantics the incoherent subsets are considered independently one from the other, one of the best values for Ph_{ps} on the whole subset is equal to the value of the domain expert validated partition.

4.4.1 Local Semantics Algorithm for Several Criteria

To find all best partitions values on a criteria graph according to local semantics, we first need to identify incoherent (and coherent) parts with a connected components algorithm (complexity $\mathcal{O}(m \log n)$).[3] Then, for the coherent part and each of the at most $n/2$ incoherent parts,[4] we execute the algorithm of global semantics (of complexity $\mathcal{O}((k + 1)^c * m \log n)$. So, the complexity in the worst case of the local

[3] with n vertexes and m edges.

[4] since an incoherent part contains at least two edges between two vertexes.

semantics algorithm for several criteria is: $\mathcal{O}(n * (k+1)^c * m \log n)$ (please see Algorithm 3).

Algorithm 3 BestPartitionsValuesForLocalSemantics

Require: $\mathbb{C}$, a criteria set on an objects set $\mathbb{O}$; $\mathbb{G}_\mathbb{C}$ criteria graph of $\mathbb{C}$
Ensure: set of best partitions values with respect to $\mathbb{C}$ on $\mathbb{O}$.
1: best partitions values set $bestV = \{\}$;
2: Partition $P_a = ref(\mathbb{VP})$;
3: set of graphs: $Gparts = \{\}$;
4: **for** each incoherent class $\mathbb{I} \in P_a$ **do**
5: add $incoherentPart(\mathbb{G}_\mathbb{C}, \mathbb{I})$ to $Gparts$;
6: **end for**
7: add $coherentPart(\mathbb{G}_\mathbb{C}, Gparts)$ to $Gparts$;
8: apply algorithm 2 on each graph in $Gparts$;
9: $bestV = \{$ best partition for $\mathbb{G}_\mathbb{C}$: best partition for each graph in $Gparts\}$;
10: **return** $bestV$;

5 Evaluation

We have experimented the algorithms on 133 SUDOC data subsets related to 133 random (on a list of common names and surnames) appellation. For each appellation, we select the associated SUDOC data subset as follows:

- each authority record which has a close appellation is selected;
- for each selected authority record, linked bibliographic records (up to 100 upper limit) are selected;
- for each link between a selected bibliographic record and a selected authority record, we construct a contextual authority.

We measure the execution time in nanoseconds on each 133 Sudoc data subsets selected for algorithms 2 and 3. The 133 appellations generated between 1 and 349 contextual authorities each. The number of criteria we considered is 6 with a 84 closeness value sets (between 0 and 6 values per criterion).

We used a Intel(R) Core(TM) i7-2600 CPU 3.40 GHz based PC with 4 GB of RAM running Windows 7 64 Bit with a Java 1.6 implementation. The execution times are shown on Figs. 2 and 3.

As seen in Fig. 2 the execution time for global semantic algorithm is fast (less than one second even for 349 contextual authorities). This is an acceptable result since, according to the domain experts, it is very rare for an appellation to have more than 350 books authored. However, we notice that the execution time is irregular. This is due to the fact that since we have too many conflicting comparison values (causing incoherences) we are in the worst case scenario thus augmenting the execution time. Please note that this case is not necessarily dependent on the number of contextual

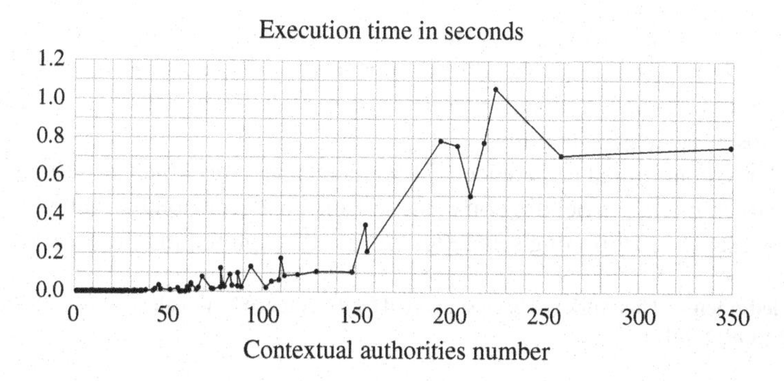

Fig. 2 Execution time for global semantics algorithm

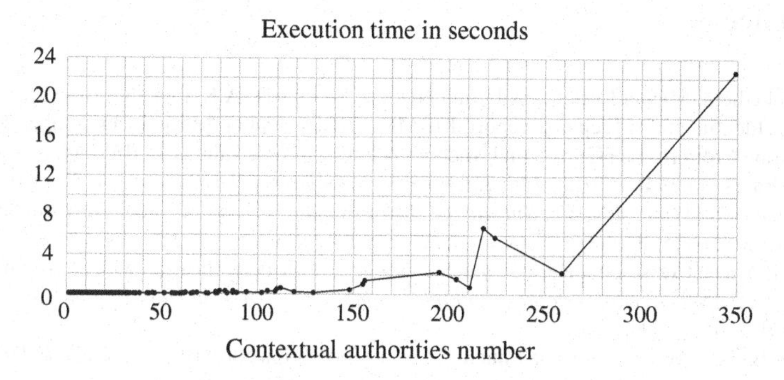

Fig. 3 Execution time for local semantics algorithm

authorities considered. We are currently investigating the relation between conflicting values and number of contextual authorities in the SUDOC by means of sampling.

In Fig. 3 the execution time for local semantics algorithm is depicted. Even if the local semantics returns better qualitative results, the execution time is much longer than the global semantics algorithm. The results are not acceptable for a large number of contextual authorities with a lot of incoherence (as seen in the case of the 349 records with a time of 22 s) but we hope to be able to better understand the SUDOC data in order to show that such cases are extremely rare. Such analysis of the SUDOC data, as mentioned before, constitutes current ongoing work.

6 Conclusion

In this paper we presented two partitioning semantics based on non-numerically valued criteria. The partitioning semantics were introduced due to a main feature of our system and namely that we want to keep the symbolic values of the criteria as much as possible (as opposed to aggregation techniques that reduce them to numerical values for manipulation). We explained the need of such semantics in the case of our application and explained how the two semantics yield different results on a real world example. We also shown than those semantics are scalable on most of real Sudoc subsets selected randomly.

Similar to conditional preferences [7], we need to decide which criterion value to improve over another depending on context. Links between conditional preferences and the presented partitioning semantics have to be explored in a future work.

Acknowledgments This work has been supported by the Agence Nationale de la Recherche (grant ANR-12-CORD-0012).

References

1. Michel Chein, Michel Leclère, and Yann Nicolas. SudocAD: A Knowledge-Based System for Object Identification. Technical report, LIRMM, INRIA Sophia Antipolis, December 2012.
2. Madalina Croitoru, Léa Guizol, and Michel Leclère. On Link Validity in Bibliographic Knowledge Bases. In *IPMU'2012: 14th International Conference on Information Processing and Management of Uncertainty in Knowledge-Based Systems*, volume Advances on Computational Intelligence, pages 380–389, Catania, Italie, July 2012. Springer.
3. Arvind Arasu, Christopher Ré, and Dan Suciu. Large-scale deduplication with constraints using dedupalog. In *Proceedings of the 25th International Conference on Data Engineering (ICDE)*, pages 952–963, 2009.
4. Nikhil Bansal, Avrim Blum, and Shuchi Chawla. Correlation clustering. In, *MACHINE LEARNING*, pp. 238–247, 2002.
5. Léa Guizol, Madalina Croitoru, and Michel Leclère. Aggregation semantics for link validity: technical report. Technical report, LIRMM, INRIA Sophia Antipolis, http://www.lirmm.fr/~guizol/AggregationSemanticsforLinkValidity-RR.pdf, 2013.
6. A. Guénoche. Partitions optimisées selon différents critères: évaluation et comparaison. *Mathématiques et sciences humaines. Mathematics and social sciences*, (161), 2003.
7. Craig Boutilier, Ronen I. Brafman, Carmel Domshlak, Holger H. Hoos, and David Poole. Cp-nets: A tool for representing and reasoning with conditional ceteris paribus preference statements. *Journal of Artificial Intelligence Research*, 21:135–191, 2004.

AI Applications

Efficient Interactive Budget Planning and Adjusting Under Financial Stress

Peter Rausch, Frederic Stahl and Michael Stumpf

Abstract One of the most challenging tasks in financial management for large governmental and industrial organizations is Planning and Budgeting (P&B). The processes involved with P&B are cost and time intensive, especially when dealing with uncertainties and budget adjustments during the planning horizon. This work builds on our previous research in which we proposed and evaluated a fuzzy approach that allows optimizing the budget interactively beyond the initial planning stage. In this research we propose an extension that handles financial stress (i.e. drastic budget cuts) occurred during the budget period. This is done by introducing fuzzy stress parameters which are used to re-distribute the budget in order to minimize the negative impact of the financial stress. The benefits and possible issues of this approach are analyzed critically using a real world case study from the Nuremberg Institute of Technology (NIT). Additionally, ongoing and future research directions are presented.

1 Introduction

Planning is the process of managing an organization to reach specific goals. 'Budgets are the formal expression of plans, goals and objectives that cover all aspects of operations for a designated time period' [12]. They are used to transform the plans

P. Rausch (✉) · M. Stumpf
Nuremberg Institute of Technology Georg Simon Ohm, Keßlerplatz 12,
90489 Nuremberg, Germany
e-mail: peter.rausch@th-nuernberg.de

M. Stumpf
e-mail: michael.stumpf@th-nuernberg.de

F. Stahl
University of Reading, School of Systems Engineering, PO Box 225, Reading
RG6 6AY, UK
e-mail: F.T.Stahl@reading.ac.uk

M. Bramer and M. Petridis (eds.), *Research and Development in Intelligent Systems XXX*,
DOI: 10.1007/978-3-319-02621-3_28,

of decision-makers into quantitative terms to be able to measure progress and to control the organization. Due to the close link between multiple plans and budgets P&B processes are usually very time-consuming and costly [1].

Furthermore, decision-makers have to make assumptions in advance. They have to take uncertain conditions into account, such as macro economic developments and estimate planning parameters. Because of this, plans can quickly become obsolete [1] and adjustments to plans and budgets have to be made to accommodate new conditions. Especially, external forces with unforeseen negative impacts on the organization's goals can put the system under stress and trigger a rescheduling of plans and budgets during the fiscal planning period. Stress can be seen as 'an external force operating on a system, be it an organization or a person' [2]. In this paper we use the term 'financial stress' to refer to a situation, where external conditions change and stress is induced on the organization's environment in a negative way. Levine defined the process of managing organizational change towards lower levels of resource consumption and organizational activity as cutback management [3]. He created a typology covering causes of public organizations' decline and examined sets of tactics and decision rules to manage cutbacks [4]. For the research presented in this paper we assume that the decision-maker cannot influence the external environment and adapts to the new situation by changing the internal system, i.e. by adding strain to the system. The authors of [2] define strain as 'change in the state of the internal system which results from this external stress'. In our case, strain is added to the system by a reduction of available resources, namely cutting budgets. A simple approach would be an 'across-the-board' ('a-t-b') budget cut, which cuts all programs of an organization's budget equally. However, [12] state that this approach may result in disastrous consequences for certain programs and that decision-makers should evaluate budget reductions carefully for individual programs.

Besides, efficient P&B methods are necessary [10]. Considering the resource intensive P&B process, it is desirable to have a high degree of automation to avoid unnecessary consumption of resources. Yet, it has to be anticipated that decision-makers want to keep control and may not accept fully automated solutions [13]. In the following sections, we are investigating how our fuzzy linear approach copes with the above mentioned issues and financial stress situations. For that purpose our previous P&B model, which we presented in [8] is described in Sect. 2 using a case study. Section 3 examines different ways how our P&B model can adapt to financial stress. First, the previous model's limits are determined and fuzzy stress constraints are introduced as an extension to handle high budget cuts. In addition results of the enhanced approach are provided. Section 4 evaluates the presented approach in terms of its benefits and possible issues, and the approach is compared to other budgeting techniques. Ongoing and future work is presented in Sect. 5. Summary and concluding remarks are provided in Sect. 6.

2 FULPAL (Fuzzy Linear Programming Based on Aspiration Levels): Previous Work and Case Study

This section highlights P&B using our existing FULPAL (FUzzy Linear Programming based on Aspiration Levels) approach, see [8]. It is based on the idea that fuzzy sets can help to manage uncertainties and to create more realistic P&B models, by incorporating uncertain conditions. In [8] fuzzy constraint borders were used. It was proposed to describe the fuzzy parameters by membership functions μA of the following type: $X \rightarrow [0,1]$ which assign a membership value to each element of a set. Piecewise linear membership functions are recommended [8] in order to reduce the modelling effort. A discussion of different types of membership functions can be found in [11].

An example of a budget for the sum of all support expenses, represented by a fuzzy parameter, is illustrated in Fig. 1a. The corresponding membership function reflects the planner's assessment that a sum of support expenses between € 200,000 and € 210,000 is perfectly justifiable. The sum must neither fall below € 180,000 nor exceed € 230,000. Despite some issues, the boundaries of € 180,000 and € 230,000 would be still defensible.

The approach we use also takes into consideration that the human decision-makers may not be able to consider and evaluate the entire solution space prior to the P&B process. Our approach is based on the theory of aspiration adaption and interactive planning iterations for managing this issue. It allows the user to enter fuzzy parameters for the constraint borders. The parameters are summarized in Table 1. An example for modeling $\sim; \leq$-constraint[1] borders is illustrated in Fig. 1b, where the upper bound b_i is fully acceptable and the assigned membership value is 1.0. However, $(b_i + d_i)$ must not be exceeded by all means, where the assigned membership value is 0.0 or an ε-value which is very close to 0.0.

FULPAL computes the minimal and maximal values for the goal once all coefficients and constraint borders are entered. Next the decision-maker defines aspiration levels concerning the restrictions and the goal. In case he does not want to set individ-

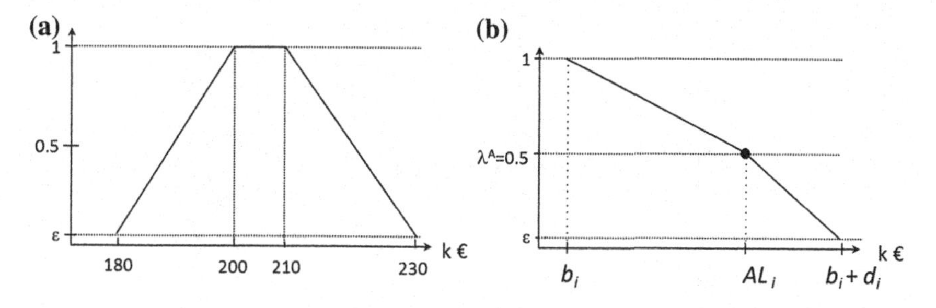

Fig. 1 **a** A planner's perspective of a budget with fuzzy parameters [8]. **b** An example for modeling $\sim; \leq$ inequations with fuzzy constraint borders [8]

[1] $\tilde{\leq}$ denotes fuzzy less or equal.

Table 1 List of FULPAL parameters used and their meaning

Parameter	Meaning
b_i	acceptable upper bound for fuzzy constraint i; assigned membership value is 1.0
$(b_i + d_i)$	upper bound for fuzzy constraint i which must not be exceeded; assigned membership value is 0.0 or an ε-value which is very close to 0.0
AL_i	aspiration level for fuzzy constraint i; assigned membership value is 0.5
λ^A	assigned membership value for the aspiration levels; equals 0.5
λ^*	achieved result: maximized minimal degree of satisfaction concerning all constraints (membership values of the results for constraints) and the goal (membership value of the achieved goal)

ual aspiration levels, he can use default values generated by FULPAL. The process of generating aspiration levels is based on ideas presented in [11]. The aspiration levels are assigned to the membership value $\lambda^A = 0.5$. Internally, the software approximates piecewise linear membership functions, computes the optimal degree of satisfaction λ^* and determines the unknowns. λ^* represents the maximized minimal degree of satisfaction concerning all constraints and the goal. In order to compute λ^*, the membership values of the achieved results are used, namely the membership values of the results for constraints and the membership value of the achieved goal. Details of this process can be found in [9, 11]. If $\lambda^* > \lambda^A$, then a solution has been found that satisfies the aspiration levels of the decision-maker. Therefore, he can now either take the solution, or further adapt some of his aspiration levels as illustrated in Fig. 2.

If $\lambda^* < \lambda^A$ then no acceptable solution can be found and the decision-maker may modify at least one aspiration level. The solution space can be explored iteratively, until an acceptable solution is found.

The following example is based on a real world budgeting problem of the NIT. Due to confidentiality reasons we changed the budgets and the related categories slightly. Assume an academic department of the NIT has to supply an 'official' budget plan to the management for the next academic year. This comprises distributing a

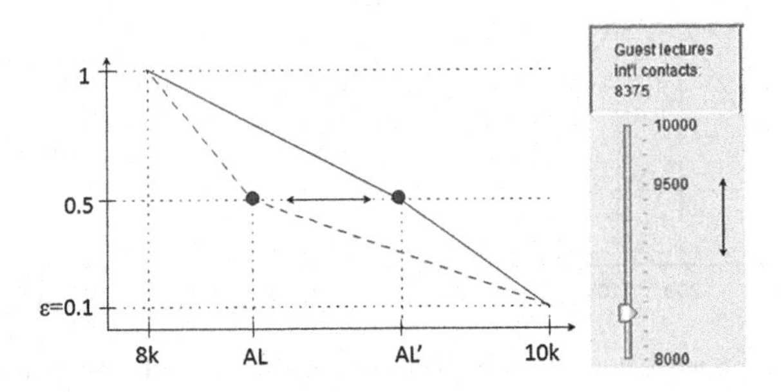

Fig. 2 Adjusting the aspiration level (AL) in order to explore the solution space iteratively by using the FULPAL GUI [8]

certain amount to sub-budgets. If discrepancies become apparent during the planning period, then resources need to be reallocated, while keeping the total budget constant. Further we assume that the whole budget of € 300,000 needs to be spent. Due to the uncertain environment and the long planning horizon, many planning parameters can change after the planning process is finished and the budgets have already been assigned. For instance, it is almost impossible to determine the exact amount of money needed for excursions and trainings at the beginning of the planning period. For a full discussion of the example we refer to [8]. Here we just highlight some constraints in a little more detail. In Table 2, the constraints are clearly described. As already mentioned, b_i is definitely within the limit, but $(b_i + d_i)$ should certainly not be exceeded. For example regarding 'Staff expenses (temporary)', C2a represents the minimal assignment to the staff expenses because existing contracts have to be fulfilled, hence $x1 \geq 165{,}000$. It is a so-called crisp constraint. Additionally, C2b makes sure that expenses up to $b_2 =$ € 170,000 are definitely acceptable, but amounts greater than $(b_2 + d_2) =$ € 180,000 are certainly not tolerable. Thus, for C2b we get $x1 \tilde{\leq} [170, 000; 180, 000]$. As displayed in Fig. 1b, the set of valid expense levels is represented very well by b_i. So, a membership value of 1.0 is assigned. On the other hand $(b_i + d_i)$ characterizes the set of valid expense levels poorly. So, the membership value is 0.0 or ε, which means very close to 0.0. The constraints C3a to C11a and C3b to C11b can be interpreted similarly. After careful consideration the decision-maker fixed the weights w_i in Table 2 to express the categories' importance for the university. The weights are used as coefficients within an objective function and the goal of the decision-maker is to maximize this objective function:

$$z(x_1, x_2, \ldots, x_{10}) = 0.6x_1 + 0.7x_2 + 0.9x_3 + 0.6x_4 + 0.9x_5 + 0.8x_6 + 0.7x_7 + 0.8x_8 + 0.8x_9 + 0.7x_{10} \quad (1)$$

Table 2 Model constraints (in k €) [8]

Constraint No.	Coefficients	Constraints	CXa	CXb		
			c_i	b_i	$b_i + d_i$	w_i
C2a/C2b	x_1	Staff expenses (temporary)	165	170	180	0.6
C3a/C3b	x_2	Assistants, student assistants	20	30	34	0.7
C4a/C4b	x_3	Mentoring program	4.5	6	7	0.9
C5a/C5b	x_4	Contracts for work labour	5	8	9	0.6
C6a/C6b	x_5	Guest lectures and international contacts	5	8	10	0.9
C7a/C7b	x_6	Excursions and trainings	5	10	15	0.8
C8a/C8b	x_7	IT-projects	7	9	12	0.7
C9a/C9c	x_8	Labs (hard- and software)	8	9	10	0.8
C10a/C10b	x_9	Library and online lecture notes	25	30	35	0.8
C11a/C11b	x_{10}	Equipment special labs	23	28	33	0.7
C1	x_1 to x_{10}	Sum of all expenses		300		

C1 is a crisp equals-constraint and makes sure that all money is spent

Table 3 Experimental results (amounts in k €) [8]

Iteration	λ^*	$x1$	$x2$	$x3$	$x4$	$x5$	$x6$	$x7$	$x8$	$x9$	$x10$
1	0.667	165.00	30.00	6.50	5.00	9.00	12.50	7.00	9.50	32.50	23.00
2	0.433	165.00	33.40	6.85	5.00	8.00	10.00	8.90	9.85	30.00	23.00
3	0.528	165.00	32.83	6.71	5.00	9.42	10.00	8.33	9.71	30.00	23.00

Solving the fuzzy linear programming system leads to the results shown in the first row of Table 3, denoted as iteration 1.

For iteration 1 we retrieved the result $\lambda^* = 0.667$ which is greater than $\lambda^A = 0.5$. The surplus of his aspiration levels encourages the decision-maker to start a new iteration with higher claims. Thus, some aspiration levels were raised in iteration 2. The corresponding solution provided a λ^* less than 0.5, which means no viable solution could be found. Then, an aspiration level was relaxed, and finally a suitable solution was found in iteration 3.

A shortcoming of the suggested approach is that it assumes that the available budget is stable over the planning period. However, there may be situations of financial stress, where budgets are slashed during the planning period, for example the elimination of student fees. In such a scenario the results of the planning period would not be consistent with the optimized result of the FULPAL model which was determined before; in fact it may be more useful to reassign the scarce resources and to minimize the negative impacts of such a situation. Section 3 discusses extensions of the model that are able to minimize the impact of financial stress.

3 Incorporating Financial Stress Awareness in FULPAL Using Fuzzy Financial Stress Constraints

In a series of experiments we allow to adjust model parameters due to changes during the planning period. We will analyze the impacts for decision-makers who are in charge of planning and budgeting. The experiments are based on the case study presented in Sect. 2.

3.1 Spending Freeze to Determine the Model Limits

For the first iteration under stress, denoted as iteration 4, we assume that the government declares a spending freeze at the beginning of the fiscal year. 10 % of the initially available amount for all expense types cannot be spent. One 'traditional' approach to manage such an issue would be an 'a-t-b' budget cut. However, cutting all budgets by 10 % can violate CXa-type constraints. Additionally, it is quite likely

Table 4 Experimental results (amounts in k €)

Iteration	λ^*	$x1$	$x2$	$x3$	$x4$	$x5$	$x6$	$x7$	$x8$	$x9$	$x10$
3	0.528	165.00	32.83	6.71	5.00	9.42	10.00	8.33	9.71	30.00	23.00
4	< 0.5	No acceptable solution found									
5	0.778	165.00	20.00	6.33	5.00	5.33	5.00	7.00	8.33	25.00	23.00

that necessities in terms of the management's goals and the academic department would not be addressed either. Using the FULPAL approach we simply change constraint C1 and assign a general budget of € 270,000. For this purpose, we change the constraint border of C1 and leave all aspiration levels untouched. As result of iteration 4 no solution with λ^* greater than 0.5 can be found. In the following iteration 5 constraint 6b is tightened, because guest lectures and international contacts are not needed to run the department's services, invitations can be reduced and expenses can be cut for a short period without too many negative impacts. b_6 is reduced to € 5,000 and the corresponding aspiration level is reduced from € 9,500 to € 5,750. Constraint 6b represents the upper limit for this expense category.

Table 4 compares the results of iteration 5 with the last satisfying iteration, which fulfilled the decision-maker's aspiration levels, before the budget was cut (iteration 3). It is obvious that the budget cut is not distributed proportionally and that the weights of the goals are considered except the category 'guest lectures/international contacts' and categories with lower constraints determined by c_6 which cannot be reduced anymore.

In iteration 6 we add more stress to the system by reducing the budget to € 267,500 which is the minimum amount of money needed to avoid violations of the CXa constraints. The result of this iteration is not really surprising. The only solution which can be found is the necessary minimum amount of money for all categories.

3.2 Fuzzy Stress Constraints to Handle Limit Violations

In the next step (iteration 7) we reduce the available sum for all categories further. We assume that the government decides to abolish study fees and that the available budget drops down to € 250,000. Therefore, € 250,000 is assigned for the sum of all expenses (C1). The current FULPAL model does not provide any solution, since the minimum assignment constraints (CXa) cannot be fulfilled anymore. To address this issue modifications of the model become necessary. Compared with the available budget in iteration 1 the system has to digest a reduction of roughly 16.67 % for the total available sum of funds. One way to handle this could be an a-t-b budget cut of 16.67 %. Nevertheless, such a budget cut for all expense categories would not result in a reasonable budget reassignment because certain minimal budgets for some categories, like temporary staff expenses, are needed to maintain the university department's business and to fulfil legal requirements. Additionally, it is recommended to modify our model in a way that some flexibility in terms of a goal-oriented assignment of resources is still possible and an interactive budgeting considering the decision-maker's preferences is maintained.

Instead of reducing all initial lower bounds for the expenses (see c_i-values, Table 2) 16.67 %, we propose to cut the lower bounds by 33.33 %. This way it is possible to cut expenses for certain categories, which can be temporarily reduced, more in favour of other important categories which should not be affected too much. This general approach saves effort and helps to maintain flexibility. Depending on the results of the model the percentage can be varied, of course.

Furthermore, it is assumed that the stress situation for the system does not allow too much flexibility in terms of spending amounts which are close to the old upper limits $(b_i + d_i)$ which represent the set of valid expense levels poorly. So, the idea is to substitute the greater equal (CXa) and the lower equal (CXb) constraints by equal constraints with reduced upper limits compared with the initial situation. The intention is to maintain as much flexibility as possible by means of this new type of constraints which we call 'fuzzy stress constraints'. Deviations from an interval of the appropriate level of expenses can decrease the total level of satisfaction. The model has to cover that for some expense types a minimum amount of money is needed to run the department's services and/or some money has already been spent. On the other hand the budget cut has to be respected on the aggregated level.

For iteration 8 the model is modified according to Table 5.

$\tilde{}$; bs denotes a fuzzy set of the so-called L-R-type [11] which is depicted in Fig. 1a. It represents the fuzzy border for the fuzzy stress constraint by means of a piecewise linear membership function, see Fig. 3. $\underline{bs}_i^{\varepsilon}$ and $\overline{bs}_i^{\varepsilon}$ are the lower and

Table 5 Modification of the FULPAL P&B model (amounts in k €)

Constraint No.	Coefficients	Constraints	$\underline{bs}_i^{\varepsilon}$	$\underline{bs}_i^{1}$	$\overline{bs}_i^{1}$	$\overline{bs}_i^{\varepsilon}$	w_i
C1	x_1 to x_{10}	Sum of all expenses	250	250	250	250	
C2	x_1	Staff expenses (temporary)	110	137.50	154.21	170	0.6
C3	x_2	Assistants, student assistants	13.33	16.67	18. 69	30	0.7
C4	x_3	Mentoring program	3	3.75	4.21	6	0.9
C5	x_4	Contracts for work labour	3.33	4.17	4. 67	8	0.6
C6	x_5	Guest lectures and international contacts	3.33	4.17	4.67	8	0.9
C7	x_6	Excursions and trainings	3.33	4.17	4.67	10	0.8
C8	x_7	IT-projects	4.67	5.83	6.54	9	0.7
C9	x_8	Labs (hard- and software)	5.33	6.67	7.48	9	0.8
C10	x_9	Library and online lecture notes	16.67	20.83	23.37	30	0.8
C11	x_{10}	Equipment special labs	15.33	19.17	21.50	28	0.7

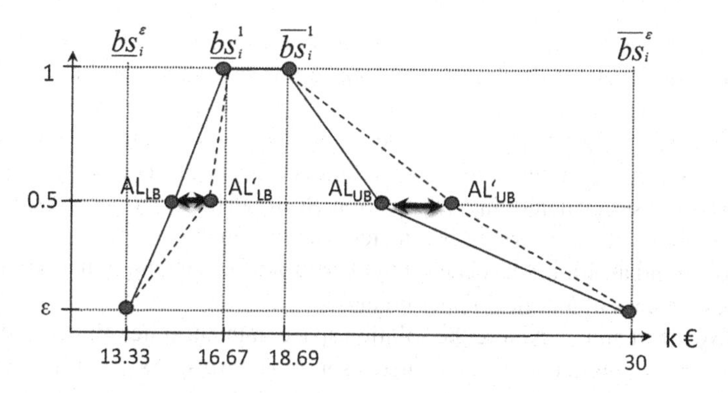

Fig. 3 Constraint border of the fuzzy stress constraint C3

upper borders of the fuzzy set which describes the appropriate budget for expense type i. Similar to $(b_i + d_i)$ in the old model, $\underline{bs}_i^{\varepsilon}$ and $\overline{bs}_i^{\varepsilon}$ represent the set of valid expense levels poorly. The assigned membership values of $\underline{bs}_i^{\varepsilon}$ and $\overline{bs}_i^{\varepsilon}$ are ε which is very close to 0.0 if a decision-maker is not able to provide a value for the zero level of the membership function. The interval $[\underline{bs}_i^1, \overline{bs}_i^1]$ includes values representing the appropriate amount of expenses very well, and the assigned membership values equal 1.0. So, we get a set of fuzzy equal constraints which replaces the old CXa and CXb constraints:

$$x_i^{\sim}\ ; =^{\sim}; bs_i \tag{2}$$

The $^{\sim}; bs_i$ -parameters for our next iteration are listed in Table 5. For the constraints C2 to C11 we suggest the following parameters: As explained at the beginning of this section we intend to maintain flexibility and choose 2/3 of the initial minimum levels (see c_i-values, Table 2) for the $\underline{bs}_i^{\varepsilon}$-values. The $\underline{bs}_i^1$-values represent the level of expenses which would be the result of a 16.67 % budget cut of the initial minimal expense levels (see c_i-values, Table 2). It has to be checked, if budget cut of 16.67 % is not too much for single categories considering the actual amount of expenses or certain constraints. In this case the parameters for the concerning categories have to be adapted as an exception. Cases like this become more likely if the regarded period progresses. In our case all $\underline{bs}_i^1$-values are the result of a 16.67 % budget cut of the initial minimal expense levels. It is assumed that the decision-maker would also accept higher levels of expenses for individual categories which could also be regarded as an appropriate budget. To determine appropriate $\overline{bs}_i^1$ parameters we assume that the percentage of the last budget cut (€ 250,000/€ 267,500) multiplied by the old c_i values would also describe an appropriate budget very well under these circumstances. Finally, the absolute upper limits $\overline{bs}_i^{\varepsilon}$ have to be fixed. Due to the critical financial situation for all expense categories, we take the acceptable upper bounds of iteration 6 (see Table 2, b_i-values), where the membership value was 1.0, as absolute limits $\overline{bs}_i^{\varepsilon}$ with a membership value of 0.0 or ε.

Figure 3 shows an example of a fuzzy stress constraint border. For each constraint two aspiration levels can be defined. AL_{LB} represents the aspiration level for the lower bound and AL_{UB} for the upper bound. Together with the parameters which were described before, a piecewise linear membership function is approximated. When a solution is computed and a constraint is critical, just one of two aspiration levels of the constraint determines the solution. So, it can be an option for the decision-maker to modify the concerning parameter. The left double arrow illustrates this if the lower bound aspiration level determines the solution, and the right double arrow illustrates the case of a critical upper bound.

To save effort and to explore the solution space in the next iteration the decision-maker chooses automatically generated aspiration levels. As already mentioned before, the generation of the aspiration levels is based on an approach presented in [11]. These automatically chosen aspiration levels are closer to the absolute bounds than to the acceptable bounds, thus increasing the solution space. Being 'generous' in such a way makes it more likely to find a satisfying solution. Depending on the results, the aspiration levels can be tightened in another iteration if λ^* surpluses 0.5.

3.3 *Results of the Fuzzy Stress Approach*

In the first iteration of the new model (iteration 8), we get $\lambda^* = 0.667$ as result. Subsequently, in iteration 9 the decision-maker raises some aspiration levels for the constraints. In the current situation, expenses for assistants or student assistants (C3), the library (C10) and special equipment (C11) can be postponed. Hence, a lower level of expenses would be desirable for this period from the decision-maker's point of view. So, the corresponding aspiration levels concerning the upper bounds are reduced as much as possible, namely to $\overline{bs}_i^1$.

The modified model provides the results shown in Table 6 (iteration 9). $\lambda^* = 0.557$ is greater than 0.5 which means that all aspiration levels are fulfilled and an acceptable solution for the rescheduling problem is found.

If we compare the results with an a-t-b approach which cuts all budgets equally the decision-maker has achieved a much better result. It takes the decision-maker's internal knowledge, the goal weights and organizational/legal requirements into account. For instance, in our case cutting staff expenses is related with serious implications and they cannot be reduced significantly. An atb approach based on the results of iteration 3, see Table 4, would have resulted in € 137,500 (= € 250,000 /€ 300,000 * € 165,000), see Table 6 (line atb), which would not be acceptable. By means of the modified FULPAL approach we get a higher allocation of about € 145,660 which can

Table 6 Experimental results of the modified FUPAL model (amounts in k €)

Iteration	λ^*	$x1$	$x2$	$x3$	$x4$	$x5$	$x6$	$x7$	$x8$	$x9$	$x10$
9	0.557	145.66	18.69	5.40	3.61	6.89	8.22	8.18	8.49	23.37	21.50
a-t-b		137.50	27.36	5.59	4.17	7.85	8.33	6.94	8.09	25.00	19.17

be enforced and is also acceptable. On the other hand less critical or less important categories, like assistants ($x2$), are strained with a relatively high budget cut.

4 Evaluation and Comparison of FULPAL with Traditional Budgeting Methods

By using fuzzy sets to represent planning parameters, the validity of the planning parameters lasts longer compared with other approaches and the quality of the resulting plans and budgets is higher. Considering the resource-intensive P&B process this is beneficial, since a rescheduling of plans and budgets has to be done less often. It was shown that the initial FULPAL model is 'stress resistant' to a certain limit. Traditional P&B approaches require a lot of effort, see [8], for rescheduling if budget cuts occur. In contrast to these approaches, the modified FULPAL P&B model allows an interactive reassignment of resources according to the decision-maker's preferences.

Traditional budgeting methods are described in [8]. To compare FULPAL with these approaches a brief summary of the most popular methods and tools will be given. Usually, the P&B process is executed by using a mixed approach. Strategic targets and guidelines for the planning period are determined top–down in the beginning. Afterwards, different operational objectives are defined for the business units. In the next step, a bottom–up planning is executed. Usually revisions are needed to fix final budgets [8]. In case of financial stress during the fiscal period, for instance because of budget cuts, new more or less resource consuming iterations become necessary.

A common approach is to start a new traditional P&B iteration. This would cause a lot of effort in comparison to the FULPAL approach. Another solution for financial stress can be an a-t-b approach. This minimizes effort, but has several disadvantages from our point of view. An a-t-b approach would be a reasonable response if the present mix of assigned budgets perfectly reflects the desired mix of designated services. This is an unrealistic assumption in many cases [5], and efficient subunits are penalized more, since they do not have inefficient parts to optimize [3]. Additionally, a-t-b cuts do not take 'the needs, production functions, and contributions of different units' into account and may prove too inflexible to small specialized units [4]. As we have shown in the last section, FULPAL avoids these issues. To a certain degree uncertainty is anticipated, and financial stress can be 'digested' without much effort.

In [8] we have also discussed modern software applications including P&B functions to address the above mentioned issues. It was shown that tasks like data input and process support, support for models and methods, as well as integration aspects can be covered quite well [8]. Other IT functions provide analysis support, and scenarios can be simulated. But that kind of support does not allow an explicit management of uncertainty. So, by means of these IT functions the management of financial stress is suboptimal. In contrast to FULPAL, they only support the 'best' deterministic model representation. Like shown before, FULPAL helps to manage budget cuts more flexible and to save effort in such a scenario. Our fuzzy approach demonstrates the ability to embed the decision-maker's knowledge and uncertainties in the P&B process. Thus plans are more robust and the models are much more resistant against future changes of parameters during the remaining planning period. Additionally, the interactivity of the approach helps decision-makers to understand the solution space which may lead to more elaborate results and a higher level of acceptance. But as stress on the system grows, like in iteration 7 of our example, it can become necessary to modify the FULPAL P&B model which is related with effort.

Regarding the advantages and the quality of the results of the presented approach, the effort for a modified fuzzy model, as shown in Sect. 3.3, is reasonable. The effort needed for modeling the membership functions depends on the model's size and can be challenging for bigger systems. Nevertheless, traditional negotiation-based approaches also consume many resources and are usually related with a lot of time-consuming discussions. Besides, even if crisp parameters of traditional approaches would be evaluated carefully the quality of the corresponding results is quite poor, since decision-makers are usually not able to provide such accuracy. To save some effort for the set-up of fuzzy P&B models and to support the parameter determination process, instruments, like enhanced regression analysis approaches, which are discussed in [7], could be used.

5 Ongoing and Future Work

The presented approach is successfully applied to budgeting problems at the NIT. However, there are several ideas for improvements. Ongoing work comprises the 'fuzzyfication' of the weights of budget categories. This will allow a more realistic approach to model uncertainties of the decision-maker on the importance of certain budget categories and thus allows to model statements such as 'the weight for the expenses for equipment of special labs is about 0.7'. It is to be determined how this modification to the approach can deal with financial stress.

For the future work we have planned to develop a hierarchical FULPAL system that enables P&B on higher levels in complex organisations. For example a university is usually divided into faculties and these faculties are divided into several departments or schools. Budgets are usually planned on faculty or higher levels and then passed through to departments who then in turn plan their sub-budgets. A hierarchical FULPAL system would allow the departmental budget planners to define weights and fuzzy constraints on their local departmental levels, and also allow the faculty planner to specify weights and fuzzy constraints on the faculty level. This would enable the decision-maker to optimize the P&B for a faculty as a whole as well as taking local departmental aspirations into account. This, together with stress parameters as proposed in this paper, would create a powerful P&B system for adjusting budgets on a higher level under financial stress.

6 Conclusions

P&B is one of the most challenging tasks in financial management, especially in times of global crisis, increasingly scarce resources and volatile markets combined with an uncertain environment in many areas. In our previous research we have proposed and evaluated a promising fuzzy approach that allows optimizing the P&B process interactively. In this paper we have explored the deployment of the fuzzy approach beyond the initial planning stage and analyzed the impact of financial stress occurring during the fiscal period. It was shown that budget cuts can be managed quite well to a certain degree. In cases of drastic budget cuts financial stress cannot be managed anymore. Therefore, we have developed an extension that handles these scenarios by means of fuzzy stress parameters which are used to re-distribute the budget in order to minimize the negative impacts of financial stress.

The presented case study shows that some effort was necessary to modify the fuzzy model. Although, simple 'a-t-b' budget cuts could save some effort, it is obvious that a unified percentage cut for all expenses would result in non-acceptable and non-enforceable reassignments. Considering this aspect, some effort is not avoidable in case of high budget cuts. Evaluating the results of the modified fuzzy approach, our research shows that it takes the decision-maker's internal knowledge, the management's preferences and other requirements into account. So, a high level of acceptance is achieved and the results are practical. If further reassignments are required, then necessary adjustments can be managed with a relatively low consumption of resources. Once a modified fuzzy model is established an acceleration of rescheduling processes can be reached.

In summary, the presented work shows that financial stress situations can be handled efficiently by means of an interactive fuzzy approach. It is intended to further develop the presented FULPAL model to a hierarchical fuzzy system that enables P&B on different levels in complex organisations.

References

1. Barrett, C.R.: Planning and budgeting for the agile enterprise. A driver-based budgeting toolkit. Elsevier, Amsterdam (2007).
2. Hall, D.T., Mansfield, R.: Organizational and individual response to external stress. Administrative science quarterly 16(4), 533 (1971).
3. Levine, C.H.: More on cutback management: Hard questions for hard times. Public administration review 39(2), 179–183 (1979).
4. Levine, C.H.: Organizational decline And cutback management. Public administration review 38(4), 316–325 (1978).
5. Levine, C.H.: Police management in the 1980s: From decrementalism to strategic thinking. Public administration review 45(Special), 691–700 (1985).
6. Raudla, R., Savi, R., Randma-Liiv, T.: Literature review on cutback management. COCOPS Workpackage 7 Deliverable 1 (2013). http://www.cocops.eu/wp-content/uploads/2013/03/COCOPS_Deliverable_7_1.pdf. Accessed 21 May 2013.

7. Rausch, P., Jehle, B.: Data Supply for Planning and Budgeting Processes under Uncertainty by Means of Regression Analyses. In: Rausch, P., Sheta, A.F., Ayesh, A. (eds.) Business Intelligence and Performance Management. Advanced Information and Knowledge Processing, pp. 163–178. Springer London (2013).
8. Rausch, P., Rommelfanger, H., Stumpf, M., Jehle, B.: Managing Uncertainties in the Field of Planning and Budgeting - An Interactive Fuzzy Approach. In: Bramer, M., Petridis, M. (eds.) Research and Development in Intelligent Systems XXIX, pp. 375–388. Springer (2012).
9. Rausch, P.: HIPROFIT. Ein Konzept zur Unterstützung der hierarchischen Produktionsplanung mittels Fuzzy-Clusteranalysen und unscharfer LP-Tools. Peter Lang Verlag Frankfurt a.M. et al. (1999).
10. Remenyi, D.: Stop IT project failure through risk management. Computer weekly professional series. Butterworth Heinemann, Oxford (1999).
11. Rommelfanger, H.: Fuzzy Decision Support-Systeme. Entscheiden bei Unschärfe, 2nd edn. Springer, Berlin (1994).
12. Shim, J.K., Siegel, J.G.: Budgeting basics and beyond, 3rd edn. Wiley & Sons, Hoboken, N.J (2009).
13. Tate, A.: Intelligible AI Planning - Generating Plans Represented as a Set of Constraints. Artificial Intelligence Applications Institute (2000). http://www.aiai.ed.ac.uk/oplan/documents/2000/00-sges.pdf. Accessed 22 May 2013.

'The First Day of Summer': Parsing Temporal Expressions with Distributed Semantics

Ben Blamey, Tom Crick and Giles Oatley

Abstract Detecting and understanding temporal expressions are key tasks in natural language processing (NLP), and are important for event detection and information retrieval. In the existing approaches, temporal semantics are typically represented as discrete ranges or specific dates, and the task is restricted to text that conforms to this representation. We propose an alternate paradigm: that of *distributed temporal semantics*—where a probability density function models relative probabilities of the various interpretations. We extend SUTime, a state-of-the-art NLP system to incorporate our approach, and build definitions of new and existing temporal expressions. A worked example is used to demonstrate our approach: the estimation of the creation time of photos in online social networks (OSNs), with a brief discussion of how the proposed paradigm relates to the point- and interval-based systems of time. An interactive demonstration, along with source code and datasets, are available online.

1 Introduction

Temporal expressions communicate more than points and intervals on the real axis of unix time—their true meaning is much more complex, intricately linked to our culture, and often difficult to define precisely. Extracting the temporal semantics of text is important in tasks such as event detection [12].

We present a technique for leveraging big-data to capture the *distributed temporal semantics* of various classes of temporal expressions (the term *distributed* began

B. Blamey (✉) · T. Crick · G. Oatley
Cardiff Metropolitan University, Western Avenue, Cardiff CF5 2YB, UK
e-mail: beblamey@cardiffmet.ac.uk

T. Crick
e-mail: tcrick@cardiffmet.ac.uk

G. Oatley
e-mail: goatley@cardiffmet.ac.uk

M. Bramer and M. Petridis (eds.), *Research and Development in Intelligent Systems XXX*,
DOI: 10.1007/978-3-319-02621-3_29,

to appear in the context of automatic thesauri construction during the 1990s [8]). Our approach models the inherent ambiguity of traditional temporal expressions, as well as widening the task to infer semantics from quasi-temporal expressions not previously considered for this task.

In Sect. 2, we discuss how existing work has overlooked the distributed semantics issue, followed by an outline of our approach in Sect. 3. In Sect. 4, we describe a technique for mining a distributed definition from photo metadata downloaded from the Flickr service. Examples are shown in Sect. 5, with a discussion of cultural nuances we find. In Sect. 6 the changes to the SUTime framework are described. Section 7 shows an example use of the system, for determining the creation time of Facebook photos—highlighting how the approach facilitates incorporation of a prior probability. Section 8 is a brief discussion of how the approach relates to the point- and interval-based systems of time used in AI, followed conclusions in Sect. 9, with directions for future work in Sect. 10.

2 Related Work

Research into temporal expressions has generally focused on their detection and grammatical parsing. Traditionally, systems used hand-coded rules to describe a formal grammar. Popular frameworks using such an approach include: HeidelTime [14], GUTime [11], with the more recent SUTime [6], part of the Stanford CoreNLP framework,[1] considered to be a state-of-the-art system, as measured on the TempEval-2 dataset [16].

Consistent with general trends in NLP, more statistical approaches have become popular, where grammars are built through the analysis of large corpora. An example is the development of *grammar of time expressions* [2], to concisely model complex compositional expressions.

Whether hand-coded or machine learnt rules are used, the terminal set of the grammar is generally the months, dates, days of the week, religious festivals, public holidays, usually with an emphasis on the culture of the authors. SUTime can be configured to use the JollyDay[2] library, which contains definitions of important dates for many cultures—but even in this case the definitions are restricted to the model of discrete intervals. This approach is the most natural and simplest approach to the mathematical modelling of time, used throughout natural science. Work such as that of Allen [1] is often cited as a philosophical underpinning of this model.

Such an approach is useful for describing the physical world with mathematical precision, but is a poor means of describing the cultural definition of temporal language. In cases where it is difficult to assign specific date ranges, the advice is to leave alone:

[1] http://nlp.stanford.edu/software/corenlp.shtml

[2] http://jollyday.sourceforge.net/index.html

> Some expressions' meanings are understood in some fuzzy sense by the general population and not limited to specific fields of endeavor. However, the general rule is that no VAL is to be specified if they are culturally or historically defined, because there would be a high degree of disagreement over the exact value of VAL. [7, p. 54]

An advantage of using this restricted, well-defined vocabulary is that it facilitates numerical evaluation of parsing accuracy, and performance can be compared with standard datasets, such as those from the TempEval series [15, 16].

However, this emphasis on grammar has resulted in the research community overlooking the meaning of the terminals themselves. An exciting development is the approach of Brucato et al. [5], who, noting the maturity of tools developed for the traditional tempex task, widen the scope by to include so-called *named temporal expressions*. First, they created a list of NTEs by parsing tables containing temporal expressions in manually selected Wikipedia articles, merging the results with the JollyDay library. This list is used to train a CRF-based detector, which in turn was used to find completely new NTEs, such as sporting events. However, they note the difficulties of learning definitions for the newly-discovered NTEs:

> ...it is difficult to automatically learn or infer the link between "New Year's Day" and 1st January, or the associations between north/south hemisphere and which months fall in summer... [5, p. 6].

They resort to TIMEX3, a traditional, discrete interval representation. In this paper we present an approach to constructing distributed definitions for temporal expressions, to hopefully overcome this issue.

Clearly, looking for temporal patterns in datasets is not a new task, many studies have observed that Twitter activity relating to some topic or event can peak at or near the corresponding real-world activity. Indeed, work closely related to our approach detects topics through periodicity [18], as compared with existing approaches of Probabilistic Latent Semantic Analysis (PLSA), and Latent Dirichlet Allocation (LDA). Our goal here is to demonstrate how such data can be viewed as a distributed *definition* of the expressions, and that this definition can be incorporated into temporal expression software by choosing a suitable probability density function.

3 A Distributional Approach to Defining Temporal Expressions

We pursue a distributional approach for two reasons: firstly, a distributed definition can capture a more detailed cultural meaning. Examples from our study show that these common temporal expressions are often associated with instances outside their official, or historical definition. We find distributions have a range of skewness and variance, some with more complex patterns exhibiting cultural ambiguity; we discuss specific detailed examples in Sect. 5.

Secondly, our approach allows a much larger range of expressions to be considered as temporal expressions. Under the current paradigm, phrases need to be associated with specific intervals or instances in time. Religious festivals and public holidays can be resolved to their official meaning, but this is not possible for expressions where no single such definition exists. Indeed, there are many expressions that have consistent temporal meaning, without any universal official dates. See Sect. 5 for a discussion of examples in this category, such as "Freshers' Week" and "Last Day of School" (Fig. 1).

In our theoretical model, we define time $t \in \mathbb{R}$. A temporal expression S is represented by a function $f(t)$, which is a probability density function for the continuous random variable T_s (Sect. 8 for interpretation of this random variable). For our purposes here, we define a p.d.f. $f(t)$ simply as:

$$P(T_s \geq t_1, T_s \leq t_2) = \int_{t_1}^{t_2} f(t)dt \quad (1)$$

it follows that:

$$\int_{-\infty}^{\infty} f(t)\, dt = 1 \quad (2)$$

and

$$f(t) \geq 0 \, \forall t \quad (3)$$

In practice, we work over a smaller, finite date range, suitable for the context. For this paper, we consider a single 'generic year', and focus on handling temporal expressions with date-level granularity.

4 Mining the Definitions

Photographs uploaded to the photo-sharing site Flickr,[3] used in numerous other studies, have been used as the basis for our definitions. The Flickr API was used to search for all photos relating to each term uploaded in the year 2012.[4] Metadata was retrieved for each matching photo,[5] the 'taken' attribute of the 'dates' element is the photo creation timestamp (Flickr extracts this from the EXIF [3] metadata, if it exists).

Our aim for using an online social network as a data source was to build culturally accurate definitions. A photo-sharing service was used because the semantics of the photo metadata would be more closely associated with the timestamp of the photo than would be the case for a status message. Tweets such as "getting fit for the summer", "excited about the summer", "miss the summer", etc, do not reveal a specific definition of the word or phrase in question. Conversely, a photo labelled "Summer", "Graduation" or such like, indicates a clear association between the term

[3] http://www.flickr.com

[4] Using the endpoint described at: http://www.flickr.com/services/api/flickr.photos.search.html.

[5] Using the endpoint described at: http://www.flickr.com/services/api/flickr.photos.getInfo.html.

and the time the photo was taken. Measuring this association on a large scale yields a distributed definition—literally a statistical model of how society defines the term.

For our initial system, we consider only temporal expressions for which we expect the pattern to repeat on an annual basis. To some extent, this obviates some of the error in the photo timestamps, inevitably originating from inaccurate camera clocks, and timezone issues. A similar approach should be suitable for creating definitions at other scales of time. A range of phrases were chosen for the creation of a distributed definition, some that are commonly used for such purposes, and other more novel examples taken from Facebook photo album titles (Sect. 5).

Having collected a list of timestamps for each term, we needed to find a probability density function to provide a convenient representation, and smooth the data appropriately. An added complication is that mapping time into the interval of a single year creates an issue when trying to fit, say, a normal distribution to the data. The concentration of probability density may lie very close to one end of the interval (e.g. "New Year's Eve"), which means we cannot neglect contributions from peaks of probability density that lie in neighbouring years in such cases.

Generally, the timestamps were arranged in distinct clusters, so we computed frequencies for 24-h intervals, and then attempted to fit mixture models to the data, using the expectation-maximization process, implemented in the Accord.NET scientific computing framework [13]. Initial attempts used a mixture of von Moses distributions, a close approximation to the *wrapped normal distribution*, the result of wrapping the normal distribution around the unit circle. We had difficulties reaching a satisfactory fit with this model, so instead we used a mixture of normal distributions, adapted to work under modulo arithmetic (using the so-called *mean of circular quantities*). Hence, the probability density greater than ± 6 months away from the mean is neglected, for each normal distribution in the mixture. With standard deviations typically in the region of a few days, this is reasonable.

After mixed results using k-means clustering to initialize the model, we settled on a uniform arrangement of normal distributions. A uniform distribution was also included to model the background activity level—without this, because the normal distributions had standard deviations of just a few days, fitting was disrupted by the presence of many outliers.

After fitting, normal distributions with a mixing coefficient of less than 0.001 are pruned from the model (as our primary goal is to generate a terse semantic representation). As discussed in Sect. 10, the inclusion of asymmetric distributions in the mixture, to better model the some of the distributions in the data is an obvious avenue for future investigation.

5 Discussion of Fitted Distributions

For **Bonfire Night** (Fig. 2) (5th November, United Kingdom) the primary concentration is near the primary date, but with more variance than is with the case with **April Fools' Day**. A number of other distributions in the fit have between 1–2 % mixing

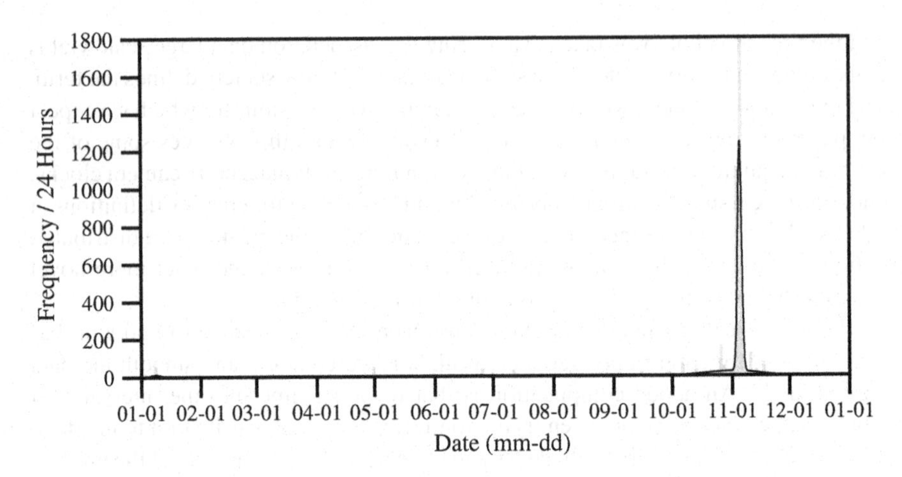

Fig. 1 Distribution of "Bonfire Night"

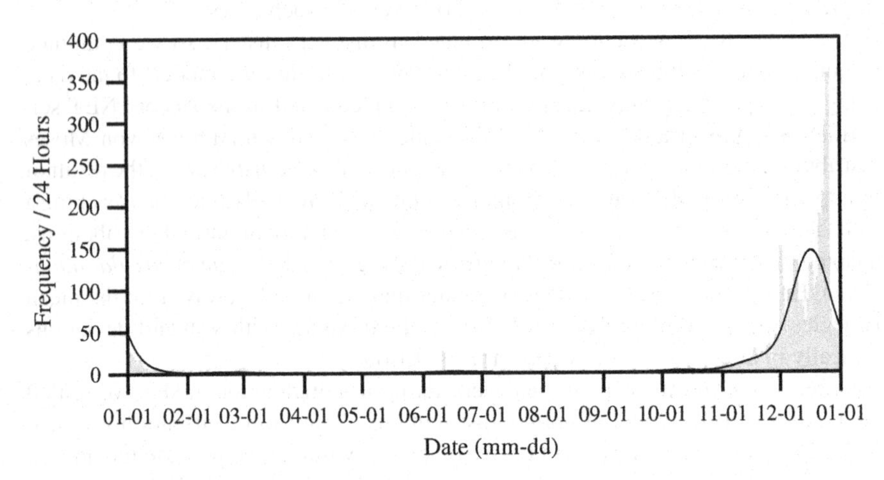

Fig. 2 Distribution of "Christmas"

coefficient, with means at 8th January (possibly relating to the solemnity of John the Baptist on 16th January), 26th June (Midsummer's Eve, 23rd June is popular for bonfires in Ireland), and 2nd May (Bonfires are popular in Slavic Europe on 1st May).

Christmas (Fig. 3) (commonly 25th December) starts early, with 10 % of the probability density contributed by a distribution with a mean of 20th November. In cultures using the Julian calendar, Christmas is celebrated on 7th January and 19th January, perhaps explaining some of the probability density we see in January. In the case of **New Years' Eve**, more than 92 % of the probability mass is centred around 31st of December. We find a normal with mean of January 15th, possibly relating to the Chinese New Year on (23rd January, 2012).

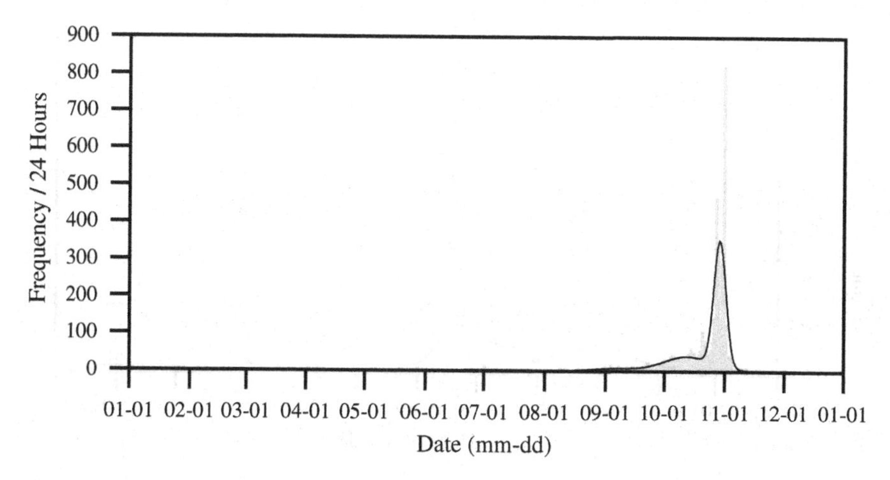

Fig. 3 Distribution of "Halloween"

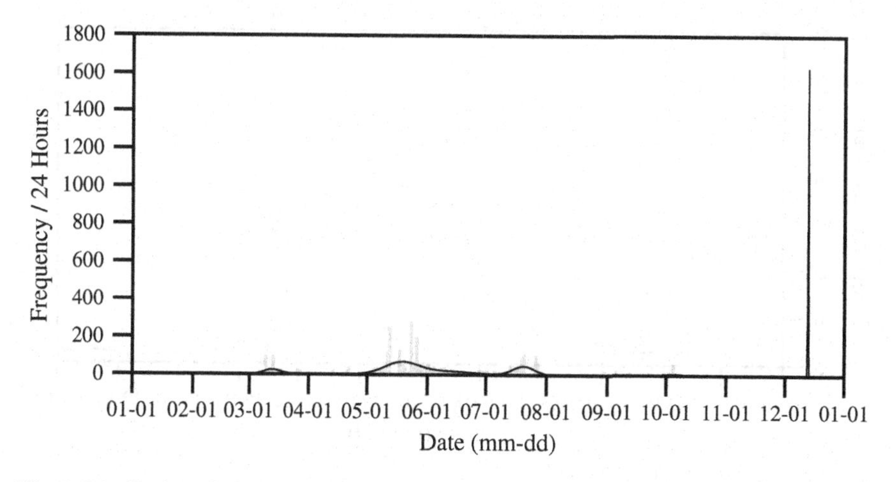

Fig. 4 Distribution of "Graduation"

The data for **Halloween** (Fig. 4) clearly has a skewed distribution about its official data, 31st October—we see much more activity in the preceding weeks, with activity rapidly dropping off afterwards. A similar distribution is exhibited in the case of **Valentines' Day** on (14th February). A limitation of our work is that we did not include asymmetric distributions in our mixture model; such distributions are fitted to a cluster of normal distributions with appropriately decaying mixture coefficients.

Freshers' Week is a term used predominantly in the UK to describe undergraduate initiation at university, usually in September or October. With the obvious differences between educational calenders between regions and institutions, a complex pattern is unsurprising. In the case of **Last Day of School** (Fig. 1), assuming a precise date range in the general case is clearly impossible. Many universities have multiple

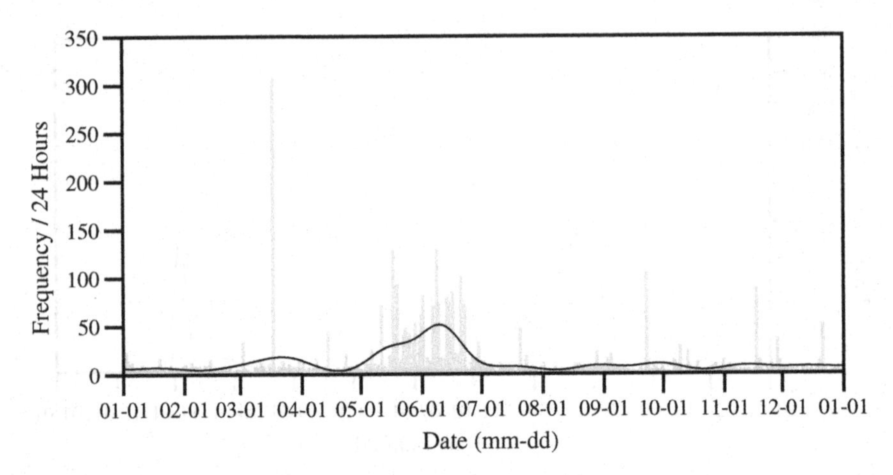

Fig. 5 Distribution of "Last Day of School"

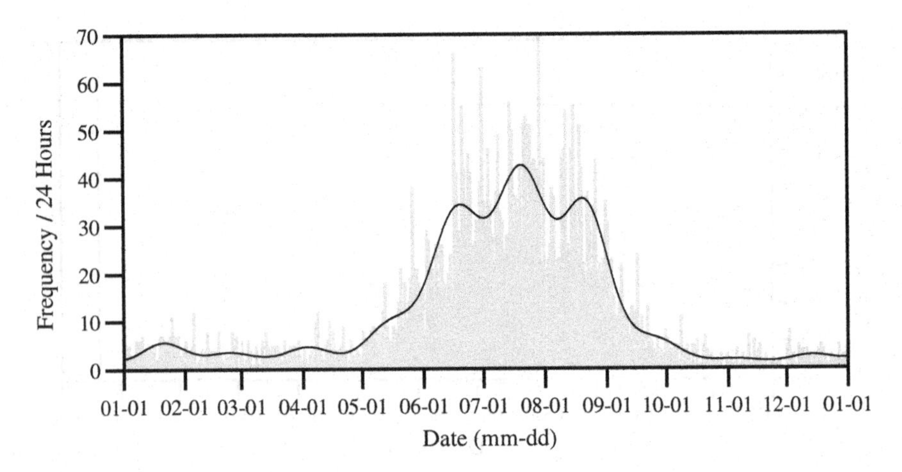

Fig. 6 Distribution of "Summer"

Graduation (Fig. 5) ceremonies a year, with loose conventions on dates, reflected in the clustering of the data.

Definitions of seasons show a significant bias toward northern hemisphere definitions, to be expected with the bias towards English language. However, we do see density at the antipodal dates in each case, modelled by normal distributions of appropriate means. For example, **Winter** has a distribution with a mean of 13th July, with a mix coefficient of 1 %, with a similar phenomenon in the other samples. It is clear from the distribution of the data that all season terms we studied are used year-round. The data for **Summer** is shown in Fig. 6, exhibiting a similar antipodal peak.

Some degree of background "noise" was present in many of the examples: the mixing coefficient was typically in the region of 2 %.

6 Modifications to SUTime

When modifying the SUTime framework [6], our aim was to preserve the existing functionality, as well as implement our distributed approach. A number of Java classes are used to represent the parsed temporal information, our key modification was to augment these classes so that they stored a representation of a probability distribution alongside their other fields. Modifications were then made to the grammar definitions to ensure that instances of these classes were associated with the appropriate probability distributions upon creation, and updated appropriately during grammatical composition.

In more detail, we begin with the core temporal classes. Where appropriate, we added a class field which could optionally hold an object representing the associated probability distribution. Effectively, this object is a tree whose nodes are instances of various new classes: AnnualNormalDistribution, AnnualUniformDistribution, (as the leaves of the tree), and those representing either a Sum or Intersection (i.e. multiplication) as the internal nodes. When no distributed definition was available (e.g. when parsing "2012"), a representation of the appropriate discrete interval is used as a leaf.[6] These classes implemented a method to return an expression string suitable for use in gnuplot (visible in the online demo[7])—with the two internal nodes algebraically composing the expressions returned by their children in the obvious way. Generation of an alternative syntax, or support for numerical integration could be implemented as additional methods. We also introduced a new temporal class, to represent a temporal expression which does not have a non-distributed definition (such as "Last Day of School"), for which composition is possible under the distributed paradigm, but which uses a dummy implementation under the traditional paradigm.

Secondly, it was necessary to make a number of changes to the grammar definitions—these files control how instances of temporal classes are created from the input text, and also how the instances of these classes are combined and manipulated based on the underlying text. After fitting the mixture models to the Flickr data (Sect. 4), definitions were generated in the syntax used by SUTime. Rules defining the initial detection of these expressions were updated so that the probability distributions were included, and modified to allow misspellings and repeated characters that we found in Facebook photo album names (common to online social networks [4]). We introduced rules to detect our new temporal expressions, and assign their distributed definitions. Other modifications were made to adapt the grammar to our domain of photo album names, relating to British English date conventions, and to

[6] $tm\_year(x)$ and related gnuplot functions were useful for this [17, p. 27].

[7] http://benblamey.name/tempex

support temporal expressions of the form 'YY. The rules for temporal composition were largely unchanged, as they are expressed in terms of the temporal operators defined separately.

SUTime defines 17 algebraic operators for temporal instances (e.g. THIS, NEXT, UNION, INTERSECT, IN). Facebook photo album names tended to contain mostly absolute temporal expressions (none of the form "2 months", or "next week"), and it was only necessary to modify the INTERSECT operator. In the distributed paradigm, intersecting two temporal expressions such as "Xmas" and "2012" is simply a case of multiplying their respective probability density functions. The existing implementation of the operator is unaffected. Adaptation of 'discrete' operators such as PREV and NEXT into the distributed paradigm presents an interesting problem, and is left for future work. All that remained was to include an expression for the final probability density functions in the TimeML output.[8]

7 A Worked Example

In tasks such as event detection, it is useful to know the time that a photograph was taken. In Facebook, the EXIF metadata is removed for privacy-related reasons [9], and the API does not publish the photo creation time (as is the case with Flickr). In Facebook, photo album titles tend to be rich in temporal expressions, and an album title such as "Halloweeeeennnn!" should indicate the date the photo was originally taken, even if it wasn't uploaded to Facebook until later. The usual technique would be to parse the temporal expression and resolve it to its 'official' meaning—in this case, October 31st.

In Fig. 4, we see that some of the probability density for "Halloween" actually lies before this date—peaking around the 29th (although the effect is greater with "Christmas", Fig. 3). Having represented the temporal expression as a probability density function, we can combine it with a prior probability distribution, computed as follows. The photo metadata collected from Flickr (Sect. 4), contains an upload timestamp[9] in addition to the photo creation timestamp. We define the *upload delay* to be the time difference between the user taking the photo, and uploading it to the web. Figure 7 shows the distribution (tabulated into frequencies for 24-h bins), plotted with a log-log scale. Taking y as the frequency, and x as the upload delay in seconds, the line of best fit was computed (with gnuplot's implementation of the Levenberg-Marquardt algorithm) as:

$$log(y) = a\ log(x) + b \tag{4}$$

with:

---

[8] We introduced an 'X-GNUPlot-Function' attribute on the TIMEX3 element for this purpose.

[9] The time when the photo was uploaded to the web, shown as the 'posted' attribute of the 'dates' element, see: http://www.flickr.com/services/api/flickr.photos.getInfo.html.

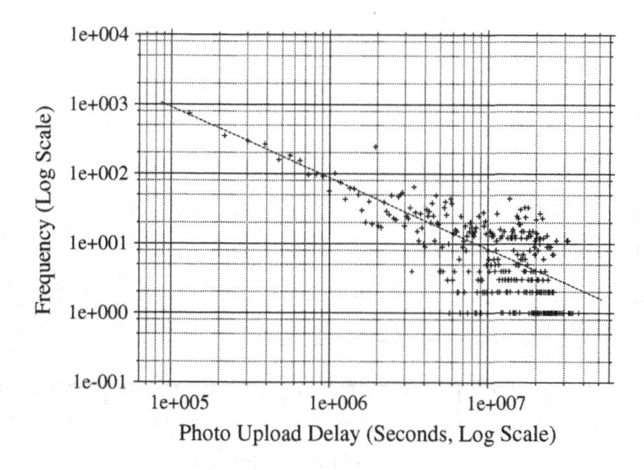

Fig. 7 Distribution of the *Upload Delay*, estimated from Flickr photo metadata

$$a = -1.0204 \tag{5}$$

$$b = 18.5702 \tag{6}$$

We can then use this equation as a prior distribution for the *creation* timestamp of the photo, by working backwards from the *upload* timestamp, which is available from Facebook.[10] Figure 8 shows this prior probability, the distribution for "Halloweeeeennnn!", and the prior distribution obtained by multiplying them together, respectively scaled for clarity. The resulting posterior distribution has much greater variance than what would have resulted from simply parsing the official definition of October 31st, accounting for events being held on the surrounding days, whilst the application of the prior probability has resulted in a cut-off and a much thinner tail for earlier in the month.

8 Interpreting $S \sim T_s(t)$

In Sect. 3, we discussed the association (denoted by $\sim$) between the temporal expression S and the continuous random variable $T_s(t)$. Detailed discussion is beyond the scope of this work, but we briefly outline a few interpretations:

1. S represents some unknown instant: $S \sim t_s$. $T_s(t)$ models $P(t = t_s)$.
2. S represents some unknown interval: $S \sim I_s = [t_a, t_b]$. $T_s(t)$ models $P(t \in I_s)$.
3. The meaning of S is precisely $S \sim T_s(t)$, and only by combining T_s with additional information can anything further be inferred. Particular instances or time intervals may have cultural or historical associations with S, it may be possible to recognize their effect on T_s. But T_s itself is the pragmatic interpretation of S.

[10] See the 'created_time' field at: https://developers.facebook.com/docs/reference/api/photo/.

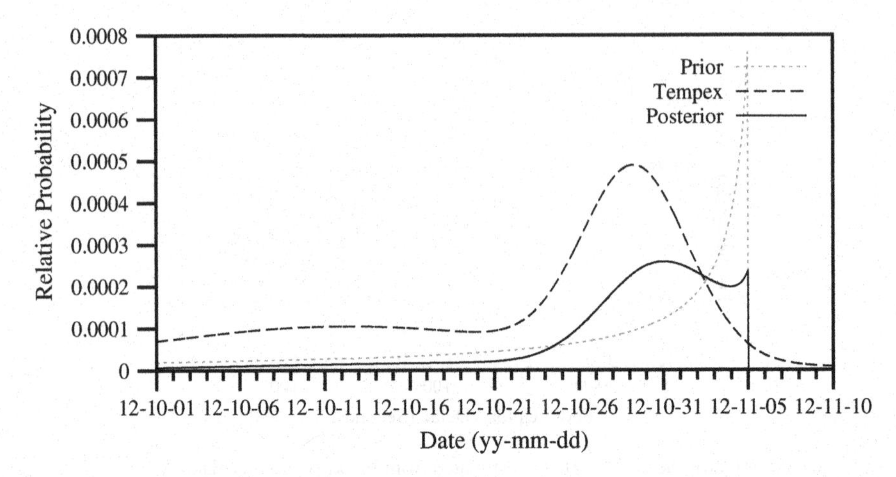

Fig. 8 Computation of the posterior probability distribution for the creation time of the photo, from the prior probability, and the distribution associated the temporal expression "Halloweeeeennnn!"

There is extensive discussion relating to the models underlying (1) and (2) in the literature, and one can construct various thought experiments to create paradoxes in either paradigm. By constructing our probability distribution by modelling time as $\mathbb{R}$ which means we are undeniably using the classical point-based model of time, rather than the interval-based model [1]. However, the situation is a little more subtle: employing a probability density function only allows computation of the probability associated with an arbitrary *interval* (see Eq. 1). For a continuous random variable, the probability of any particular instance is zero by definition; something which is arguably more akin to an interval-based interpretation of time. So, associating the temporal expression with a p.d.f. means that the theoretical basis of the point-based system of time is retained, whilst the mathematics restricts us to working only with intervals. Whether this dual-nature obviates the dividing instant problem [10], requires a more rigorous argument, is something we leave to future work.

The thrust of our contribution is to suggest that temporal expressions in isolation are intrinsically ambiguous[11] (interpretation 3). We argue that such expressions cannot be resolved to discrete intervals or instants (without loss of information), and attempts to do so are perhaps unnecessary or misguided. In some cases, it may be desirable to defer resolution, perhaps to apply a prior probability (as in Sect. 7).

[11] The "weekend", and precisely when it starts, is a good example of this. Readers will be able to imagine many different possible interpretations of the word.

9 Conclusions

In Sect. 2 we have discussed how existing work focuses grammar and composition. Recent work to widen the task [5] requires methods (such as ours) for assigning meaning to these expressions. We've noted how the usual approach of representing meaning as discrete intervals limits the scope of the temporal expression task.

Our main contribution is a proposal for an alternative *distributed* paradigm for parsing temporal expressions (Sect. 3). The approach has several advantages:

- It is able to provide definitions for a wider class of temporal expressions, supporting expressions where there is no single official definition.
- It captures greater cultural richness and ambiguity—arguably a more accurate definition.
- It facilitates further processing, such as the consideration of a prior probability, as demonstrated with an example in Sect. 7.

Secondly, we have demonstrated a technique for mining definitions from a large dataset, and statistically modelling the results to create a distributed definition of a temporal expression (Sect. 4).

Thirdly, we have adapted a state-of-the-art temporal expression software framework to incorporate the distributed paradigm, allowing some of the temporal algebraic operators to be implemented as algebraic operators (Sect. 6). An online demonstration, datasets and source code, and figures omitted for brevity are available at http://benblamey.name/tempex.

10 Future Work

We hope to extend the work by modelling semantics at alternative scales, consider a wider range of expressions, including durations—which means expanding support for SUTime's temporal operators. Alternative, asymmetric distributions could be included in the mixture model, with an appropriate algorithm to determine initial parameters, to achieve a better fit to some of the distributions we found. Furthermore, we intend to develop a framework for evaluating the distributional approach against the existing approaches, and explore the philosophical issues discussed in Sect. 8 in greater depth.

References

1. Allen, J.F.: An interval-based representation of temporal knowledge. In: Proceedings of the 7th International Joint Conference on Artificial Intelligence (IJCAI'81), pp. 221–226. Morgan Kaufmann (1981)
2. Angeli, G., Manning, C., Jurafsky, D.: Parsing time: Learning to interpret time expressions. In: Proceedings of the 2012 Conference of the North American Chapter of the Association for Computational Linguistics: Human Language Technologies, pp. 446–455 (2012)

3. Association, C.I.P.: Exchangeable image file format for digital still cameras: Exif version 2.3 (2010). http://www.cipa.jp/english/hyoujunka/kikaku/pdf/DC-008-2010_E.pdf. English Translation.
4. Brody, S., Diakopoulos, N.: Cooooooooooooooollllllllllllll!!!!!!!!!!!!!! using word lengthening to detect sentiment in microblogs. In: Proceedings of the 2011 Conference on Empirical Methods in Natural Language Processing, pp. 562–570. Association for Computational Linguistics, Edinburgh, Scotland, UK. (2011). http://www.aclweb.org/anthology/D11-1052
5. Brucato, M., Derczynski, L., Llorens, H., Bontcheva, K., Jensen, C.S.: Recognising and interpreting named temporal expressions (2013). http://derczynski.com/sheffield/papers/named_timex.pdf
6. Chang, A.X., Manning, C.: SUTime: A library for recognizing and normalizing time expressions. In: Proceedings of the 8th International Conference on, Language Resources and Evaluation (LREC'12) (2012)
7. Ferro, L., Gerber, L., Mani, I., Sundheim, B., Wilson, G.: TIDES 2005 standard for the annotation of temporal expressions (2005)
8. Grefenstette, G.: Explorations in automatic thesaurus discovery. Springer (1994)
9. James, J.: How facebook handles image exif data (2011). http://windowsitpro.com/blog/how-facebook-handles-image-exif-data
10. Ma, J., Knight, B.: Representing the dividing instant. Comput. J. 46(2), 213–222 (2003). http://dblp.uni-trier.de/db/journals/cj/cj46.html#MaK03
11. Mani, I., Wilson, G.: Robust temporal processing of news. In: Proceedings of the 38th Annual Meeting on Association for, Computational Linguistics, pp. 69–76 (2000)
12. Ritter, A., Mausam, Etzioni, O., Clark, S.: Open domain event extraction from twitter. In: Proceedings of the 18th ACM SIGKDD international conference on Knowledge discovery and data mining, KDD '12, pp. 1104–1112. ACM, New York, NY, USA (2012). DOI 10.1145/2339530.2339704. http://doi.acm.org/10.1145/2339530.2339704
13. Souza, C.R.: The Accord.NET Framework (2012). http://accord.googlecode.com
14. Strötgen, J., Gertz, M.: HeidelTime: High quality rule-based extraction and normalization of temporal expressions. In: Proceedings of the 5th International Workshop on Semantic, Evaluation (SemEval'10), pp. 321–324 (2010)
15. UzZaman, N., Llorens, H., Allen, J.F., Derczynski, L., Verhagen, M., Pustejovsky, J.: TempEval-3: Evaluating events, time expressions, and temporal relations. In: Proceedings of the 7th International Workshop on Semantic, Evaluation (SemEval'13), pp. 1–9 (2013)
16. Verhagen, M., Pustejovsky, J.: Temporal processing with the TARSQI toolkit. In: Proceedings of the 22nd International Conference on, Computational Linguistics, pp. 189–192 (2008)
17. Williams, T., Kelley, C.: gnuplot 4.6 (2013). http://www.gnuplot.info/documentation.html
18. Yin, Z., Cao, L., Han, J., Zhai, C., Huang, T.: LPTA: A probabilistic model for latent periodic topic analysis. In: Proceedings of the 11th IEEE International Conference on Data Mining (ICDM'11), pp. 904–913. IEEE Computer Society (2011)

Genetic Programming for Wind Power Forecasting and Ramp Detection

Giovanna Martínez-Arellano and Lars Nolle

Abstract In order to incorporate large amounts of wind power into the electric grid, it is necessary to provide grid operators with wind power forecasts for the day ahead, especially when managing extreme situations: rapid changes in power output of a wind farm. These so-called ramp events are complex and difficult to forecast. Hence, they introduce a high risk of instability to the power grid. Therefore, the development of reliable ramp prediction methods is of great interest to grid operators. Forecasting ramps for the day ahead requires wind power forecasts, which usually involve numerical weather prediction models at very high resolutions. This is resource and time consuming. This paper introduces a novel approach for short-term wind power prediction by combining the Weather Research and Forecasting—advanced Research WRF model (WRF-ARW) with genetic programming. The latter is used for the final downscaling step and as a prediction technique, estimating the total hourly power output for the day ahead at a wind farm located in Galicia, Spain. The accuracy of the predictions is above 85 % of the total power capacity of the wind farm, which is comparable to computationally more expensive state-of-the-art methods. Finally, a ramp detection algorithm is applied to the power forecast to identify the time and magnitude of possible ramp events. The proposed method clearly outperformed existing ramp prediction approaches.

1 Introduction

Wind power has become one of the renewable resources with the strongest growth over the last years [1]. However, a high penetration of wind power can lead to instability in the electric grid due to the intermittency and variability of the wind.

G. Martínez-Arellano (✉) · L. Nolle
Nottingham Trent University, Clifton Lane, NG11 8NS Nottingham, UK
e-mail: giovanna.martinezarellano@ntu.ac.uk

L. Nolle
e-mail: lars.nolle@ntu.ac.uk

M. Bramer and M. Petridis (eds.), *Research and Development in Intelligent Systems XXX*, 403
DOI: 10.1007/978-3-319-02621-3_30,

Hence, a wind power forecast is essential for grid operators when planning the operation of the electric grid. This is commonly known as the Unit Commitment (UC) and Economic Dispatch (ED) problems.

The UC problem refers to deciding which units of a power system will be turned on to generate, in a more cost-effective way, the amount of power that is demanded. If a system is on permanently, it will be more efficient than if it is used intermittently [2]. ED is the process of deciding what the individual power outputs should be of the scheduled generating units at each time-period. A certain amount of unit backup is also considered, because of the possibility of a break down of a unit at any time. The backup units need to be able to cover for any scheduled unit being unavailable. In power markets, UC and ED practices are carried out as follows: 12 to 36 h before the physical dispatch of power (day-ahead planning), energy producers report to the grid operator how much power and reserve they are able to provide for the next day. The grid operator will then run a unit commitment algorithm to decide which providers will better satisfy the estimated demand. After defining which units will be operating the next day, an economic dispatch process will be carried out to determine the amount of power each generator will be producing and the schedule of each unit. Wind power forecasting models are nowadays used by some grid operators for UC and ED, however, existing tools need to be improved to be able to handle extreme situations related to wind power generation [3]. These extreme situations may be related to specific meteorological events, such as cold fronts or high pressure levels, which can produce drastic increases or drops in the level of power production of a wind farm [4]. These sudden increases or drops, also known as *ramp events*, may happen from a range of minutes to a couple of hours and the amount of the change in the power output may vary according to the needs of the grid operator. In this paper, ramp events are referred to as changes from one to a couple of hours due to the interest of ramp forecasts in the day-ahead market. An accurate wind power ramp forecast for the day-ahead can be used to estimate more accurately the necessary unit back up.

To predict wind power ramps for the day ahead, first a wind power forecast on the short-term (up to 48 h into the future) is needed. Short term wind power forecasting usually relies on numerical weather prediction models (NWP) due to their capacity to capture the atmospheric flow. Meteorological models represent the atmospheric flow by a set of physical equations, which govern the behavior of the dynamics and thermodynamics of the atmosphere. These models can run globally or at specific regions with different grid resolutions. Models such as High Resolution Limited Area Model (HIRLAM), WRF, ETA and Global Forecast System (GFS) have been used for short term forecasting. In [12], different post-processing methods are proposed to improve forecasts from the Consortium of Small Scale Modeling (COSMO) model. These methods included a Kalman filter approach [13], ANN approach [14], bias correction methods and a combination of these. In [15], a WRF model is implemented together with the Kalman filter method for wind speed and wind power forecasting in a wind farm in China. Kalman filter approaches have also been applied in [16] and [17], as post-processing tools for correcting the bias of WRF wind speed predictions, reducing significantly the size of the training set, compared to ANN based methods.

The majority of these approaches require to run the numerical weather prediction model at a very high resolution, which is time and resource consuming. In [14] a neural network approach is used to perform the final downscaling from the MM5 mesoscale model to the observation sites avoiding the execution of the numerical model at high resolutions. However, neural networks behave as black boxes , which do not provide internal information about the model that was found. They also need a significant amount of training data to ensure generalisation. The same forecasting model used in [14] was implemented in [18] replacing the ANN with a support vector machine (SVM) approach. In [19], NWP are used for wind power ramps forecasting. NWP forecasts from two sources are used to improve the characterisation of the timing uncertainty of the ramps. In [20] NWP ensembles are used for wind power ramp predictions. The ensembles are used to improve the timing error of ramp forecasts. However, running numerical models for ensemble forecasting is computationally expensive. According to [21], the timing or *phase error* of numerical models can also be addressed taking into account a wider area of the NWP grid, not only the closest point to the observation site, due to the misplacement errors that the numerical model may have.

In this paper, a short-term wind power forecasting approach that combines the WRF-ARW model with Genetic Programming (GP) is presented. GP is used as a final downscaling procedure, predicting the total power output of a wind farm located in Galicia, Spain, which is located in a semi-complex terrain, and avoiding the execution of high resolution forecasts. Additionally, a ramp detection algorithm is applied to identify the ramp events within the forecasted horizons. The rest of the paper is organised as follows: the next section presents in more detail the numerical weather prediction models used in this study and explains in detail the proposed approach for wind power forecasting. Section 3 presents the experiments carried out, results and discussion. Finally, Sect. 4 presents the conclusions and future work.

2 Short-Term Wind Power Prediction

In order to predict the wind power output of a wind farm at each hour for the next day, two major steps are carried out in the proposed approach. First, the prediction of the wind speeds in a close point to the park, which is achieved by the use of the numerical weather prediction model WRF-ARW. Secondly, wind speed predictions and wind power observations at the park are used to train a GP based algorithm and obtain a model which best represents the relationship between the wind speed obtained from the numerical model and the output power produced by the wind park. The model found can be applied to new available wind speed forecasts to predict the power output of the wind park.

For this study, data from the experimental wind farm, Sotavento, located in Galicia, Spain, was used [22]. This wind park is composed of 24 wind turbines of 9 different models (each model with a different power curve) and has a total nominal power of 17.56 MW. The wind farm provides open access to wind speed and wind power observations. The wind speed observations are obtained by an anemometer situated in the park at 45 m height. The day-ahead electricity market requires of hourly forecasts to plan the commitment and dispatch of the power units. For this reason, mean hourly wind speeds from the park where used.

2.1 Wind Speed Prediction Using the WRF-ARW Model

WRF-ARW is a non-hydrostatic limited area model from the National Center of Atmospheric Research (NCAR) in the USA [9]. Similar to GFS model, WRF-ARW solves a system of differential equations that represent the dynamics of the atmospheric flow, except it does not take into account the ocean-land interactions. This mesoscale model uses a series of parameters to set the initial state and boundary conditions, as well as the parameters to determine how the execution will be carried out. In this research, the initial state was generated using GFS forecasts and terrestrial data. GFS runs four times a day, at 00Z, 06Z, 12Z and 18Z (UTC time). At each run, it produces low resolution forecasts. This means the entire globe is divided into a grid of, usually $1° \times 1°$, producing forecasts at each of the intersection points of the grid. Each run predicts up to 16 days in advance with a three-hour time step. For this study, the runs from 06Z were used, taking into account that forecasts need to be provided to the operator before noon for the day ahead. From each run, only the first 48 h of the complete forecasted horizon were considered. For each GFS run, a WRF-ARW run was executed, producing a higher resolution forecast in time and space. Those values forecasted for the next day (19–42 h into the future), as shown in Fig. 1, are the values of interest in this study.

WRF-ARW, as other mesoscale models, allows nesting. This means the model can run at different resolutions or domains; one contained into the other, were the inner domains have a higher resolution in a smaller area. Figure 2 shows the domain settings that were used. As shown in the figure, the model was set to run in two domains. The first domain, which covers a major part of Spain, has a resolution of 30×30 km and results from the first integration of the WRF model from the GFS grid (111×78 km). The second domain, which is centered on Galicia, the area of interest, has a resolution of 10×10 km and is obtained by a second model integration that uses the first domain as boundary conditions. A third domain was considered, but

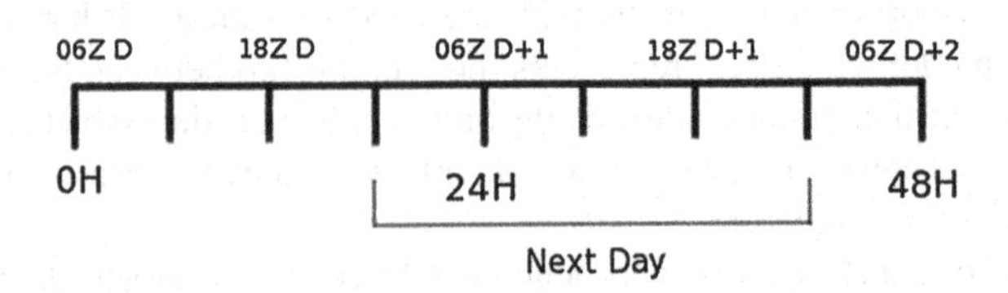

Fig. 1 WRF-ARW 48 h horizon in a 06Z run started at day D. The next day forecast corresponds to those values 19–42 h into the future

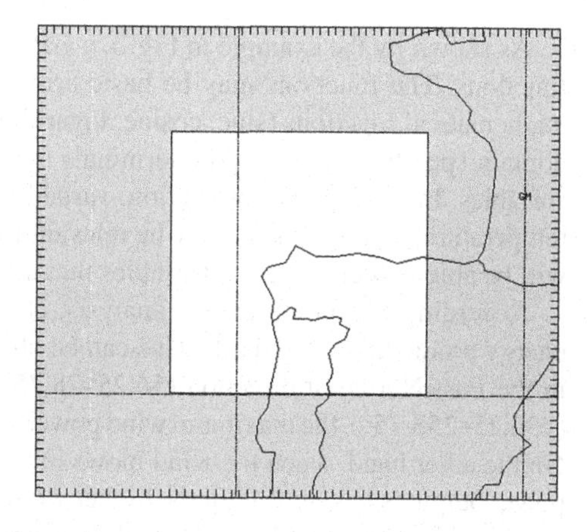

Fig. 2 WRF-ARW two domain setting. The second domain (*white area*) is centered in the point of interest, which is Galicia, Spain

the computational cost of running at 3 km resolution was considered too high. From the mode output, wind speed and wind direction forecasts from the closest point of the grid to the location of the wind farm will be considered for power prediction.

2.2 Wind Power Prediction Using Genetic Programming

Genetic Programming [23] is a biologically inspired computation technique based on the evolution of individuals over time, through events such as crossover and mutation, which progressively refines them into better individuals. In GP, a population of programs (in a binary tree layout) is evolved, each program representing a set of instructions to solve a specific problem. GP, like nature, is a stochastic process, which cannot guarantee to find the global optimum but it is that randomness which can lead it to escape local optima, which deterministic methods may be captured by [24].

Symbolic regression via GP is a non-parametric, non-linear regression technique that looks for an appropriate model structure and model parameters as opposed to classic regression that assumes a certain model structure and estimates the optimal parameters that best fit a given sample of data [25]. In a previous work [26], GP was used as a final downscaling step from the mesoscale model to the location of the observation site for wind speed forecasting. Results obtained showed the ability of the GP approach to model the relationship between the forecasts of the WRF-ARW model and the observations at specific sites. Wind speed forecasts can be easily converted to wind power using the power curve provided by the wind turbine manufacturer. However, the performance of the wind turbines depends on the characteristics of the place where they are located. For this reason, the use of local wind power observations to estimate the power output of each turbine is suggested [27]. This paper revisits and extends the GP downscaling approach from [26] as a wind power prediction technique. While the input to the algorithm stays the same (numerical model predictions), the output changes to the total wind power produced by the wind farm.

As shown by the example in Fig. 3, a GP tree is formed by a set of terminals and functions. The functions may be basic arithmetic operators {+, −, *, /}, standard mathematical functions (sine, cosine, logarithmic, exponential), logical functions or domain-specific functions. The terminals may be constants or any problem-related variables. In wind power prediction, variables such as wind speed, wind direction, temperature, among others, may be relevant for the problem. The evolution process will be able to identify those variables that are relevant for the model.

According to some previous analysis, changes in the wind direction affect the energy production of the park. This can be observed in Fig. 4. When the wind blows in the East Northeast direction (56.25–78.75°) or in the West Southwest direction (236.25–258.75°), the maximum wind power output is achieved for high wind speeds. On the other hand, when the wind blows in the North direction (348.75–11.25°) the power output is lower even for the higher wind speeds. For this reason, it is important to take into account both wind speed and wind direction as inputs to the GP.

As any other Machine Learning (ML) technique, achieving good generalization is one of the most important goals of the GP approach. Failure to generalize, or overfitting, happens when the solution performs well on the training cases but poorly on the test cases. The most common approaches to reduce overfitting consist in biasing the search towards shorter solutions. The parsimony pressure technique [25], penalises the fitness of a program according to its complexity (number of nodes in the tree), reducing the probability for it to be selected in future generations for crossover.

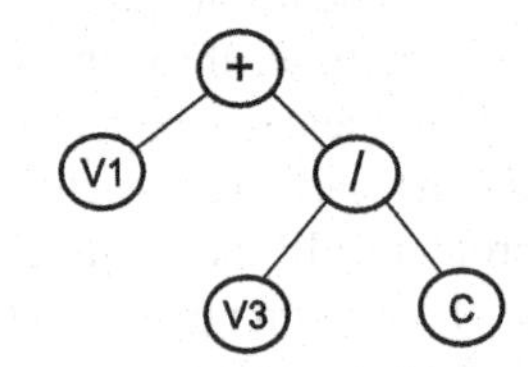

Fig. 3 Example of a tree expression of the program $v_0 = v_1 + v_3/C$

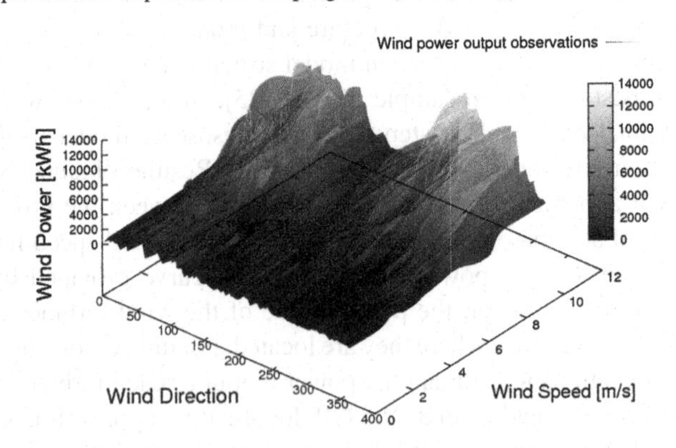

Fig. 4 Relationship between wind speed, wind direction and the power output at Sotavento for three months, January to March, 2012

The mathematical representation of the penalisation function used is shown on Eq. 1. The first term of the fitness equation is the sum of the errors between the obtained output (new forecast) and the desired output (observations) in the test set (test set size = s). The second term is the complexity factor, where t is the number of nodes of the GP tested and k is a trade-off weight that allows to control the level of pressure of the complexity factor. A small value of k (e.g. $k = 0.1$) would be translated into low complexity pressure, and higher values of k (e.g. $k = 1.0$) will add a strong pressure to the penalisation.

$$f = \frac{1}{s}\sum_{i=0}^{s} e(i) + k\left(\frac{(t^2 log_2(t))}{s}\right)^{\frac{1}{2}} \quad (1)$$

The GP algorithm is implemented as follows: First, an initial population p of n trees of different sizes (not larger than a fixed depth) is created, selecting random combinations of variables and operators. The fitness of each tree is calculated based on the RMSE between the output given by the model on each training point and the real observation plus the penalisation function (Eq. 1). Once the tree with the best fitness is identified, the algorithm iterates x number of generations. At each iteration, a new population is created by copying the best tree so far and the rest of the individuals by means of selection, crossover and mutation operators. The new population replaces the previous population and the best tree of the new population is identified. This process is repeated until the specified number of generations has been reached. The algorithm will output the best tree on the training set and the best tree obtained on the test set.

3 Experimental Results and Discussion

To assess the performance of the proposed approach, a set of experiments were carried out as follows: First, to create the training, test and validation sets, four months of day-ahead WRF-ARW forecasts were produced, from January to April 2012 (not consecutive days, due to missing global model data). Observations from the same period of time were obtained from Sotavento. Therefore, for every hourly wind speed and direction forecast, a wind power observation was associated. The first three months were used as a training/testing set and 15 non consecutive days from the fourth month as a validation set. Data pre-processing was performed to remove the *outlier* points in the training/testing set. These outliers correspond to wind power observation errors caused possibly by wind turbine maintenance or shut downs related to strong winds or other meteorological events. To remove the outliers, first each data point in the data set was classified according to its cardinal wind direction (16 categories). All the points that fell in a direction category were approximated by a curve and those falling outside the curve by certain percentage were removed. The remaining data, was divided randomly in 80 % for training (942 data points) and 20 % for testing (236 data points). The reason for removing the

Table 1 Fixed GP parameters used for the experiments

Runs	50
Population	1,000
Generations	100
Crossover operator	Standard subtree crossover, probability 0.9
Mutation operator	Standard subtree mutation, probability 0.1,maximum depth of new tree 17
Tree initialisation	Ramped Half-and-Half, maximum depth 6
Function set	+,−, *, / log, exp
Terminal set	v1, d1 and random constants
Selection	Tournament of size 20
Elitism	Best individual always suvives

outliers is to avoid the algorithm to be guided by these outliers, and evolving into models that are not a good representation of the complete set.

Once the data sets were pre-processed, the experiments were carried out as follows. First, to identify the best pressure parameter for the algorithm, 50 runs per pressure value were executed. At each run, the training and test sets were selected randomly. Additional parameters that were set for these runs are shown in Table 1.

Figure 5 shows that at higher pressures, the correlation in the test set decreases, however, the correlation on unseen data is better. This means that if low pressure is applied, the models tend to get more complex and over trained.

According to the results obtained, values of $k = 2.0$ achieved the best correlation in unseen data, keeping the models at a low complexity. In terms of RMSE (Fig. 6), it can be observed that a complexity of approximately 100 is enough to achieve the best results. After the ideal value of the pressure parameter was identified ($k = 2.0$), the

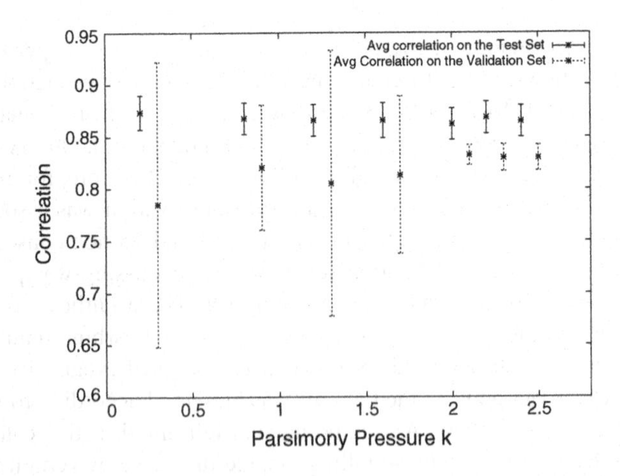

Fig. 5 Average correlation of the best model to the test and validation sets and standard deviation in 50 runs applying different pressure parameter values

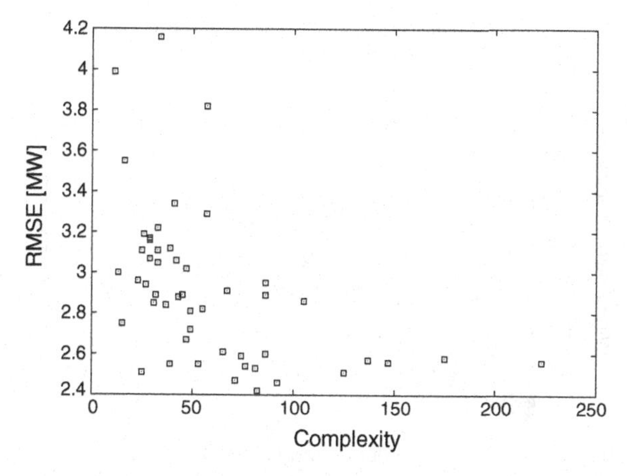

Fig. 6 Relationship between complexity and RMSE in 50 runs using the pressure parameter $k = 2.0$

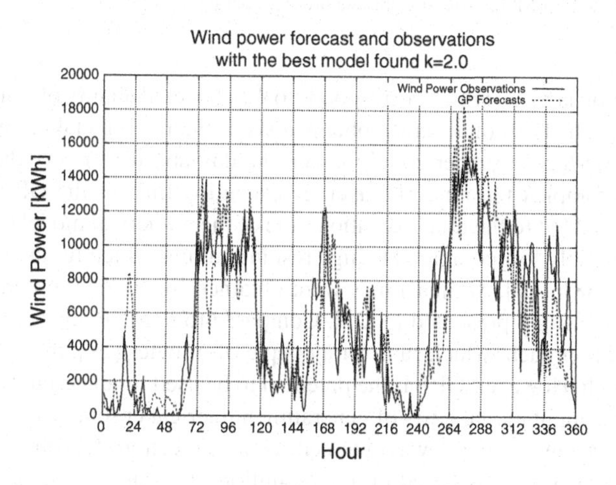

Fig. 7 Best Model found applied to unseen data (April 2012) and the real power output at Sotavento. Pressure parameter $k = 2.0$

best model in 50 runs using this pressure value was selected to forecast wind power for 15 days in April 2012. Results are shown in Figs. 7 and 8. It can be observed that the global trend is well captured but in short time steps, e.g. day 15 (hour 336–360 in Fig. 7), the high peaks are not well captured, in comparison to other peaks in the rest of the forecasted days. In addition, it can be seen that in some of the forecasted days (hours 144–168), there is a time misplacement, which is associated with a weakness of NWP modeling to forecast synoptic events like cold fronts and high or low pressure systems in the adequate position [21]. This misplacement could be addressed by taking into account a wider area from the mesoscale grid instead of taking only the closest point. According to the average RMSE obtained, the accuracy

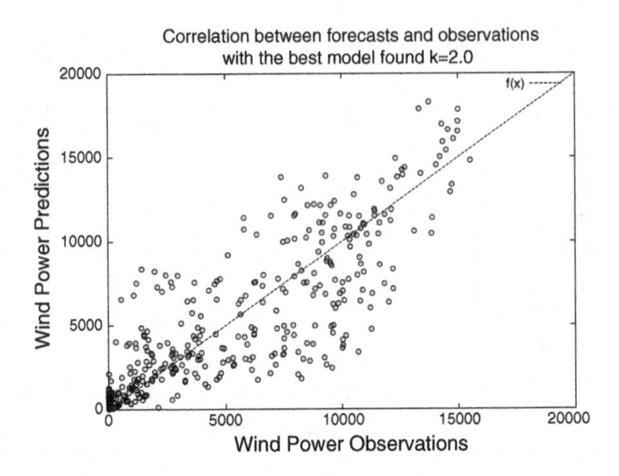

Fig. 8 Correlation of the forecasted wind power and real power output from Sotavento on April 2012, applying the best Model found using the pressure parameter $k = 2.0$

of the prediction is 87 %, which corresponds to the expected quality of state-of-the art methods [20]. The level of accuracy obtained is considered high taking into account that forecasting local weather conditions on the day ahead is a very difficult task, specially on complex terrains. The average execution time of the GP approach is 2,856 s which added to the total execution time of the numerical model (approx. 3 h) is still considerably low compared to high resolution runs which for up to 5 domains could run for 48 h [29]. Similar times were obtained by the GP approach for very short term forecasting presented in [28]. Compared to non-GP approaches, e.g. the Kalman filter approach, which runs for a couple of minutes [29], the time could be considered high. However, GP has the potential to be used for uncertainty estimation avoiding the computational costs of generating numerical ensembles.

Once a wind power forecast was obtained, a ramp detection algorithm was applied. The approach presented is based on the definition of a ramp event as a change in power output in a short period of time, specifying both percentage of change and window size. The approach proceeds as follows:

1. The percentage p of the amount of change in power output and the maximum window size w are set.
2. Using a 1 hour window, the power signal is analysed to identify any increase (ramp-up) or decrease (ramp-down) by p percent. As soon as a ramp is detected, this is identified as the start time of the ramp and the end of the ramp will be identified in the following hours as soon as the ramp changes direction.
3. In the final set of ramps identified, if any two are overlapped, they are considered as one ramp, and the start and end times are readjusted. Then, go to step 5.
4. The size of the window is increased one hour and the power signal re analysed to identify any increase or decrease by p percent with the new window size. Start and end times are identified like in step two.

5. If the window size has not reached its maximum then go to step 4, else, check any overlaps and output the identified ramps.

The algorithm was applied on both forecasted and real power output of the farm to identify the false and true forecasted ramps. The time window was set to 5 h, according to [19]. The percentage of change was set to 30 % due to a very small number of real events of higher change. Of the 14 ramps observed, 8 of them where forecasted accurately in direction (ramp up or down) and with a *phase error* less than ±12 h. This time period of association is the maximum time difference between the timing of the forecast and the observed ramps, according to [19], that can maintain realistic connection between the forecast and the observed event. Events further apart can be representing totally different events. The total ramp accuracy and ramp capture percentage were calculated using Eqs. 2 and 3.

$$ramp\,capture = \frac{true\,forecasts}{true\,forecasts + missed\,ramps} \tag{2}$$

$$forecast\,accuracy = \frac{true\,forecasts}{true\,forecasts + false\,forecasts} \tag{3}$$

True forecasts are those forecasted ramps that are associated with an observed ramp; false forecasts are those ramps that were forecasted but did not occur; finally, missed ramps are those observed events that were not forecasted. Comparing to results obtained by a recent study in the U.K. and U.S. [19] in terms of accuracy and capture for the day-ahead, GP delivers competitive results. The accuracy obtained by [19] in the U.K. is 59.1 % and in the U.S. 57.8 %, compared to 61 % obtained with the GP approach. In terms of ramp capture, the GP approach obtained 57 % compared to the 35 % and 29.6 % obtained in [19]. However, the validation set used in this study presented a small number of observed ramps. A larger set including other times of the year would determine more accurately the effectiveness of the ramp forecasts. Figure. 9 shows a true forecast of a ramp up that occurred on the 9th of April. Although the observations indicate that the ramp started around 17:00 hrs, the forecast was able to model this event one hour later, from 18:00 hrs to 02:00 hrs the next day.

The method proposed is sensitive to the time window size and percentage of change, which reduces the detection of events to a certain type. To improve the characterisation of these events, it is of interest to employ a signal filtering step eliminating the "false" ramps from the signal to produce a signal were relevant changes of any type can be detected. In addition, ramps where identified in time, duration and magnitude. However, there is no estimation of the uncertainty of the events. An approach presented in [20] uses numerical weather prediction ensembles to predict the uncertainty in the timing of the event. The ensemble forecasts are obtained by initializing the numerical model with slightly different initial states. Evolving these systems is computationally expensive. As an alternative to numerical ensembles, two approaches can be implemented with GP. The first one is based on

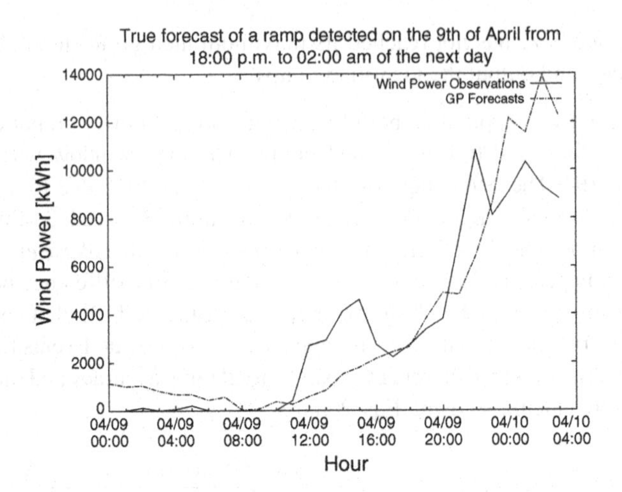

Fig. 9 True ramp forecasted on April 9th, 2012

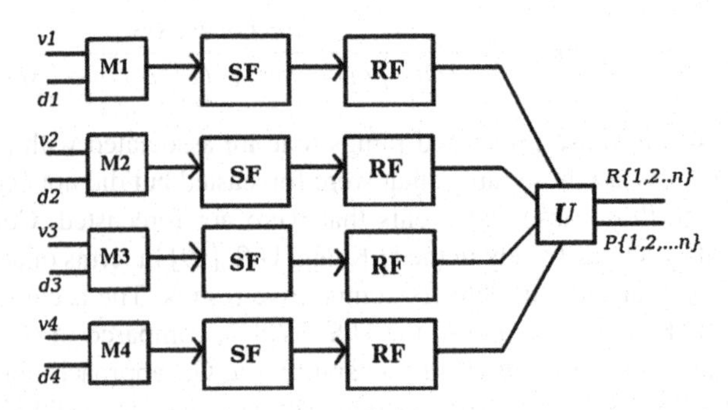

Fig. 10 Model proposed for uncertainty estimation of ramp events. For each pair v_i, d_i a model is obtained through the GP approach. Each forecast obtained is filtered in a second step to detect ramp events in a third step. Finally, the predictions are combined to estimate the uncertainty of each ramp event

using not only the best, but a set of the best models obtained by the GP. If a high number of models indicate the occurrence of a ramp event, then the event will be considered as highly probable. Another approach is to obtain several models using different *neighbour* points (close to the location of the wind farm) of the mesoscale grid. This can, in fact, address the misplacement error that the numerical model could produce and be used as a way to estimate the uncertainty of the event by analysing how many neighbour points of the grid are predicting the same event. If a high number of neighbours are indicating the presence of a ramp, this can be interpreted as a high probability of the event to occur at that time. The proposed approach is shown in Fig. 10.

Figure 11 shows an example of a false forecast on the 4th of April. Forecasts obtained using four close points to the park (v_1, v_2, v_3, v_4) were used. It can be

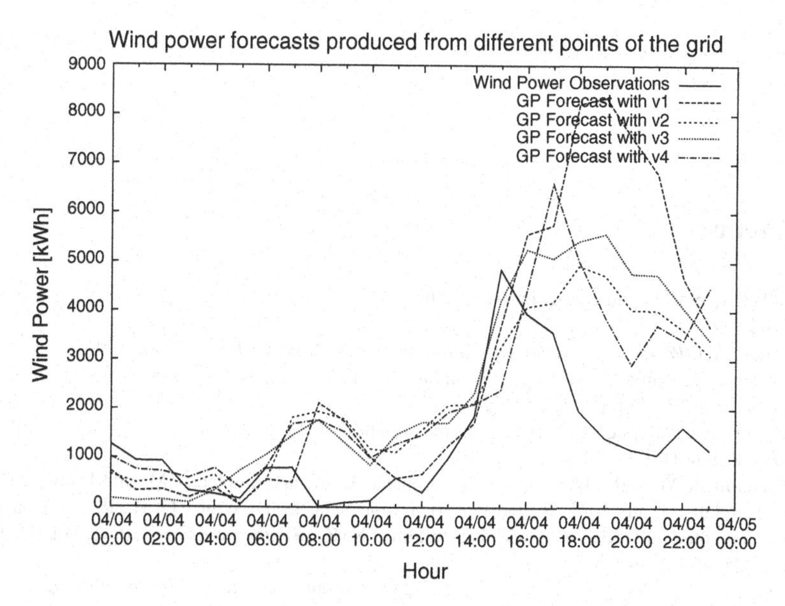

Fig. 11 Wind power forecast using different points of the grid

observed that $v1$ forecasts a high peak (ramp up) some hours after the original event and with a higher magnitude. Neighbor point $v4$, on the other hand, is able to predict the peak closer to the observation in both time and magnitude. This finding shows the *phase error* that can occur with numerical models. If neighbor points are considered, more certainty of the occurrence of the event could be obtained. In this case, the neighbor points do not indicate a high peak, therefore, the ramp event modeled with $v1$ can be considered less probable. Estimating the uncertainty of a ramp event with this approach can reduce the computational effort of ensemble forecasting approaches.

4 Conclusions and Future Work

In this paper, a novel approach for wind power forecasting and ramp detection is proposed. One advantage of this approach, compared to other state-of-the-art methods, is the reduction of computational effort by using GP as a downscaling procedure, delivering competitive results for day-ahead forecasting. In addition, two alternative methods for the estimation of the uncertainty of a ramp event are proposed. Preliminary results show the potential of taking into account other points of the grid for estimating the uncertainty of an event. Further investigations in this matter will be conducted to determine the adequate size of the area to be explored and to determine the potential of the approach compared to ensemble forecasting. Additionally, further experimentation will be carried out using data from different times of the year to analyse the quality and accuracy of the method under different meteorological events. So far, the ramp detection algorithm looks at changes in specific size windows, which restricts the ramp detection to certain events. Fourier and Wavelet transforms will be explored to better characterise these events without restricting the time window.

References

1. WWEA, World wind energy half-year report 2012, Tech. rep.*World Wind Energy Association*, 2012.
2. Wood, A.J. *Power Generation, Operation and Control*. Wiley-Interscience, 1996.
3. Ferreira, C., Gama, J., Matias, L., Botterud, A. and Wang, J. A Survey on Wind Power Ramp Forecasting, Tech. rep. *ARL, DIS-10-13*, 2010.
4. Pinson, P. Catalogue of complex extreme situations.*Technical Report, EU Project SafeWind, Deliverable Dc1.2*, 2009.
5. Kanamitsu, M. and Alpert, J.C. and Campana, K. A. and Caplan, P.M. and Deaven, D.G. and Iredell, M. and Katz, B. and Pan, H. L. and Sela, J. and White, G. H., Recent Changes Implemented into the Global Forecast System at NMC, *Weather and Forecasting*, Vol. 6, 1991, pp. 425–435.
6. Landberg, L., Short-term prediction of the power production from wind farms, *Journal of Wind Engineering and Industrial Aerodynamics*, Vol. 80, 1999, pp. 207–220.
7. Landberg, L., Short-term prediction of local wind conditions, *Journal of Wind Engineering and Industrial Aerodynamics*, vol. 89, 2001, pp. 235–245.
8. Constantinescu, E.M., Zavala, E.M., Rocklin, M., Sangmin Lee, Anitescu, M., A computational framework for uncertainty quantification and stochastic optimization in unit commitment with wind power generation, *IEEE Transactions on Power Systems*, vol. 26, 2011, pp. 431–441.
9. Skamarock, W.C. and Klemp, J.B. and Dudhia, J. and O. Gill, D. and Barker, D. M. and Wang, W. and Powers, J. G., A Description of the Advanced Research WRF Version 2, *AVAILABLE FROM NCAR*, Vol. 88, 2001, pp. 7–25.
10. Alexiadis, MC., Dokopoulos, PS., Sahsamanoglou, H., Manousaridis, IM., Short-term forecasting of wind speed and related electrical power, *Solar Energy*, Vol. 63(1) 1998, pp. 61–68.
11. Lazić, L. and Pejanović, G. and Živković, M., Wind forecasts for wind power generation using Eta model, *Renewable Energy*, Vol. 35, No. 6, 2010, pp. 1236–1243.
12. Sweeney, C. P. and Lynch, P. and Nolan, P. Reducing errors of wind speed forecasts by an optimal combination of post-processing methods, *Meteorological Applications*, doi.10.1002/met.294, 2011.
13. Sweeney, C.P. and Lynch, P., Adaptative post-processing of short-term wind forecasts for energy applications, *Wind Energy*, doi.10.1002/we.420, 2010.
14. Salcedo-Sanz, S. and Ortiz-García, E. G. and Portilla-Figueras, A. and Prieto, L. and Paredes, D, Hybridizing the fifth generation mesoscale model with artificial neural networks for short-term wind speed prediction.*Renewable Energy*, Vol. 34, No. 6, 2009, pp. 1451–1457.
15. Zhao, P. and Wang, J. and Xia, J. and Dai, Y. and Sheng, Y. and Yue, J. Performance evaluation and accuracy enhancement of a day-ahead wind power forecasting system in China, *Renewable Energy*, Vol. 43, 2012, pp. 234–241.
16. Delle Monache, L. and Nipen, T. and Liu, Y. and Roux, G. and Stull, R. Kalman Filter and Analog Schemes to Postprocess Numerical Weather Predictions, *Monthly Weather Review*, Vol. 139, No. 11, 2011, pp. 3554–3570.
17. Cassola, F. and Burlando, M., Wind speed and wind energy forecast through Kalman filtering of Numerical Weather Prediction model output, *Applied Energy*, Vol. 99, 2012, pp. 154–166.
18. Salcedo-Sanz, S. and Ortiz-García, E. G. and Pérez-Bellido, A. M. and Portilla-Figueras, A. and Prieto, L., Short term wind speed prediction based on evolutionary support vector regression algorithms, *Expert Systems with Applications*, Vol. 38, No. 4, 2011, pp. 4052–4057.

19. Greaves, B., Collins, J., Parkes, J., Tindal, A. Temporal forecast uncertainty for ramp events. *Wind Engineering*, Vol. 33, No. 11, 2009, pp. 309–319.
20. Bossavy, A., Girard, R. and Kariniotakis, G., Forecasting ramps of wind power production with numerical weather prediction ensembles, *Wind Energy*, Vol. 16, No. 1, 2013, pp. 51–63.
21. Cutler, N.J., Outhred, H.R., MacGill, I.F., Kepert, J.D., Characterizing future large, rapid changes in aggregated wind power using numerical weather prediction spatial fields. *Wind Energy*, Vol. 12, No. 6, 2009, pp. 542–555.
22. Sotavento Galicia Experimental Wind Farm, sotaventogalicia.com, accesed on 29 April, 2013.
23. Koza, J.R., *Genetic Programming: on the programming of computers by means of natural selection*, MIT Press, 1992.
24. Poli, R. and Langdon, B. and McPhee, N. F., *A field guide to genetic programming with contributions by J. R. Koza*, Published via http://lulu.com and freely available at http://www.gp-field-guide.org.uk, 2008
25. Kotanchek, M. E. and Vladislavleva, E. Y. and Smits, G. F., *Genetic Programming Theory and Practice VII*, Springer US, 2010.
26. Martinez-Arellano, Giovanna, Nolle, Lars and Bland, John. Improving WRF-ARW Wind Speed Predictions using Genetic Programming. SGAI Conf. 2012: 347–360.
27. Tindal, A., Johnson, C., LeBlanc, M., Harman, K., Rareshide, E. and Graves, A. Site-especific adjustments to wind turbine power curves. AWEA Wind Power Conf., 2008.
28. Vladislavleva, E., Friedrich, T., Neumann, F. and Wagner, M., Predicting the Energy Output of Wind Farms Based on Weather Data: Important Variables and their Correlation. Renewable Energy, Vol. 50, 2013, pp. 236–243.
29. Louka, P., Galanis, G., Siebert, N., Kariniotakis, G., Katsafados, P., Pytharoulis, I., Kallos, G., Improvements in wind speed forecasts for wind power prediction purposes using Kalman FIltering. Journal of Wind Engineering and Industrial Aerodynamics. Vol. 96, No. 12, 2008, pp. 2348–2362.

Automated River Segmentation Using Simulated Annealing

N. Richards and J. M. Ware

Abstract A simulated annealing based algorithm is presented for segmenting a river centre line. This process is required for the purposes of river symbolization that is often required when generalizing (simplifying) a large scale map to produce a map of smaller scale. The algorithm is implemented and then tested on a number of data sets. A gradient descent based alternative is also implemented. Simulated annealing is shown to produce significantly better results.

1 Introduction

The problem addressed in this paper falls into the category of research termed automated map generalization (i.e. the process automatically reformatting a map as a necessary part of reducing its display scale). Useful collections of automated map generalization research papers can be found in [1–6].

On large-scale (detailed) maps, river networks are usually represented as a series of connected polygons. At reduced scale (less detail) these polygons are typically replaced by centre lines. These centre lines are themselves likely to require line simplification. In addition, when displaying centre lines (either on paper or electronically), the question of symbolization needs to be addressed (i.e. what thicknesses should be applied to the centre line?). This is not a straightforward decision since line thickness along the length of the river centre line will not be uniform, but will rather vary according to some criteria—typically based on river width. This sequence of operations is illustrated in Fig. 1.

N. Richards · J. M. Ware (✉)
University of South Wales, Pontypridd, CF37 1DL, UK
e-mail: mark.ware@southwales.ac.uk

N. Richards
e-mail: nigel.richards@southwales.ac.uk

M. Bramer and M. Petridis (eds.), *Research and Development in Intelligent Systems XXX*,
DOI: 10.1007/978-3-319-02621-3_31,

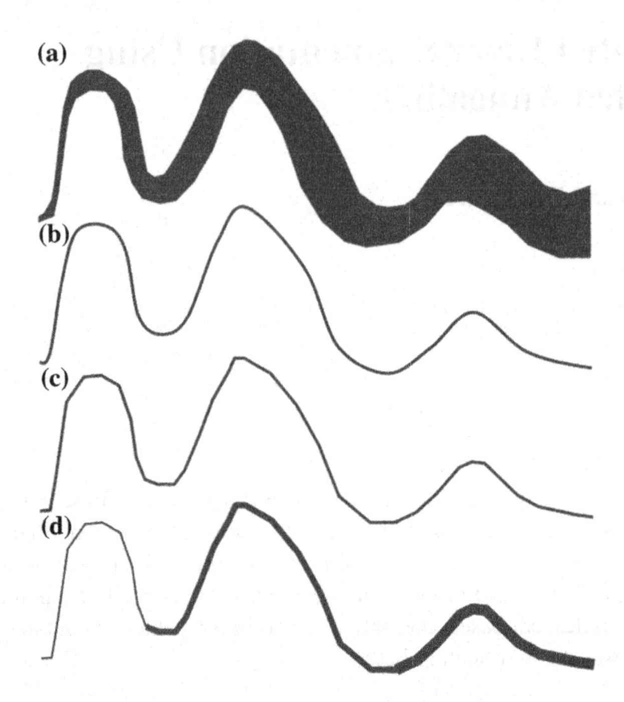

Fig. 1 Generalization of a river polygon. **a** Original polygon. **b** Centreline. **c** Simplified centreline. **d** Symbolised centreline

The problem addressed in this paper is that of automating the process of river centre line symbolization (Fig. 1d). The authors' are not aware of any previous automated solution to this problem. When performed manually, a cartographer will usually split the river centre line into a fixed number of segments. The location of each split (and hence the length of each segment) will be decided according to predefined criteria. Each segment is then symbolized (i.e. allocated a line thickness), again in accordance with predefined criteria. Consider the following example criteria (typical of those used by national mapping agencies):

- River segment width >10 metres (m) shown by single line, 2 mm width;
- River segment width between 7 and 10 m shown by single line, 1 mm width;
- River segment width between 4 and 7 m shown by single line, 0.5 mm width;
- River segment width < 4 m shown by single line, 0.25 mm width;
- Line thickness of single lines does not decrease downstream.

Automating the river segmentation process is non-trivial since, even though the width of a river generally increases when moving from upstream to downstream, this increase will not be monotonic. Therefore, identifying just a single location along a river at which its width equals a particular value is likely to be not possible. Automation therefore requires the use of algorithms that can produce good solutions given the initial predefined criteria, or rules. One approach to developing these algorithms is to consider the problem in terms of a search space, for which there is a set of possible solutions. There are many algorithms that can be used to find an optimal solution. This paper focuses on designing and evaluating a simulated annealing (SA) solution. SA is chosen since it has been applied successfully to previous map generalization problems (e.g. [7]) and has been adopted as the map generalization optimizer by ESRI, the leading commercial GIS software provider [8]. A gradient descent (GD) solution is also provided for bench-marking purposes.

2 Generating a River Centre Line from its Polygon Representation

At large scales, river networks are usually represented digitally by collections of polygons, as is the case with Ordnance Survey MasterMap data [9]. Each polygon is represented by a series of connected vertices, with the position of each vertex being defined by an (x, y) coordinate pair (see Fig. 2). At reduced scale river polygons are typically represented by polylines (i.e. they have been collapsed down to a centre line). This collapse process is illustrated in Fig. 1a and b.

Automated collapse of a polygon to a polyline has been the focus of much previous research. The approach adopted in this work is similar to that used in [9] and [10], and involves implementing algorithms that:

- Triangulate the polygon (i.e. decompose the polygon into a set of triangles);
- Produce an ordered chain of triangle medial axis edges, which forms the centre line approximation.

In particular, this work makes use of a constrained Delaunay triangulation algorithm [11] as the basis for triangulating the polygon (Fig. 3). The medial axis (centre line) is generated by connecting the midpoint of each internal triangle edge to its adjacent midpoints (Fig. 4). Full details of this process, which is not straightforward, can be found in [12].

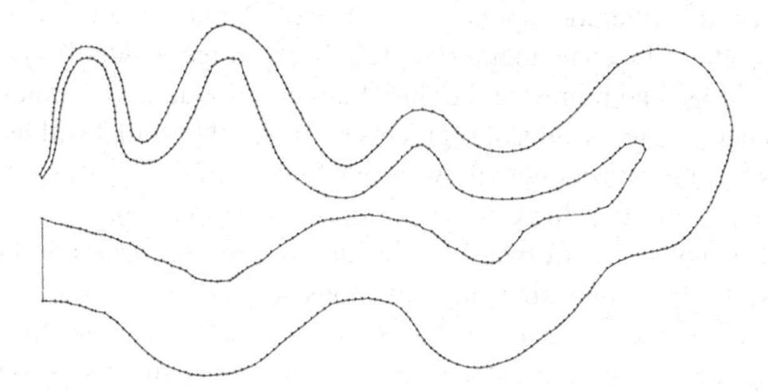

Fig. 2 Polygon representation of a river

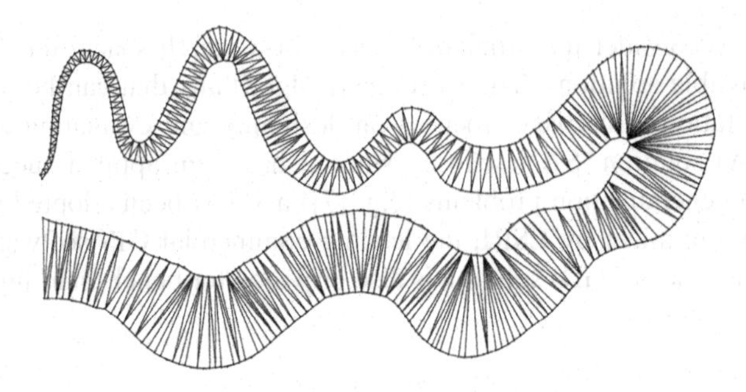

Fig. 3 Constrained Delaunay triangulation of the river polygon

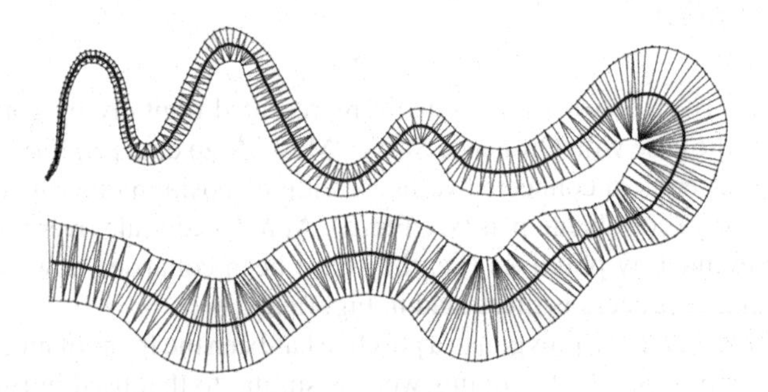

Fig. 4 Centre line produced from constrained Delaunay triangulation

3 River Segmentation

A more formal definition of the river segmentation problem now follows. Consider a polygon P representing a river, and its associated centreline C, which is made up of n vertices $(v_1, v_2, \ldots., v_n)$. It can be assumed that C has been derived from P using an appropriate triangle-based technique. The goal here is to divide C into k segments or divisions $(d_1, d_2, \ldots., d_k)$, with the segments having associated average widths $W = (w_1, w_2, \ldots., w_k)$. The average width w_i of a division d_i represents the average width of the corresponding section of P, and is calculated by finding the average height of its constituent triangles. The average widths W are referred to as the target widths, and represent the initial target criteria, or constraints. It makes sense to consider average widths of segments as opposed to actual widths at splitting points because of the non-monotonic way in which a river increases (or decreases) in width when moving in a downstream (or upstream) direction.

Dividing C is not straightforward. For example, consider a simple segmentation that gives rise to D_A, consisting of k divisions $(d_{A1}, d_{A2}, \ldots., d_{Ak})$ of (approximately) equal number of edges (note that some other measure could have been used for controlling this initial segmentation, such as having segments of roughly equal length), with average widths $W_A = (w_{A1}, w_{A2}, \ldots., w_{Ak})$. In order for D_A to meet initial criteria, then W and W_A would need to be equal. W and W_A are compared by calculating the value $|W - W_A| = |w_1 - w_{A1}| + |w_2 - w_{A2}| + \dots + |w_k - w_{Ak}|$. Typically, W and W_A will not be equal (i.e. $|W - W_A| > 0$). The problem is therefore to automatically reposition the splitting points to produce a segmentation D_X such that $|W - W_X| = 0$. Finding a set of split positions that results in $|w_i - w_{Xi}| = 0$ for all divisions is likely to prove impossible; the problem is therefore redefined as trying to find a set of positions that minimises width error. The width error $\hat{W}_Y$ is the value $|W - W_Y|$ associated with a segmentation D_Y.

Consider the situation where C is divided into 5 initial divisions (segments) resulting in D_A, made up from $(d_{A1}, d_{A2}, \dots d_{A5})$ where $d_{A1} = (v_1, v_2, \dots, v_a)$, $d_{A2} = (v_a, v_{a+1}, \dots, v_b)$, $d_{A3} = (v_b, v_{b+1}, \dots, v_c)$ and so on. Now consider a slightly changed segmentation D_B made up from $(d_{B1}, d_{B2}, \dots d_{B5})$ where $d_{B1} = (v_1, v_2, \dots, v_{a-1})$, $d_{B2} = (v_{a-1}, v_a, \dots, v_b)$, $d_{B3} = (v_b, v_{b+1}, \dots, v_c)$ and so on. That is, D_B is derived from D_A by removing a single vertex (in this case v_a) from one segment (in this case d_{A1}) and adding to another (in this case d_{A2}) resulting in the new segments d_{B1} and d_{B2}. In other words, the split location between a pair of adjacent divisions has moved by one vertex in one direction or the other (Fig. 5). It is now possible to compare the width errors $\hat{W}_A$ and $\hat{W}_B$ associated with segmentations D_A and D_B. Potentially, one of the segmentations will have a smaller error than the other. For example, if $\hat{W}_B < \hat{W}_A$, then it follows that D_B is a better segmentation than D_A (at least, that is, in terms of the target criteria being used). In other words, the reallocation of a vertex has resulted in an improved solution.

If the location of centre line splits is restricted to coincide with vertex locations only, then, based on standard combinatorial theory, the total number of alternative segmentations of C into k segments is:

$$\mathrm{N}(S) = (n-2)!/((n-2)-(k-1))!(k-1)!) \tag{1}$$

Note that in Eq. 1 $(n - 2)$ is used instead of n since vertices v_1 and v_n cannot be considered as splitting locations. Furthermore, it follows that there are $k - 1$ split vertices, each of which joins 2 adjacent divisions. The set of possible solutions S can be regarded as a search space, and the problem of segmentation can be redefined as that of finding the segmentation D_Z in S with smallest width error. One strategy might be to generate and evaluate all possible combinations. However, this is not feasible for realistic data sets. For example, consider the situation where $n = 1000$ and $k = 5$; this would result in over 40 billion possible segmentations. An alternative strategy for finding D_Z is therefore required.

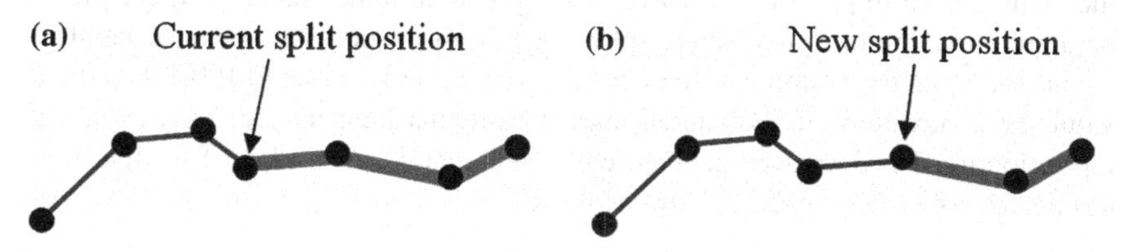

Fig. 5 Moving a vertex to an adjacent segment. **a** Original split position. **b** New split position

4 A Simulated Annealing Solution

One method of finding good solutions in a search space is to adopt an iterative improvement algorithm. There has been a vast quantity of research carried out in this area. A wide range of algorithms have been developed and applied to an even wider range of application domains. The technique used here is simulated annealing.

4.1 Simulated Annealing Algorithm

Algorithm 1 represents a SA solution. The algorithm accepts an initial centreline segmentation $D_{initial}$, which is immediately designated as being the current solution $D_{current}$. Next a random successor D_{new} is generated. This is achieved by selecting a split vertex at random and randomly relocating the associated split position to either of two adjacent vertices (see Fig. 5). If the repositioning results in a smaller width error (i.e. $\hat{W}_{new} < \hat{W}_{current}$), then the new split location is retained ($D_{current} \leftarrow D_{new}$). If, however, the repositioning fails to reduce width error (i.e. $\hat{W}_{new} \geq \hat{W}_{current}$), then the new split location is either discarded (i.e. $D_{current}$ is retained) or accepted ($D_{current} \leftarrow D_{new}$). The decision is based on the value P, which represents the probability of D_{new} being accepted. P in turn is dependent on two values ΔE (the change in width error) and T (the current annealing temperature)—as given by Eq. 2.

$$P = \mathrm{e}^{-\Delta \mathrm{E}/\mathrm{T}} \quad (2)$$

High values of T will encourage poor moves to be accepted. The idea therefore is to start off with a relatively high initial T; this has the effect of biasing the value of P towards higher values, thus encouraging poorer solutions (large ΔE) to be accepted more often (thus allowing a wide exploration of the search space). As the optimization progresses T is reduced, and poorer solutions will be rejected more often, although split location repositioning resulting in small ΔE might still sometimes be accepted. Split location repositioning continues until stopping conditions are met (e.g. a line segmentation is found that satisfies the constraint criteria or a meets target criteria is found or a pre-defined maximum number of iterations have elapsed).

It should be noted that the success or failure of SA implementations is influenced greatly by the choice of annealing schedule. The approach used here is based around the method used in [13]. This involves setting T to an initial value τ. At each temperature a maximum of ωk object split repositioning (successful or unsuccessful) are allowed. So, for example, if there are 5 divisions and ω is set to 1000, then there would be a maximum of 5000 iterations at a particular temperature. After every ωk repositioning, T is decreased geometrically such that $T_{new} = \lambda T_{old}$. The algorithm terminates when T_{new} reaches a lower limit ψ.

```
function SimulatedAnnealing
        input: D_initial, Schedule, Stop_Conditions
        D_current←D_initial
        T←initialT(Schedule)
        while NotMet(StopConditions)
                D_new←RandomSuccessor(D_current)
                ΔE← (Ŵ_current - Ŵ_new)
                if ΔE >0 then D_current←D_new
                else
                        P = e^(-ΔE/T)
                        R=Random(0,1)
                        if (R<P) then D_current←D_new
                endif
                T←UpdateT(Schedule)
        endwhile
Return(D_current)
```

Algorithm 1: Simulated annealing

4.2 Cost Function

In order to determine the quality of the current state a cost function is used to compare the target segment widths against those of the current state. The cost C, in the first instance, therefore corresponds to the width error, and is given by Eq. 3.

$$C = |W - W_{current}| = |w_1 - w_{current1}| + |w_2 - w_{current2}| + \dots + |w_k - w_{currentk}| \tag{3}$$

A potential problem here (which was identified during experiments) is that solutions with relatively low overall cost can be found, but these solutions may contain individual segments with relatively high cost. In order to distribute cost evenly across segments, the cost function is therefore modified by adding in segment width error standard deviation $\sigma_{current}$ (Eq. 4).

$$C = |W - W_{current}| + \sigma_{current} \tag{4}$$

Note that segment widths are calculated by taking the average of the heights of triangles that lie within a segment.

4.3 Results

A bespoke Java application was developed specifically to solve the river problem using the SA algorithm. Initial experiments have been carried out using a simple river polygon, Fig. 4, which shows the corresponding polygon triangulation and river centre line. This centre line is made up from 490 vertices. The goal here is to divide the centre line into 5 divisions (d_1, d_2, d_3, d_4, d_5), with associated target average widths W = (170.0, 130.0, 90.0, 50.0, 10.0). These widths represent screen units, but could just as easily correspond to appropriate mapping units. Figure 6 shows the initial segmentation (where each segment contains roughly equal number of edges) and corresponding initial segment widths of (156.71, 158.56, 105.12, 60.94, 22.67), giving an initial overall width error $\hat{W}_{initial} = 86.95$. Using Eq. 1, it follows that the total number of possible segmentations = 2,334,078,990.

A number of experiments were carried out using different parameter values to determine the settings that produced the best result. The initial temperature setting τ was varied from a very high initial temperature to a low temperature, coupled with different rates of cooling λ and iterations per temperature reduction ω. Experiments showed that at low initial temperatures (less than 4) the best solution could not be found, even if a very slow cooling rate was used. In these instances, it is probable that at low temperatures the SA algorithm was not accepting enough poor perturbations to explore the search space in enough depth. At very high temperatures the best solution could be found but a high number of iterations was also required as many more poor perturbations were being accepted due to the initial high temperature: This allowed a greater exploration of the search space, but required a long cooling period to find the best solution. It was found that a balance between the two temperature extremes produced the optimal result. In all cases a minimum of $\omega = 3000$ as necessary to allow enough iterations to explore the search space in sufficient depth to find the best solution.

The best result found produced an output with width error = 17.24, which was arrived at after the algorithm ran for a total of 765,000 iterations. This was produced using the following parameters: $\tau = 10.0$, $\lambda = 0.9$ and $\omega = 3000$. The algorithm terminated when the final temperature ψ reached a value of 0.05. The final solution shown in Fig. 7 took a total time of 15 s to produce. During the first 170,000 iterations significant cost improvement was achieved in stages, followed by small cost reductions up to 765,000 iterations, which produced the final width error value of 17.24. Figure 8 shows the performance in terms of a graph plotting width error (cost) over iterations.

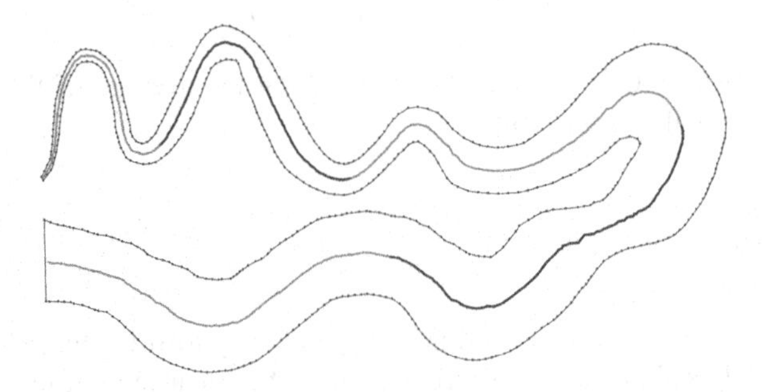

Fig. 6 Initial segmentation of the river into 5 parts shown without triangulation

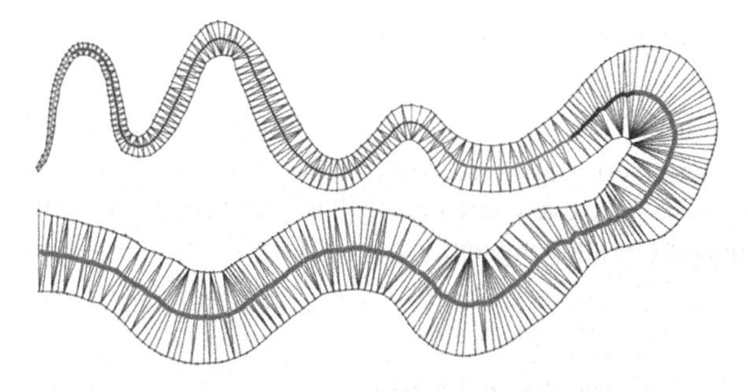

Fig. 7 Segmentation after application of simulated annealing algorithm

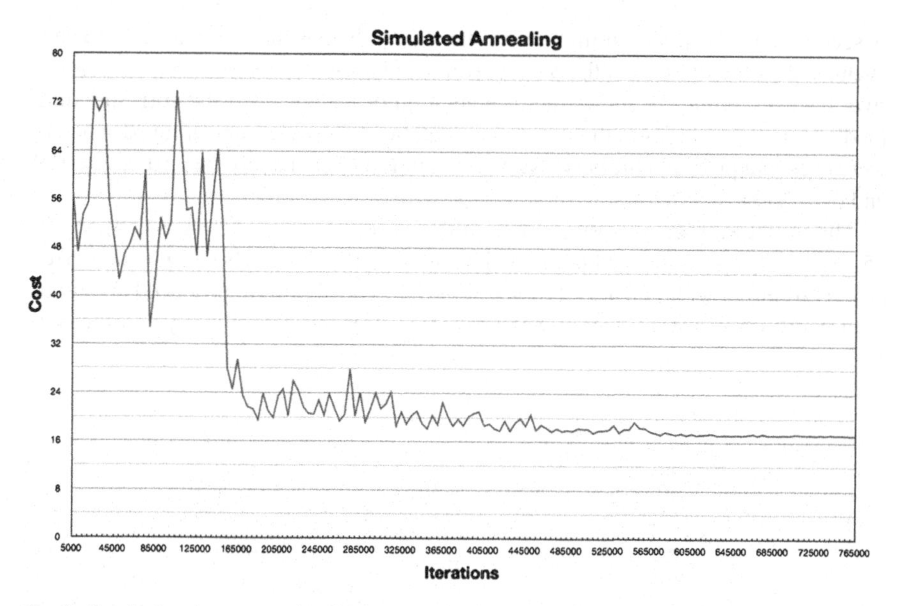

Fig. 8 Graph showing cost against iterations for simulated annealing

The graph shows that during the first 170,000 iterations the algorithm accepts a high number of poor (negative) moves. This is due to the initial high temperature allowing for a greater degree of search space exploration. Significant cost savings are made during this time. As the temperature cools the graph shows less activity representing a lower acceptance of negative moves. Many more iterations were required to find the optimal result. As with all probabilistic optimisation approaches, there is no guarantee that the global optimum solution has been found. That is, it is likely in this case that 17.24 represents the best possible solution, but this is not known for certain.

4.4 Gradient Descent Comparison

For the purpose of initial comparison, a GD solution has also been implemented. This achieved a width error of 61.18. This result justifies the use of more advanced techniques (such as SA) to overcome the local optimum problem. Figure 9 shows the resulting segmentation following application of GD.

4.5 SA Applied to the River Thames

A second data set, representing part of the River Thames, has been used to further evaluate the effectiveness of the SA algorithm and provide a comparison to the initial river data set. Figure 10 shows the triangulation of the river and the corresponding medial axis; Fig. 11 shows the initial segmentation. The same annealing parameters were used that provided the best result for the initial river, namely $\tau = 10.0$, $\lambda = 0.9$ and $\omega = 3000$.

The initial average segment widths in this case are $W = (86.84, 81.54, 61.27, 35.58, 21.14)$ and associated target widths of $W = (100, 70, 50, 30, 15)$, which gives an overall initial width error of $\hat{W}_{initial} = 50.77$. The algorithm produced an output with width error = 8.63, which was arrived at after the algorithm cycled through

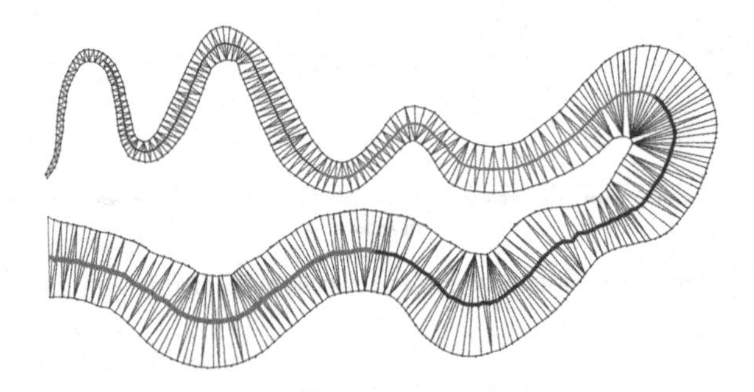

Fig. 9 Segmentation after application of gradient descent algorithm

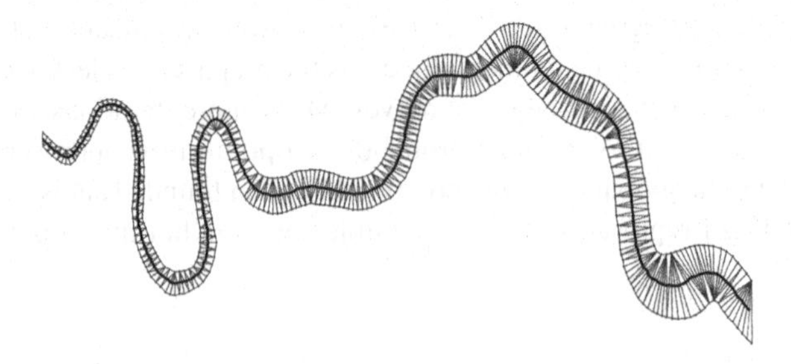

Fig. 10 River Thames triangulation and medial axis

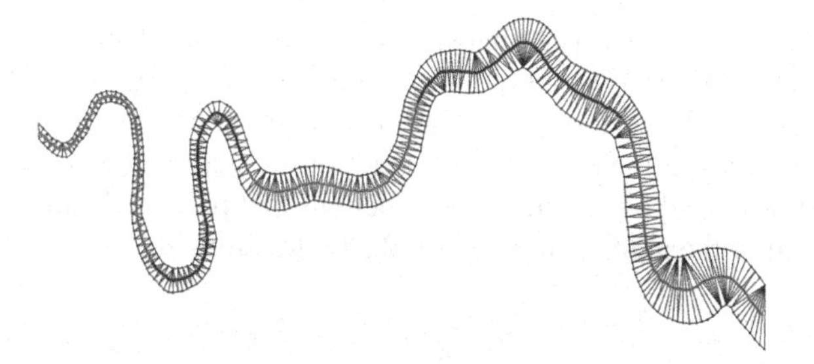

Fig. 11 River Thames triangulation and initial segmentation

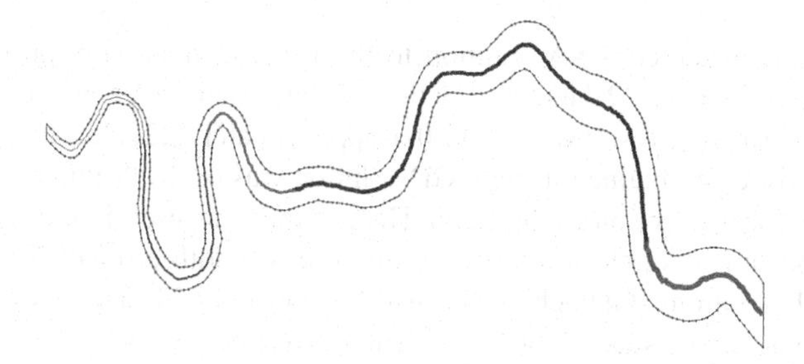

Fig. 12 River Thames solution using SA this time shown without triangulation

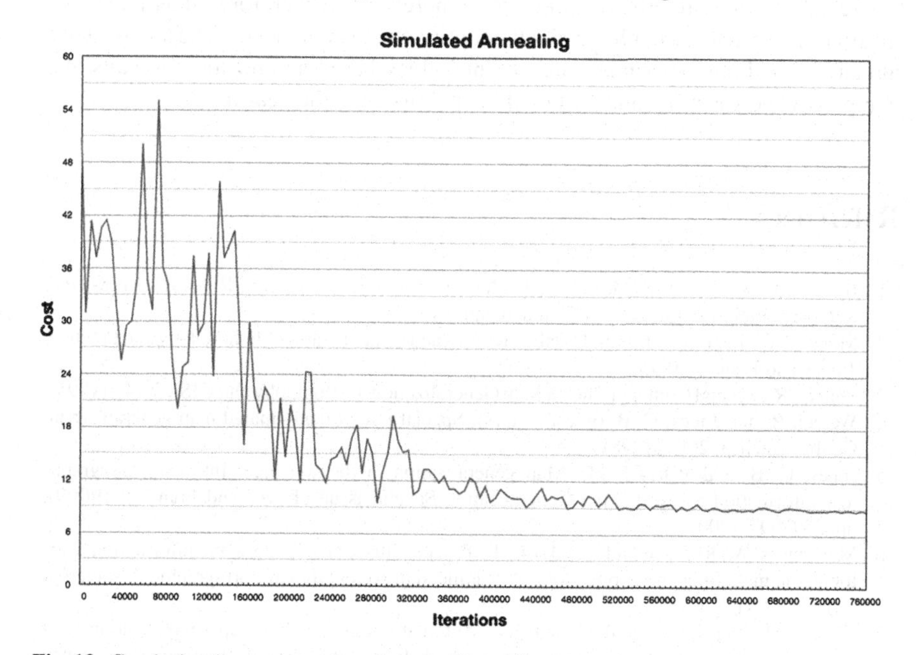

Fig. 13 Graph showing cost over iterations for River Thames

a total of 765,000 iterations. The final solution show in Fig. 12 took and a total time of 12 seconds to produce. The GD solution achieved a width error of 17.31.

Figure 13 shows the performance in terms of a graph plotting width error (cost) over iterations. During the first 250,000 iterations a significant cost improvement was achieved, followed by smaller cost reductions that produced a final width error value of 8.63 after the algorithm ran for 372,381 iterations.

5 Conclusions

This paper has presented a SA solution to the river segmentation problem. This solution has been tested on 2 data sets. For the first data set, the SA algorithm produced a best width error cost result of 17.24 after approximately 227,352 iterations. The best width-error cost obtained using a GD solution was 61.18. Similar results were also obtained for the section of the River Thames that was used in the second case. The SA algorithm was able to reduce an initial cost width error of 50.77 to 8.63 after 372,381 iterations. GD achieved a width error of 17.31. The SA approach is therefore deemed to be successful in segmenting river centerlines.

It is noted that the segmentation problem is conceptually straightforward, in that the search space is relatively simple to comprehend. It therefore lends itself to a meaningful comparison of SA with other optimization approaches. The authors have already carried out experiments using Ant Colony based optimization. Results of a comparison between SA and ACO will be published in the near future.

References

1. Buttenfield, B. P. and McMaster, R. B.: Map Generalization: Making Rules for Knowledge Representation, Longman Science and Technical (1991).
2. Müller, J. C., Lagrange, J.-P. and Weibel, R.: GIS and Generalization Methodology and Practice, Taylor and Francis (1995).
3. Weibel, R.: Special Content Issue on Map Generalization, Cartography and GIS', 22(4), (1995).
4. Weibel, R. and Jones, C. B. (guest editors): Special issue on automated map generalization, GeoInformatica, 2(4), (1998).
5. Jones, C. B. and Ware, J. M.: Map generalisation in the web age, Int. J. of Geographical Information Science: Map Generalisation Special Issue (Taylor and Francis), 19(8/9), pp. 859–997, (2005).
6. Mackaness, W., Ruas, A. and Sarjakoski, L. T.: Generalisation of geographic information: Cartographic modelling and applications. A volume in the International Cartographic Association series, 282 pages (2007).
7. Ware, J.M., Jones, C.B. and Thomas, N.: Automated cartographic map generalisation with multiple operators: a simulated annealing approach, Int. J. of Geographical Information Science (Taylor and Francis), 17(8), pp. 743–769, (2003).
8. Punt, E. and Watkins, D.: User directed generalization of roads and buildings for multi-scale cartography, Proceedings of 13th Workshop of the ICA commission on Generalisation and Multiple Representation, Zurich, (2003).

9. Regnauld, N. and Mackaness, W.: Creating a hydrographic network from its cartographic representation: A case study using Ordnance Survey MasterMap data, Int. J. of GIS Geographical Information Science (Taylor and Francis), 20(6), pp. 611–631, (2006).
10. Jones, C. B., Bundy, G. L1. and Ware, J. M.: Map generalisation with a triangulated data structure, Cartography and, GIS, 22(4), pp. 317–331, (1995).
11. Chew, L.P.: Constrained Delaunay Triangulations, Algorithmica, 4, pp. 97–108, (1989).
12. Richards, N.: An Evaluation of the Usefulness of Ant Colony Optimization to Solving Automated Map Generalization Problems, PhD thesis, available from the British Library, (2012).
13. Christensen, J., Marks, J. and Shieber, S.: An empirical study of algorithms for point-feature label placement, ACM Trans. on Graphics, 14(3), pp. 203–232, (1995).

Short Papers

A Multiagent Based Framework for the Simulation of Mammalian Behaviour

Emmanuel Agiriga, Frans Coenen, Jane Hurst and Darek Kowalski

Abstract A Mammalian Behaviour Multi-Agent Based Simulation (MBMABS) framework is proposed. Central to the framework is the concept of a behaviour lattice comprised of vertices representing states and edges representing possible state changes. State changes occur as a result of an agent completing some self-appointed task or as a result of some external event. Each state has one one or more predefined potential follow on states. Where there is more than one follow on state selection is made according to a weighted random selection process. The weightings are derived dynamically according to individual agent desires. The elements of the MBMABS framework are described in detail. The operation of the framework is illustrated using a case study.

1 Introduction

Computer simulations are used widely with respect to all kinds of applications [2, 5, 6]. A growing area of interest for computer simulation is animal behaviour. Animal behaviour can be perceived of as the way in which animals react to an environment as typically exhibited through movement [3, 4]. Simulations of animal behaviour are seen as desirable for a variety of reasons, the most significant

E. Agiriga (✉) · F. Coenen · D. Kowalski
Department of Computer Science, University of Liverpool, Liverpool L69 3BX, UK
e-mail: grigs@liverpool.ac.uk

F. Coenen
e-mail: coenen@liverpool.ac.uk

D. Kowalski
e-mail: darek@liverpool.ac.uk

J. Hurst
Mammalian Behaviour and Evolution Group, University of Liverpool, Liverpool L69 3BX, UK
e-mail: g.e.hurst@liverpool.ac.uk

M. Bramer and M. Petridis (eds.), *Research and Development in Intelligent Systems XXX*,
DOI: 10.1007/978-3-319-02621-3_32,

of which are: (i) once established they are inexpensive to operate, (ii) they can be used for what if style experiments without causing any permanent damage, (iii) they provide a simple mechanism for experiments to be repeated using the same set of parameters or by changing only one parameter, and (iv) they provide an excellent tool to enhance understanding of animal behaviour. The primary purpose of animal behaviour simulation is to allow behaviourologists to extend their current knowledge without needing to resort to expensive real life experimentationThe work described in this paper proposes the Mammalian Behaviour Multi-Agent Based Simulation (MBMABS) framework, a framework to support computer simulation of mammalian behaviour. The framework is founded on ideas first proposed by the authors in [1]. The basic idea is that each animal is represented by an agent. The behaviour for each agent is encapsulated in terms of a set of desires (D) and a behaviour lattice (B). These are described in Sects. 2, 3 respectively. Each agent has five main attributes: (i) a location within some environment (described by an x-y coordinate pair, (ii) a direction in which it is facing, (iii) a velocity (which may be zero indicating that it is not moving), (iv) a *state* defined by a vertex in a behaviour lattice, and (v) a set of desires. A third element of the MBMABS framework is the concept of environments (landscape) in which the agents are intended to operate. The MBMABS framework has been designed to provide a generic simulation facility that allows the inclusion of a range of desires and behaviours. The simulation operates on a iterative basis. On each iteration agents either perform some action according to their current "state" or undertake a state change. To provide for a full understanding of the proposed environment a case study is presented in Sect. 4. Some conclusions are presented in Sect. 5.

2 Desires

An agent can have any number of desires (k), goals or objectives that the agent wishes to adhere to, $D = \{d_1, d_2, \ldots, d_k\}$. Each desire has a "strength" associated with it, a number between 0.0 and 1.0. Desires are characterized as being either: (i) constant or (ii) dynamic. A constant desire is one whose strength? remains fixed throughout a simulation, while a dynamic desire is one whose strength changes (i.e. increases,reduces or remains static) with time. We model the changing strengths associated with dynamic desires using a cosine curve. A change in the character of a dynamic desire is usually associated with a state change. Typically an agent has several competing desires at a given time point in a simulation. A simple application of desires is in the selection of a direction for an agent that has decided to adopt a moving state. When an agent has decided to move it will have n directions to move in where n is defined by the number of immediate neighbour tiles into which the agent can move although some of these may feature obstructions. Thus we can identify a set of T possible tiles $T = \{t_1, t_2, \ldots, t_n\}$ where $0 \leq n \leq 8$. Note that the set T can be empty (the agent is unable to move). Each tile in T will also have a weighting associated with it calculated as the simulation progresses. Thus we have a set weightings $W = \{w_1, w_2, \ldots, w_n\}$ associated with each possible location in T indicating there desirability with respect to D.

3 Behaviour Lattice

A central feature of the MBMABS framework is the behaviour lattice. The behaviour lattice comprises: (i) a set of vertices each describing a "state" and (ii) a set of directed edges describing permitted state changes. Only certain states follow on from other states (have edges between them). Throughout a simulation each agent in the simulation is associated with one and only one vertex in the behaviour lattice at any discrete time point. State changes occur as a consequence of some event. With respect to some states there may be a number of alternative independent events that can trigger a state change. Events may be either: (i) external or (ii) internal. An external event is associated with some occurrence resulting as a consequence of the agent moving around its environment, for example encountering an obstruction or another agent. An internal event is concerned with an agent completing some self appointed task, for example changing the direction in which it is facing or timing out. Timing out is concerned with the duration whereby an agent may remain in some states; agents are assumed to be unable to remain in any one particular state indefinitely.

Although not applicable to all states, timing out is implemented using a value p (a field in each agents definition) that is set to 1.0 when the agent moves into a relevant state (vertex in the behaviour lattice). This value is then decreased according to the definition of a cosine curve, on each iteration of the simulation. With each iteration a random number r ($0.0 \leq r \leq 1.0$) is generated. If r is greater than p a state change is triggered. Thus at time 0 the probability that an agent will remain in its current state is 1.0 (definitely remain), at time N the probability that an agent will remain in its current state is 0.0 (definitely not remain); thus, as time progresses, the likelihood of a state change increases. The value of N will depend on the nature of the state under consideration.

Each vertex in the behaviour lattice will have at least two methods associated with it: (i) an action method and (ii) a state change method. The action method is used to process the current action of the agent. Three standard action methods are: moving, stopped and turning. The state change method is used to identify a follow on state and undertake any preparatory processing required before the follow on state can be commenced. Follow on states are selected in either a fixed manner or a probabilistic manner. Fixed selection occurs where, as a result of some event, there is only one possible follow on state. Probabilistic state changes occur where there are a number of competing alternative follow on states, in which case one is chosen in a probability influenced random manner whereby a weighting mechanism is used to influence follow on state selections according to current desire strengths.

4 Case Study, A Mouse in a Box Simulation

This section describes the operation of the MBMABS framework by considering a case study directed at mouse behaviour. The authors have used the proposed MBMABS framework to implement a mouse behaviour simulator. More specifically the case study considers the situation where a mouse is placed in a new environment which it is then expected to explore. The exploring is directed by a desire to explore. The environment for the mouse behaviour simulation was a simple box. This was adopted because identical boxes are used with respect to laboratory based experiments using real mice; hence the operation of simulated scenarios could be compared with similar scenarios run in "real-life". The environment E in this case comprised a set of tiles labelled using the the set $\{0, 1, 2, 3, 4, 5\}$ indicating (respectively): wall locations,corner locations, tunnel locations, choice locations (a location where we wish to enforce consideration for change) and open space. These all have significance with respect to mouse behaviour. The rest of this section is organised as follows; in Sects. 4.1, 4.2 we consider the mouse desires and the behaviour lattice for the case study. Then in 4.3 we discuss the operation of the simulation.

4.1 Mouse Agent Desires

For the purpose of the case study presented in this section it is assumed that our mice agents have only two desires: (i) a constant desire to stay close to walls and (ii) a dynamic desire to explore their environment. The preference for wall locations is a behaviour exhibited by mice called *thigmotaxis*. The desire to explore is a feature of many mammalian behaviours. In the case of mice they "know" the best (fastest and/or safest) route back to their nest site. With respect to our mice behaviour simulation the desire to explore is expressed as the desire to create a mental map of their environment (which can later be utilised). This map comprises a set of vertices and edges (and should not be confused with a behaviour lattice). The vertices are *waypoints* and the edges represent *travel lines*. Waypoints are significant locations and are currently defined as corners or choice points. The desire to explore is a dynamic desire, initially set to one, that will decrease until a new waypoint is found. If no waypoint is found the desire to explore reaches zero it will remain at zero for the remainder of the simulation or until such time as a new waypoint is found when the desire to explore will jump back to 1 before starting to decrease again.

4.2 Mouse Agent Behaviour Lattice

The behaviour lattice for the mouse behaviour case study is given in the Fig. 1. From the figure it can be observed that the behaviour lattice features seven vertices (states), they include: (i) Start, (ii) Moving Along Wall, (iii) Stopped At Wall, (iv) Moving

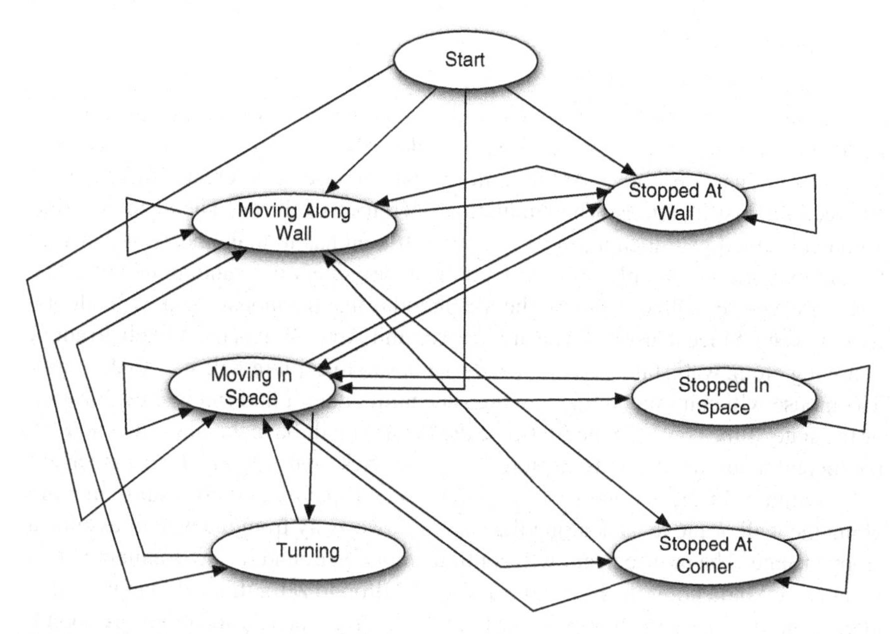

Fig. 1 Fragment of behaviour lattice for mouse case study

In Space, (v) Turning, (vi) Stopped In Space and (vii) Stopped At Corner. Each representing a particular activity which the mouse agent may be performing at a particular time T in the simulation. The meaning of each state can be derived from its nomenclature. For instance the Start state is the *current state* at the beginning of the simulation. Each of these States have one or more permissible *follow on states* (states which may be adopt whenever a relevant events occurs). The directed edges of the behaviour lattice (Fig. 1) indicate a transition from a *current state* to a *follow on state* as indicated by the direction of the arrow. Some states have several possible *follow on state.*

The mouse agent assumes the Start state as the *current state* at the beginning of the simulation (the Start state cannot exist as a follow on state). At the start of the simulation, the mouse agent will immediately select one of the three permissible follow on states (the follow on state is selected using the process described above in Sect. 3). As noted above, at the start of the simulation the dominant desire is the dynamic desire for the agent to explore its surroundings. There is also a constant desire for wall locations, so the agent is more likely to choose the Moving Along Wall state.

4.3 Operation

The simulation thus mimics the conjectured process whereby a mouse agent might build up a mental map of its environment describing *interest locations* (corners locations in this case study). At the same time the mouse agent creates links (paths) between interest locations. The simulation operates as follows. The agent starts at a predetermined gate location next to a wall. It then has to make decision whether to start exploring or simply move around the environment in a random manner. The desire to explore will be strong so the likelihood is that the mouse agent will adopt a moving state. Since it also has a desire for proximity to walls it is most likely to adopt a moving along wall state. As it proceeds the desire to explore will start to decrease. The mouse will continue in this moving state until either it finds an interest location or the state "times out". In the first case the location will be mapped (with a node in the mental map), the desire to explore will jump back up to 1.0 and the mouse agent will continue. In the second case the mouse agent may decide to resume moving along the wall or move back along the wall or move away from the wall or assume a stopped state. The mouse agent will continue moving around its environment in this manner. At some point its desire to explore will drop to zero, this will happen when after some time t no further, previously undiscovered, interest locations are found. While the desire to explore is strong the mouse agent will try to make decisions about where to go next (which states to adopt) influenced by the current state of its map.

5 Discussion and Conclusions

The MBMABS Multiagent Based Simulation framework for modelling animal behaviour has been presented. The central features of the framework are a set of desires and a behaviour lattice. The operation of the framework was illustrated using a mouse behaviour case study. Creation of the case study, and others not reported here, has demonstrated that the proposed framework also indicated that the framework readily supports the creation of such simulations. It was also easy to observe the behaviour of the simulated entities. The evaluation of the simulations was conducted with the help of animal behaviourists by comparing simulation behaviour with real behaviour. The evaluation also indicated that the MBMABS framework readily supports the addition of states and desires. It is however the case that as the number of states increase, the behaviour lattice becomes more complex and difficult to understand because the number of vertices and directed edges will also increase. For future work the research team intend to investigate more challenging animal behaviour scenarios such as nest site selection, territory guarding and threat avoidance. Experiments have been conducted using four mouse agents, however more testing environments with up to 64 mouse agents is being considered for future work.

References

1. E. Agiriga, F. Coenen, J. Hurst, R. Beynon, and D. Kowalski. *Towards Large-Scale Multi-Agent Based Rodent Simulation: The Mice In A Box Scenario*, pages 369–382. Research and Development in Intelligent Systems XXVIII. Springer, 2011.
2. MP Anderson, DJ Srolovitz, GS Grest, and PS Sahni. Computer simulation of grain growth-i. kinetics. *Acta metallurgica*, 32(5):783–791, 1984.
3. MP. Arora and C. Kanta. *Animal behaviour [electronic book] Mohan P. Arora; edited by Chander Kanta*. Mumbai India]: Himalaya Publications, House, 2009; Rev. ed, 2009.
4. R. Mathur. *Animal behaviour [electronic book] Reena Mathur*. Meerut, India: Rastogi Publications, 2009.
5. A. Newell and HA Simon. *Computer simulation of human thinking*. Rand Corporation, 1961.
6. Y. Zhang, BE Hobbs, A. Ord, and HB Mhlhaus. Computer simulation of single-layer buckling. *Journal of Structural Geology*, 18(5):643–655, 1996.

Parameter Estimation of Nonlinear Systems Using Lèvy Flight Cuckoo Search

Walid M. Aly and Alaa Sheta

Abstract Metaheursitc algorithms are used to solve hard optimization problems which can not be solved using traditional approaches within reasonable time and using feasible resources. One of the new natural inspired metaheursitc algorithms is the Cuckoo Search (CS) which is stimulated by the brood parasitism of some Cuckoo species. In this research, we explore the application of CS to solve the problem of parameter estimation of a nonlinear manufacturing process model. An industrial metal cutting system is used to examine the effectiveness of the proposed approach and also to compare CS with other metaheuristic approaches like genetic algorithm and particle swarm optimization. Results shows the high efficiency and robustness of CS when applied to the problem of parameter estimation of a nonlinear system model.

1 Introduction

Nature inspired algorithms use inexact approaches to solve computationally hard problems, they include different approaches like genetic algorithms (GAs), neural networks (NN), particle swarm optimization (PSO), Bat Algorithm, Water drops algorithm, fuzzy logic and Cuckoo Search (CS). As theses algorithms proved to be efficient, many researches investigated their usage to solve various complex industrial and engineering problems, among these problems is the system modeling problem. System modeling is the concept of representing the behavior of a system by presenting

W. M. Aly (✉)
College of Computing and Information Technology, Arab Academy for Science, Technology and Maritime Transport, Cairo, Egypt
e-mail: walid.ali@aast.edu

A. Sheta
Computers and Systems Department, Electronics Research Institute (ERI), Cairo, Egypt
e-mail: asheta66@gmail.com

M. Bramer and M. Petridis (eds.), *Research and Development in Intelligent Systems XXX*,
DOI: 10.1007/978-3-319-02621-3_33,

it in a form of a mathematical equation or set of equations. Parameter estimation of a model is a complex optimization problem that standard analytic approaches might fail to solve [7]. To estimate these parameters, search procedures like least square estimation, likelihood and instrumental variable methods [1] can be applied. These procedures aim to minimize the error between the actual model and the predicted model, although they can usually provide good results, they have no exact solution and they suffer from efficiently reduction in the presence of noise. As traditional techniques would fail to reach satisfactory solutions for the parameter estimation problem, different nature inspired algorithms have been investigated in this area; Multi-objective Genetic Algorithm (MOGA) was applied to estimate the parameters of pressure swing adsorption model [6]. PSO was used to build a model to predict the thickness and surface roughness of printed patterns in roll-to-roll printing systems which is a nonlinear system with complicated dynamics [5]. In this paper we investigate the use of Cuckoo search as a nature inspired algorithm for solving the parameter estimation problem for nonlinear manufacturing process model of the cutting tool temperature in an industrial metal cutting system.

2 Complexity of the Problem

To get a sense of the complexity of the parameter estimation of a nonlinear model problem, we defined the following hypothetical nonlinear system:

$$y(x) = 4cos(45x) - 2 * sin(10x) \tag{1}$$

Assuming that the structure of this hypothetical model is already known but without the values of the parameters, the hypothesis function $h_\theta(x)$ can be declared as follows:

$$h_\theta(x) = 4cos(\theta_0 x) - 2sin(\theta_1 x) \tag{2}$$

The role of parameter estimation is to figure out the values of θ_0 and θ_1, assuming that the exploration range for these values is from 0 to 50. The search space for minimizing the cost function $J(\theta)$ -as shown in Fig. 1 is a non convex search space. The quality of a solution proposed by traditional parameter estimation techniques for such a problem will be affected by the arbitrary staring search point and can be easily trapped in a local optima [8]. A cost function $J(\theta)$ represents the difference between the actual values of outputs and the predicted value calculated using $h_\theta(x)$. Different equations can be used as a cost function, among them are residual standard deviation and sum of square error (SSE). SSE is declared for n inputs (where $x_0 = 1$) as follows:

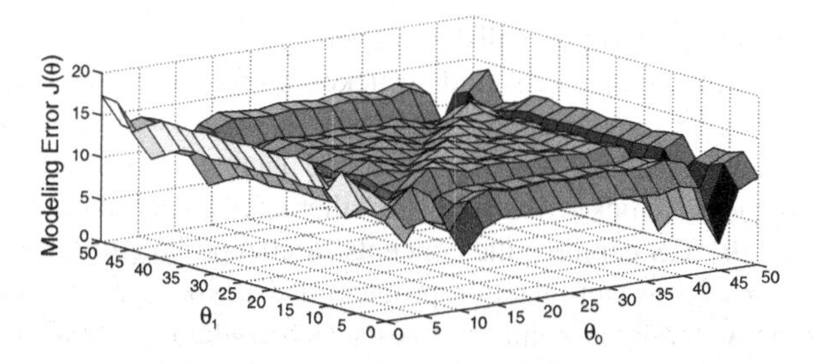

Fig. 1 Error surface for the system given in Eq. 2

$$J(\theta) := \frac{1}{m}\sum_{i=1}^{m}(h_\theta(x^{(i)}) - y^{(i)})^2 \quad (3)$$

where $J(\theta)$ is the cost function; x^i is the training data. i represents the training sample. y_i is actual known output. m is the number of training examples. The cost function $J(\theta)$ represents the calculated error of the model based on certain values of θ. θ can be estimated by solving the minimization problem of $J(\theta)$. Traditional techniques like normal equation can be applied when the size of the input vector is not large otherwise iterative techniques like gradient decent would be more efficient. Regularization modification can be applied to the cost function to prevent over fitting, where the proposed hypothesis fit the training set very well but fail to generalize to new examples.

The cost function of linear regression is always a convex function, gradient decent will always reach an optimum solution without being affected by the choice of initial solution. However, when solving the regression problems of nonlinear systems, the minimization of the cost function will produce normal equation with nonlinear parameters that does not have an exact solution, although iterative procedures like Taylor series, steepest descent and Levenberg-Marquardt methods could be applied to solve the problem, these procedures are not optimal, they suffer from slow convergence and the sensitivity to the initial solution.

To overcome the difficulties of parameter estimation of nonlinear models, Nature-Inspired Learning Algorithms [8] had been applied to solve many nonlinear modeling problem in industry [2–4]. One of these famous algorithms is the CS which shall be explored to solve the parameter estimation problem of the cutting machine.

3 Cuckoo Search

CS is a search algorithm inspired by breeding behavior of cuckoos, cuckoo breeding can be described as an act of parasitism. A cuckoo bird lay its egg in other birds nests and relay on that bird for hosting the egg. Sometimes the other bird discover that an egg is not their own, it might demolish the alien egg or just move to another nest. To protect its egg from being discovered, a cuckoo might imitate the shape, size and color of the host eggs, some cuckoos might take an aggressive action and remove other native eggs from the host nest to increase the hatching probability of their own eggs, a hatched cuckoo chick will also throw other eggs away from nest to improve its feeding share [9]. CS captures the concepts of cuckoo breeding by formulating candidate solutions for an optimization problem as Cuckoo eggs in different nests. The search starts with a fixed number of nests each contain a candidate solution to form an initial generation of solution. This generation evolve from one iteration to another while a fraction of the solutions in nests will be eliminated and replaced by new solution to model the concept of alien egg discovery in a real cuckoo world. CS depends on Lèvy flight as the random walk used to generate new solution (cuckoos) from current solution according to the following equation:

$$cuckoo_i^{(t+1)} = cuckoo_i^{(t)} + \alpha \oplus L\acute{e}vy(\lambda), \tag{4}$$

where $cuckoo_i^{(t+1)}$ is the value of the ith Cuckoo at instance t, α is the step size, usually chosen to be equal to one. λ: Lèvy distribution coefficient ($1 < \lambda < 3$). The main advantage of Cuckoo search is its simplicity. CS has fewer parameters that need to be tuned before starting the search compared with other techniques, in contrast, PSO needs tuning of mainly three parameters: Inertia weight, effect of self confidence and effect of social impact. The range of tuning parameters of PSO affect the quality of search dramatically.

4 Parameter Estimation of the Cutting Tool Temperature Model

Machining is cutting a piece of raw material into a desired form and size by a controlled process of material removal. The cutting forces of this process is converted into heat. Thus, the cutting tool suffers from very high temperature. The high temperature of the cutting tool decreases the tool life and affects the surface finish and geometrical dimensions of the product and thus the quality of the whole process [2]. The temperature computation found to depend mainly on the following three variables:

1. x_0: Depth of the cut (mm)
2. x_1: Cutting feed rate (mm/rev)
3. x_2: Cutting speed (m/min)

Moreover, An empirical model exits for cutting temperature presented in [10]. The model was declared as follows:

$$h_\theta(x) = \theta_0 * x_0^{\theta_1} * x_1^{\theta_2} * x_2^{\theta_3} \tag{5}$$

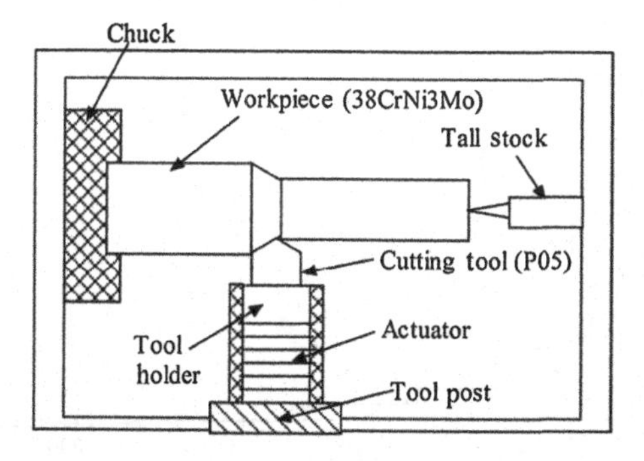

Fig. 2 Metal cutting machine

Our case study will be to estimate the parameters for the nonlinear model of cutting tool temperature. We will use the data measured experimentally for the cutting tool PO5 with the metal of the workpiece is 38CrNi3Mo, the system is shown in Fig. 2. In our case study, the cost function $J(\theta)$ that we aim to minimize will be the residual standard deviation declared as follows:

$$J(\theta) := \sqrt{\frac{\sum_{i=1}^{m} (h_\theta(x^{(i)}) - y^{(i)})^2}{n-2}} \tag{6}$$

where $J(\theta)$ is the cost function, y is actual known output and m is the number of training examples. Solving this parameter estimation problem means finding values for the model parameters which minimizes the error between outputs from the predicted model and the actual output. The data set used was provided in [10].

4.1 Experimental Results

CS succeeded in estimating the parameters of the temperature model, by the end of our experimentation the computed residual standard deviation (rsd) reached is 5.129, this result was reached using 10 nests, $beta = 1.5$ and $p_a = 0.1$. Using Genetic algorithm, Least Square and PSO rsd was 5.187, 5.695 and 5.129 respectively. Figure 3 shows the improvement of the cost of the hypothesis from iteration to another. The developed model parameters are given in Eq. 7.

$$h_\theta(x) = 1410.29 x_0^{0.032} * x_1^{0.082} * x_2^{0.158} \tag{7}$$

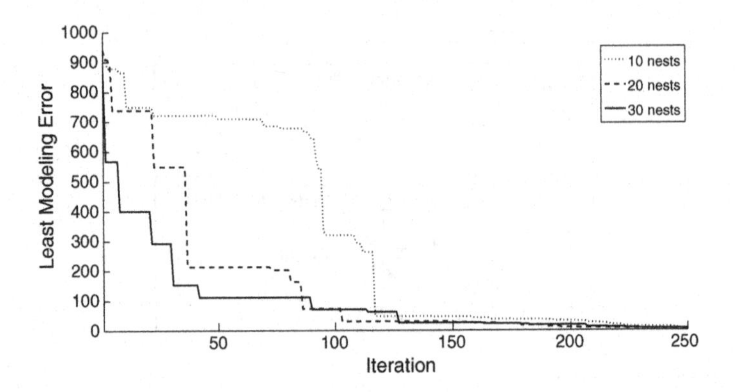

Fig. 3 Convergence of the CS evolutionary process

5 Conclusions

Cuckoo Search is a powerful search algorithm that it inspired by the breeding behavior of cuckoos. It has been used successfully in many areas. In this paper we have explored its use in solving the parameter estimation problem for nonlinear systems. This problem typically presents difficulties to traditional parameter estimation techniques which basically depend on linearizing the system in order to apply available algorithms for linear systems, also as most of these traditional techniques are based on gradient descent technique, the search might be trapped at a local optimal solution. Cuckoo Search succeeded in estimating the values of the parameters for the nonlinear model for the cutting tool temperature in a cutting machine. It is anticipated that CS will play a very important role in solving complex optimization problems.

References

1. Atanasov, N., Ichtev, A.: Closed-loop system identification with recursive modifications of the instrumental variable method. Informatica 22(2), 165–176 (2011).
2. Faris, H.: A symbolic regression approach for modeling the temperature of metal cutting tool. To appear in the International Journal of Control and Automation (2013).
3. Faris, H., Sheta, A.: Identification of the tennessee eastman chemical process reactor using genetic programming. International Journal of Advanced Science and Technology 50, 121–140 (2013).
4. Faris, H., Sheta, A., Oznergiz, E.: Modeling hot rolling manufacturing process using soft computing techniques. International Journal of Computer Integrated Manufacturing (2013).
5. Kessentini, S., Barchiesi, D., Grosges, T., Giraud-Moreau, L., Lamy de la Chapelle, M.: Adaptive non-uniform particle swarm application to plasmonic design. Int. J. Appl. Metaheuristic Comput. 2(1), 18–28 (2011).
6. Liu, Y., Sun, F.: Parameter estimation of a pressure swing adsorption model for air separation using multi-objective optimisation and support vector regression model. Expert Syst. Appl. 40(11), 4496–4502 (2013).

7. Nguyen-Van, T., Hori, N.: New class of discrete-time models for non-linear systems through discretisation of integration gains. Control Theory Applications, IET 7(1), 80–89 (2013).
8. Sheta, A., Jong, K.D.: Parameter estimation of nonlinear systems in noisy environment using genetic algorithms. In: Proceedings of the IEEE International Symposium on Intelligent, Control (ISIC' 96), pp. 360–366 (1996).
9. Yang, X.S., Deb, S.: Cuckoo search via levy flights. In: Nature Biologically Inspired Computing, (NaBIC' 2009). World Congress on, pp. 210–214 (2009).
10. Zhou, J.: GA algorithm for cutting experiment data drawing. Journal of Southwest Petroleum Institute 29(3), 1062–63 (1998).